RESTRUCTURING THE EMPLOYMENT RELATIONSHIP

Restructuring the Employment Relationship

DUNCAN GALLIE
MICHAEL WHITE
YUAN CHENG
and
MARK TOMLINSON

CLARENDON PRESS · OXFORD

This book has been printed digitally and produced in a standard specification
in order to ensure its continuing availability

OXFORD
UNIVERSITY PRESS

Great Clarendon Street, Oxford OX2 6DP

Oxford University Press is a department of the University of Oxford.
It furthers the University's objective of excellence in research, scholarship,
and education by publishing world-wide in

Oxford New York

Auckland Bangkok Buenos Aires Cape Town Chennai
Dar es Salaam Delhi Hong Kong Istanbul Karachi Kolkata
Kuala Lumpur Madrid Melbourne Mexico City Mumbai Nairobi
São Paulo Shanghai Taipei Tokyo Toronto

Oxford is a registered trade mark of Oxford University Press
in the UK and in certain other countries

Published in the United States
by Oxford University Press Inc., New York

ISBN 0-19-829441-7

Antony Rowe Ltd., Eastbourne

ACKNOWLEDGEMENTS

The principal survey on which this book is based—*The Employment in Britain Survey*—was co-funded by an Industrial Consortium, the Employment Department, the Employment Service, and the Leverhulme Trust. We are particularly grateful to Sir Bryan Nicholson and Alistair Graham for their leading roles in bringing together the Industrial Consortium.

The industrial funders were: British Nuclear Fuels; British Steel; British Telecom; Calderdale and Kirklees Training and Enterprise Council; Eastern Electricity; Institute of Personnel Management; J. Sainsbury; Local Government Management Board; London Electricity; London Regional Transport; Manweb; National Economic Development Office; National Westminster Bank; Peugeot Talbot Motor Co.; Scottish Hydro-Electric; South-Western Electricity; Southern Electric; Tesco Stores; the Boots Company; the National Grid Company; the Post Office; the Thomas Cook Group; the Wellcome Foundation; Trusthouse Forte; Tyneside Training and Enterprise Council; W. H. Smith; the Wyatt Company.

The fieldwork for the survey was conducted by Public Attitude Surveys Ltd, under the direction of Stuart Robinson, Eileen Sutherland, and Ruth Lennox. The research team worked in close liaison with PAS at all stages of the survey process: design of the sampling, piloting, briefing meetings, monitoring the fieldwork, and coding. The openness of the agency to this very close interaction and continuing discussion represented a model of good practice for this type of research. We thank them for the high quality of the work they put in throughout the survey.

Martin Range and Jane Roberts at Oxford were responsible for setting up the data on the computer, a demanding job with a data set of this size and complexity. We are also very grateful to David Firth of Nuffield College, for the very generous way in which he gave time to respond to our queries on particular aspects of data analysis. Sarah McGuigan was responsible for the final preparation of the manuscript. Our thanks too to Geoff Jones, at Nuffield College, for handling the unusually complex financial arrangements of the project.

An early version of some of the descriptive findings was published by the Policy Studies Institute under the title *Employee Commitment and the Skills Revolution*. The current book, however, represents a complete reworking of the data analysis, as well as a major extension of the themes addressed. An earlier version of material in Chapter 2 was also published in a collection of conference papers edited by Rosemary Crompton, Duncan Gallie, and Kate Purcell, *Changing Forms of Employment*, Routledge, 1996.

Finally, our thanks to the staff of Oxford University Press for their work in the preparation of this book.

Duncan Gallie   Michael White
Nuffield College, Oxford   Policy Studies Institute, London

CONTENTS

LIST OF FIGURES

LIST OF TABLES

1

Introduction

Between the early 1970s and 1990s, a rapidly changing technical and economic environment was widely seen as heralding a transformation in the character of employment relations in Britain. Economic and political change combined to produce a far-reaching restructuring of British industry. It was a period that witnessed the collapse of large sectors of traditional manufacturing industry and a marked shift towards an economy based primarily on the service industries. At the same time, heightened competitive pressures, linked to the increasingly global nature of markets and, particularly, to the process of integration of the West European economies, confronted British management with the need for major improvements in performance in terms of both cost and quality. The economic pressures for change were reinforced by political pressures. The election in 1979 of a government committed to 'new right' economic policies encouraged employers to rethink fundamentally their employment policies by giving priority to market principles and labour flexibility, while precipitating restructuring through monetary policies that provoked a particularly brutal recession.

Among the ways in which employers responded to the constraints and opportunities that came with economic and political change, four have particularly drawn attention. The first was the rapid adoption of new computerized and information technologies, which were thought to have substantial consequences for skills and work organization in both the manufacturing and the service industries. Second, there was the emergence of a new philosophy of management—Human Resource Management—that involved a direct challenge to the traditional pattern of industrial relations. Third, there was a shift away from 'standard' forms of employment contract and an increased use of 'non-standard' contracts, such as part-time and temporary work, which threatened a fragmentation and even polarization of employment conditions. And finally there was a marked increase in labour market insecurity, resulting partly from the experience of two recessions, but also from a continuous process of reorganization and manpower reduction.

While there was widespread acceptance of the potential significance of these changes, there have been sharp differences in views about their relative importance, their extent, and the nature of their impact. For some, these developments represented a fundamental rupture with the types of production system that had prevailed in the first three post-war decades; opening the way to radically changed employment relations (Bell 1974; Piore and Sabel 1984; Kern and Schumann 1987, 1992; Boyer 1988; Womack et al. 1990). The highly differentiated division of labour, the oppressive systems of control of work performance, and the rigid forms of hierarchy associated with Taylorist methods of production were giving way to a new system of production that would once more give a central role to the skills of the workforce. The employment relationship would be based increasingly on high levels of discretion for employees, the reduction of status distinctions within the workforce, and the adoption of organizational policies designed to secure the long-term commitment of employees. In effect, the traditional class distinctions in the nature of the employment relationship would be dissolved. For others, however, these changes were viewed in a very different light. They were seen rather as a means for employers to reassert their control in a period of trade union weakness and to intensify the work process, leaving intact or even reinforcing the core principles of the traditional pattern of employment relations (Elgar 1991; Hyman 1991; Berggren 1993).

One thing is clear: speculation far outran the availability of rigorous empirical evidence. The objective of this study is to examine more systematically how far in practice such changes affected the nature of work, the employment relationship, and people's experiences of work. In particular, how far did they lead to an improvement in skill and discretion in work? Did they involve a marked change in traditional patterns of control over work and in the influence employees could exercise over decision-making? What was their impact on employees' involvement in their work and on their attitudes to the organizations they worked for? And did they reduce or accentuate inequalities in the employment situation of different sectors of the workforce?

THE RESTRUCTURING OF WORK AND THE EMPLOYMENT RELATIONSHIP

New Technologies

Predictions that new forms of technology would lead to a transformation of the nature of work have abounded since the first wave of automation in the 1950s. This had seen the development in certain sectors of manufacturing industry of continuous-process technologies, in which the production process was regulated by complex feedback mechanisms and workers no longer had any direct physical contact with the product. At the time it was believed that automation would spread rapidly across the industrial scene, transforming the character of the work task, forms of work organization, and patterns of industrial organization. In practice, the adoption of manufacturing systems using highly integrated continuous-process technologies proved to be much slower than the original prophets of automation had anticipated. They were technically adapted only for fluid or crystalline products, and they required a level of expenditure on a relatively inflexible production process that could only make economic sense for a stable mass market. It was only in the 1980s, some two decades after the first wave of automation, that the development of microelectronic technology gave reality to the scenario of a technological transformation that would sweep across the broad spectrum of industry.

The theories of the implications of automation that had been formulated by industrial sociologists were remarkable not least for their highly contradictory nature. Very broadly, one can distinguish between an 'optimistic' perspective that argued that automation would lead to a marked improvement in the nature of work and a 'pessimistic' perspective that viewed it as heightening alienation in work.

The optimistic scenario suggested that the spread of computerized technologies would transform the traditional character of manual work by upskilling the work task. Whereas earlier phases of mechanization had led to an increased division of labour, involving ever-simpler and more repetitive work tasks, automation reversed this trend. It gave greater emphasis to conceptual rather than to manual skills and it tended to reduce the harshness of physical working conditions. At the same time, since the worker was no longer directly tied to a specific machine, it encouraged the growth of polyvalence or multi-skilling, which would have the effect of increasing the variety and intrinsic interest of work. Indeed, it changed the work task in a way that was likely to blur the conventional distinction between manual and non-manual work (Touraine 1955; Blauner 1964; Wedderburn and Crompton 1972). These changes would have important implications for the way work was organized. Given the new skills required of workers, traditional forms of direct supervision would give way to a decentralization of decision-making responsibilities to 'autonomous' work groups, in which employees would themselves be responsible for allocating work tasks and enforcing work norms.

Changes in the organization of work, it was argued, would be paralleled by changes in employment relations. Where responsibility was decentralized to the work group, it was essential to ensure a high level of organizational commitment in order to prevent work groups using their power to further sectional rather than organizational interests. Further, given the costs of training for employees using new and complex equipment, there was a strong incentive to provide conditions of employment which reduced turnover. The use of advanced technologies, then, would tend to be accompanied by greater job security and generally more favourable terms of employment. Organizations would seek to bind employee commitment over the longer term by providing extensive fringe benefits and developing internal labour markets that gave good opportunities for advancement up a highly stratified promotion ladder.

In sharp contrast, the pessimistic scenario saw the computerization of work as depriving workers of their ability to actively intervene in the work process, and hence as making work less skilled, more monotonous, and offering reduced possibilities for self-development. In this perspective, automation was seen as completing the process of the 'dehumanization' or 'degradation' of work that had been already inherent in the adoption of earlier mass-production techniques, in particular the introduction of assembly-line technologies (Friedmann 1946; Braverman 1974). Moreover, this was by no means confined to manual work. The development of computer technologies meant that automation could be applied every bit as thoroughly to the office (Braverman 1974; Crompton and Jones 1984). Automation would replace a skilled by a largely unskilled workforce, and this in turn would lead to the truncation of career ladders and a general decline in employment conditions.

Further, a number of studies pointed to the effect of automation in disrupting traditional forms of work group organization, increasing social isolation at work and leading to higher levels of stress (Naville 1961, 1963; Chadwick Jones 1969). The high capital costs led employers to seek maximum utilization of the equipment; this implied the extension of forms of shift work that increased the strain of work and disrupted the normal patterns of leisure and family life. Finally, since a central characteristic of automation was an acceleration of the replacement of human skills by mechanical self-regulation, it was likely to lead to a much greater sense of job insecurity.

The debate about the implications of technological change for skill gained new impetus with the rapid spread of microelectronic technology from the early 1980s. Unlike the earlier wave of automation, microprocessors provided the flexibility to transform a very wide range of production activities—design work, production, office work, distribution, and retail. For many observers, it seemed clear that a new metatechnology had arrived that could change the nature of work across both manufacturing and services industries, stimulating a new era of technical and organizational innovation (Forester 1980; Kaplinsky 1984; Freeman 1986; Bessant 1991). Surveys of employers certainly confirmed the very rapid diffusion of new applications. Whereas in 1981 only 18 per cent of establishments with more than twenty employees were using microelectronics in their products, by 1987 this had risen to 60 per cent (Northcott and Walling 1988). While much of this early implementation took the form of individual machine applications, there was also a clear growth of the potential of the new technologies to enhance the integration of work processes. Already by 1990, 50 per cent of all establishments and 87 per cent of workplaces with over 1,000 employees had introduced some form of computer networking. The very general nature of the change was also evident from the fact that it affected non-manual employees at least as much as, if not even more than, manual employees (Daniel 1987).

Yet, the core of the underlying debate remained remarkably similar to that developed in the first wave of automation. The central issue was still whether there was some inherent characteristic of the new technologies that undercut or enhanced skills (for reviews see Francis 1986; McLoughlin and Clark 1994). While some writers extended to the new context the concern with deskilling and enhanced managerial control (for example, Downing 1980; Martin 1984;

Shaiken 1984; Smith 1987; Thompson 1989), others emphasized the potential for developing conceptual skills and increasing the range of problem-solving and decision-making activities in work (Kern and Schumann 1987; Zuboff 1988; Womack *et al.* 1990; Clark *et al.* 1988). Perhaps the boldest vision of the potential for emancipation was developed by Piore and Sable (1984). In the context of a major change in product markets and the competitive environment, computer technologies brought new opportunities for low-cost customized production in small-scale and highly skilled environments. The accumulating evidence, they suggest, indicates that 'technology has ended the dominion of specialized machines over un- and semi-skilled workers, and redirected progress down the path of craft production. The advent of the computer restores human control over the production process; machinery again is subordinated to the operator' (ibid.: 261).

However, the very pervasiveness and diversity of the applications of microelectronic technology posed major difficulties for any empirical assessment of its impact. While case-study research could produce fine-grained descriptions of particular instances of work reorganization, there was little way of establishing an overall picture of the direction of change. Reviews of research in the field simply revealed the very wide range of possible developments, without being about to resolve the problem of whether some were more typical than others (Lane 1988; Martin 1988). It was clear that, in part, this reflected the fact that the mode of implementation of new technologies was heavily contingent upon the specific nature of managerial policies (Buchanan and Boddy 1983; Wilkinson 1983; Child and Loveridge 1990), which themselves could be very diverse.

The question of the principal direction of change could only be addressed by research that was deliberately designed to achieve representativity and this forms one of the central objectives of the present study. Has there been any overall trend towards either upskilling or deskilling? How exactly do skill developments affect other aspects of job quality such as the intrinsic interest of work and the degree of discretion that people have over the way the work is carried out? And how far can such developments be seen as related to the adoption of new technologies?

New Forms of Management

A second potentially important source of change in the experience of work derives from developments in managerial thinking about employee relations. Very broadly this could be characterized as the shift from an 'industrial relations' to a 'human resource management' perspective.

The traditional model of employment relations in the large firm sector had been based on collective bargaining and the joint regulation of the conditions of employment. The distinctive feature of the British pattern of industrial relations, at least in the private sector, was the significance of local workplace bargaining. The assumption was that the individual's relationship to management was mediated through the representatives of the workforce, in particular through the shop stewards. The rapid growth of shop steward organization in the 1950s and 1960s was followed by the increased formalization and institutionalization of shop floor representation in the 1970s (Clegg 1979; Batstone 1988).

The nature of work organization and the terms of employment were not subject to unilateral management prerogative, but were defined in detailed workplace agreements. Change in the organization of work was often brought about through productivity bargaining in which specific work practices were collectively 'bought out' and new working arrangements were agreed through negotiation. While the actual operation of such procedures may have been less obstructive to technological change than was often assumed (see Daniel 1987), it clearly imposed significant constraints on management's freedom to adjust work organization and in its treatment of employees (Flanders 1964; Gallie 1978). In particular, it made it very difficult for management to introduce individualized employment relations. An important example of this is that the system of joint regulation was largely inimical to individualized assessment and pay practices. The emphasis was on the 'rate for the job' rather than on the reward of individual merit. Even where there was some element of individualization in the payment system, for instance with piecework systems, this was in practice heavily constrained by informal workgroup controls (Lupton 1963).

The institutionalized power of the unions at workplace level became one of the major targets of Conservative government 'reform' from 1979. Successive waves of legislation were designed to weaken the local membership strength of the unions and to undermine the coercive power that they could wield through strike action. There was a severe haemorrhage of union membership nationally and a rise in the proportion of establishments with no union presence. While in the early 1980s the unions' influence over work organization proved remarkably resilient in those establishments where they already had a presence (Daniel and Millward 1983; Gallie *et al.* 1996), it is clear that their local bargaining strength was eroded later in the decade (Millward and Stephens 1986; Millward 1994). This provided at least a potential opportunity for British employers to reconsider the underlying pattern of relations with their employees. And the weakening of union power coincided with the emergence of a new (or at least revived) 'philosophy' of management—human resource management—which placed the emphasis very firmly on the strengthening of direct links between management and workers.

These new policies originated in the USA during the 1970s and 1980s and, at least initially, were linked to the development of non-unionized and greenfield sites (Foulkes 1980; Kochan *et al.* 1994). The specific practices that were associated with the notion of human resource management were highly diverse and of questionable coherence. At one extreme, they involved a marked individualization of the employment relationship. Instead of the fixed rate for the job applicable to all employees, payment systems were to be based on careful assessment of individual performance and significant individual merit payments. At the other, there was an emphasis on the importance of encouraging team involvement, for instance through participation in quality circles. Nonetheless, underlying this diversity, there was a common assumption that management should develop close direct links with its employees, rather than accept an indirect relationship that was mediated by elected shop floor representatives. This was intimately connected to the view that management should take a much more active role both in enhancing the quality of work performance and in building up high levels of employee commitment to the organization.

A number of factors underlay this new managerial thinking. In part, it could be seen as a belated recognition of the limits of Taylorist methods of control of work performance, which had emphasized easy substitutability of labour, tight supervision, and output-linked pay. This had been compounded by the use of Fordist assembly-line technologies, which provided mechanical control over work pace. The marked rise in industrial conflict in the 1960s and 1970s was followed, in several countries, by a serious reflection on the dysfunctional effects of traditional systems of management and work organization (for France, for instance, see Durand 1978; Martin 1994; for the USA, Walton 1972; Beer *et al.* 1984). There was accumulating evidence that these led to low levels of job involvement, weak commitment (or even hostility) to the employing organization, and a willingness to disrupt production to achieve higher financial rewards.

A shift towards a greater concern for human resources was also encouraged by the increased pace of technological change considered in the previous section. The rapid spread of computerization appeared to have introduced an era of virtually permanent technical change. The era when skills could be learned early in life either in the general education system or through apprenticeship was being replaced by one which required continuous learning. But, for this to be possible, employers would need to be able to assess the individual training needs of their employees and provide the necessary investment in the updating of skills. This implied the development of a much closer relationship between managers and employees, and much more detailed information about individual circumstances, than had been the case in the past.

But the 1980s also saw the emergence of factors that gave a new urgency to the need to obtain positive commitment, rather than merely control. Heightened international competitiveness, and particularly the striking success of the Japanese economy, brought about a new awareness of the need to reconsider techniques of management (Jurgens *et al.* 1993). Studies of Japanese transplants in North America appeared to show that Japanese organizational methods led to far higher levels of productivity even without the specific cultural conditions of Japanese society. The superiority of the Japanese model was consecrated by an MIT research programme, in the second half of the 1980s, which concluded from its study of the automobile industry that Fordism would inevitably be supplanted by the 'lean production' model of organization. Lean production, which combined the best features of both craft production and mass production, was destined to become 'the standard global production system of the twenty-first century' (Womack *et al.* 1990: 277–8; see also Kenney and Florida 1993). Whatever, its limitations in terms of research, it is clear that the study had an interested audience among employers.[1] While primarily focusing on the economies to be obtained by the systematic reduction of organizational 'slack' through the implementation of just-in-time techniques (JIT), increasing the functional flexibility of labour and decentralizing responsibilities within the organization, the study emphasized the high-skill requirements of such systems and their fragility if management failed to build up sufficient commitment in the workforce (ibid.: 100–3). Essential to the idea of lean production, then, was the view that firms would have to ensure high levels of commitment on the part of employees, finding ways of leading them to identify with the values and objectives of management.

The other aspect of the Japanese model that had implications for the importance attached to employee commitment was its emphasis on quality. This was a period when notions of 'total quality management' began to be adopted by at least some of the more progressive firms. The ideal was one of continuous improvement (*kaizen*), which would involve all levels of the workforce. At one level it was thought that this might be achieved through giving employees a sense of 'empowerment', at another it was to be brought about by ensuring much better direct communications between managers and employees. The institutional model for shop floor organization drawn from Japan was that of the 'quality circle'. While there was increased interest in quality circles in Britain in the 1980s, many of these experiments had collapsed in good part because of the indifference or even resistance of middle management (Bradley and Hill 1983, 1987). 'Total quality management', however, by making quality enhancement a central objective for managers and emphasizing a continuing initiative from all levels of management, gave such experiments a new relevance for the 1990s (Hill 1991*a*). But, as with lean production, the emphasis on total quality required a high level of motivation among employees, and therefore pushed strongly towards a reappraisal of more general motivational policies.

There has been considerable discussion of how extensive such shifts in managerial thinking have been and of how important they are for the actual practice of employment relationships. Quite apart from the residual power of local trade union organization, which made any full-scale implementation of such policies difficult to achieve (Stewart and Garrahan 1995), a number of commentators have emphasized the relatively short-term nature of British managers' perspectives whether with respect to investment, labour relations, or workforce policies (Edwards 1987; Hakim 1990). The introduction of thoroughgoing policies of

---

[1] For the reaction of German employers, see e.g. Jacobi and Hassel 1995.

individual evaluation and merit pay require a very substantial commitment of organizational resources to administer the system in a way that is likely to be seen as impartial. The objective of individualization is to increase the commitment of employees to their organizations. However, if there is little confidence in procedures, it may be seen to involve little more than favouritism, creating a sense of injustice and greater distrust of management. Even where there is a will to introduce change, there may be substantial problems in achieving a coherent relationship between different human resource management policies (Storey 1992, 1995). For instance, individualized pay systems may make it less likely that employees will happily participate in collective activities such as quality circles.

Finally, such policies are likely to be successful primarily where there is a relatively stable environment in which longer-term career planning is feasible and there is a reasonable assurance that assessed merit will be rewarded in terms of enhanced promotion opportunities. The economic environment in Britain in the 1980s and early 1990s was not favourable in this respect. The rapid alternation of recession, spectacular growth, and renewed recession could only generate a high degree of volatility in organizational plans. It was a period marked by the upheavals of large-scale mergers and by the slashing of traditional career routes as a result of the growing popularity of policies of 'delayering', which reduced the ranks of middle management. As the organizational pyramid flattened, it became decreasingly evident how management would be able to reward performance through promotion.

There is still a dearth of satisfactory evidence about the growth and prevalence of human resource management policies. The studies available suggest that employers have shown a greater interest in some components than in others. There appears to have been a substantial increase in forms of internal communication, whereas there is less sign of any rapid growth in organizational innovations such as the quality circle (Hill 1991*a*; Millward 1994: ch. 5; Wood and Albanese 1995). Moreover, such studies have only examined the formal commitment of employers to such policies, and very much less is known about how they actually impact upon employees' experience of work and the employment relationship. Of particular concern here is whether they are best viewed in their own terms as ways of increasing the dialogue between management and individual employees, improving the opportunities for employee involvement, and relating more adequately reward to individual performance or rather as an alternative mode of control of work performance that may be even more constraining in its effects than the systems it supplants. If the latter were the case, such policies could be expected to lead to marked pressures for an intensification of work rather than to any liberation from control.[2]

---

[2] A similar point has been made by sceptics about the impact of lean production on the quality of work experience (Berggren 1993).

Flexibility and the Fragmentation of Contractual Statuses

While new managerial policies designed to increase the commitment of the work-force might have significant advantages in terms of work effort and quality, they have the drawback of turning labour into virtually a fixed cost. As their pro-ponents stressed, such policies were unlikely to achieve their objectives unless employers could guarantee high levels of job security and indeed opportunities for longer-term career advancement. Yet one of the features of the changed eco-nomic situation in the decades after the oil crises of the 1970s that has been most regularly emphasized by analysts is the increased volatility of the product market and the consequent pressures on employers to adapt their costs more rapidly to fluctuations in demand. It was the attempt to assess how employers were responding to this dilemma that led to the extended and controversial lit-erature on the growth of the flexible firm.

The argument that gained greatest prominence in Britain was Atkinson's (1984, 1985, 1986) thesis that employers were meeting new market conditions by divid-ing their workforces into two distinct segments—a core and a periphery—that were regulated by very different employment conditions. The core consisted of more highly skilled workers, who benefited from high pay, relatively high job security, and generally advantageous conditions of employment. The drive to increase flexibility among these employees would largely take the form of increased functional flexibility, through polyvalence and job rotation. However, these priv-ileged conditions allocated to the core were made possible by the employment of a peripheral workforce, which had poor terms of employment and, above all, could be easily disposed of in times of economic difficulty. These were employees on non-standard contracts, in particular part-time employees, temporary workers, and the self-employed. According to one estimate, this peripheral workforce had come to represent as much as a third of the overall workforce (Hakim 1987). The key trend implied, then, by this perspective was one of the polarization of employment relations.

There has been a sustained debate about how far British employers actually came to adopt the type of strategic workforce policies outlined in the model. A major survey of employers that used employees on non-standard contracts indicated that only just over a third (35 per cent) of such employers were 'guided by some sort of manpower strategy or plan', and of these only 11 per cent operated a policy which differentiated a core and peripheral workforce (Hakim 1990; McGregor and Sproull 1991; see also Hunter and McGuiness 1991). While this may be deploying a rather demanding conception of 'strat-egy' (Procter *et al.* 1994), it scarcely suggests a tidal change in employer pol-icies. Nonetheless, there was a marked change over time in the proportion of the workforce that were employed on regular full-time contracts and this posed questions about whether there was growing inequality in conditions of employment.

Particularly significant was the expansion of part-time work, which was integrally linked to the changing gender composition of the workforce. It was fuelled by the marked increase in the labour market participation of women in the 1980s, predominantly as a result of middle-class women shortening the period they remained out of the labour force while having children (Harrop and Moss 1995). The proportion of women working full time (approximately a third of women of working age) remained very stable between the 1950s and 1990s. The expansion of women's work has been virtually entirely an expansion of part-time work. The number of female part-time jobs rose from 3.3 million in 1971 to 5.0 million in 1995 (Hakim 1996a: 63). Over the same period the proportion of employed women who were in part-time jobs rose from 38 per cent to 44 per cent in 1995.

The explanation for this increase in the prevalence of part-time contracts has been the subject of considerable controversy (Tam 1997). Some have seen it as a development that reflects more or less deliberate employer policies to obtain a lower-paid workforce. As part-time work provides insufficient pay to support a family, it leads to the recruitment primarily of female workers. It thereby produces the basis for the gender segregation of the workforce, with women primarily in female-dominated and men in male-dominated jobs. This in turn makes it difficult to establish pay comparisons and helps to contribute to the low pay of women's jobs irrespective of skill, education, and job experience. An alternative account sees part-time work as primarily benefiting employers in terms of the flexibility it provides with respect to the hours of work or the ability to dismiss employees. Part-time workers, it is suggested, will be more willing to vary their hours of work, they can be recruited to cover peak workloads, and they can be more easily dismissed in periods of economic difficulty because of their limited coverage by employment protection legislation (Beechey and Perkins 1987).

In contrast to such 'employer-centred' perspectives, it has been argued that a powerful factor behind the growth of part-time work has been women's own preferences about working hours (Hakim 1996a). In a culture where women's identity is still tightly bound up with their domestic roles, in particular with their responsibilities for child-rearing, it is suggested that they are concerned to find jobs with working hours that make it possible to reconcile their employment and family lives. Employers create part-time jobs as a way of attracting women into the workforce, while part-time female workers are likely to be inflexible rather than supple in the working hours they are prepared to accept. Even among those that accept the importance of women's own orientations to work in accounting for the growth of part-time work, however, views can differ strongly about the degree of voluntarism in such 'choices'. While some see such trade-offs between employment and family life as reflecting value choices, others argue that they reflect the constraints created by the division of work in the family or the failure of the State to provide a reasonable level of childcare provision (Dex 1985, 1988).

However, whatever the factors that have driven the expansion of part-time work, there is general agreement that it has profound implications for the nature of employment relationships. In particular, it has been shown that part-time workers are severely disadvantaged in terms of both skill development and promotion opportunities (Tam 1997). Even those that distance themselves from the more dramatic scenario of labour market polarization into a core and peripheral workforce, recognize that part-time work is associated with significant relative deprivation in employment conditions and employment opportunities. Moreover, the disadvantages suffered by part-time workers may affect employment relations more generally. Part-time workers, either through inclination or constraint, are particularly unlikely to be union members. The expansion of part-time work, then, may have been one of the factors that accounts for the decline of the traditional systems of joint workplace regulation discussed in the previous section.

While part-time work has caught attention because of its numerical importance, the issue of the increased fragmentation of the contract of employment has become particularly salient with the growth of temporary work. By the spring of 1995, there were 1.5 million temporary workers in the UK (CSO 1996: 88). While the rate of growth had been very low in the later 1980s and early 1990s, there was a marked rise in this type of employment in the mid-1990s, with the figure for 1995 some 10 per cent higher than in the previous year. In contrast to the discussions of part-time work, there is much greater consensus that this is a change in employment relations largely designed to benefit employers. While there may be occupations in which highly qualified people can use temporary contracts to play the market, the emphasis has been on the very low-skill level, the poor employment conditions, and the acute job insecurity of such work. As such the employment of temporary workers can be seen as a way in which employers can fundamentally tip the balance of power in their favour, producing a relatively cowed and docile workforce. Not only may temporary workers themselves be too frightened to join unions, but their availability may also affect the solidarity and self-confidence of the permanent workforce (Geary 1992).

The expansion of non-standard employment contracts has been seen then as generating a distinct labour market segment, in which the employment relationship is very different from that in the traditional employment model based on full-time permanent work. However, our detailed knowledge of the employment conditions of these groups remains very imperfect. To begin with, those who propound the polarization thesis tend to view the different types of non-standard contract as effectively interchangeable in terms of their function for employers and the labour market disadvantages they involve. They are essentially seen as sharing a relatively low-skilled, low-paid, and highly vulnerable labour market position. Yet existing research provides no strong evidence, of a nationally representative type, for the cumulative nature of the employment disadvantages associated with the different types of non-standard contract. If the disadvantages are not cumulative, but rather take different forms from one type of non-standard contract to another, then this would raise serious doubts about a dualistic conception of the labour market in terms of core and periphery. Second, the notion of polarization implies that there is a progressive *widening* of the gap between the employment conditions of the core and the flexible workforce over time. However, to date, no serious evidence has been advanced to establish whether or not this is the case. The type of detailed evidence across time for the full range of employment conditions that are held to define core and periphery status has simply not been available.

The Growth of Job Insecurity

Perhaps the least debatable change in employer policies from the late 1970s was a much more extensive reduction of staffing levels and, frequently, the resort to large-scale redundancies. In part as consequence, one of the most fundamental changes in the labour market in Britain, as in most other Western societies, was the return of the threat of mass unemployment. For most of the 1950s and 1960s, the unemployment rate remained less than 2 per cent of the labour force. While unemployment showed a rising trend in the early 1970s, the steep increase came after the second 'oil shock' of 1978–9. By 1983 some 12.2 per cent of the working population was out of work and the figure remained over 10 per cent until 1988. Any belief that this was a one-off adaptation to the new conditions of international trade was dispelled by a second sharp recession in the early 1990s. The UK unemployment rate again reached 10 per cent in 1992 and 1993. Moreover, this time unemployment affected not only the declining areas of traditional manufacturing in North and Central England, but also the hitherto secure and wealthy areas of the South. By the 1990s, it had come to appear that high levels of unemployment were likely to be a recurrent feature of the labour market and that the type of job security that had prevailed in the early post-war decades was unlikely to return. Clearly, this represents a fundamental change in the *de facto* nature of employment relationships.[3]

A good deal is known from existing research about the groups most affected by unemployment (White 1983, 1991; Gallie *et al.* 1993). It is clear that vulnerability differed substantially between social groups. In contrast to the situation in most West European societies, in Britain men were more exposed to unemployment than women. This was consistently the case for the 1980s and early 1990s. The sharp rise in unemployment in the early 1990s primarily affected men. Whereas the female rate peaked at 7.6 per cent in 1993, the male unemployment rate rose to as high as 12.4 per cent (CSO 1996: 94). Even with the reduction of unemployment in the following years, the sex differential remained substantially greater in 1995 than it had been at the beginning of the decade. While the risk of unemployment was particularly high among young people (those aged 16–19), men were more likely to be unemployed than women *in every age category*. Similarly, with the exception of people under 20, men were more

---

[3] Some view this growth in insecurity as part of a wider tendency to destabilization in industrial societies (see e.g. Beck 1992).

likely to be long-term unemployed than women in every age category (CSO 1996: 94).

It is also clear from cross-sectional studies in the 1980s and early 1990s that the occupational categories that suffered most from unemployment were those with lower skills, in particular low-skilled manual workers. For instance, a national survey in 1992 (White *et al.* 1996) showed that the highest proportion of the unemployed for both men and women came from those in semi- and non-skilled work. This category accounted for 45 per cent of all unemployed men and 37 per cent of all unemployed women, although it constituted only 23 per cent of men and 19 per cent of women in employment.

There are, however, a number of major issues about vulnerability to unemployment and its implications for people's careers about which the evidence is still very inadequate. To begin with, relatively little is known about the cumulative experience of unemployment across people's work lives. While there was an indisputable rise in the risk of unemployment over time (especially among men), it is less clear whether this was accompanied by a more generalized growth in job insecurity across the wider workforce. In part, this is likely to depend upon the degree of segmentation of the employment structure. The experience of unemployment can take very different forms when viewed from a work-history perspective. At one extreme, it may be very widely dispersed across the workforce, with a substantial proportion of employees experiencing unemployment at some stage in their work lives but very few people experiencing repeated unemployment or spending large amounts of time unemployed. At the other extreme, unemployment may be heavily concentrated on a small minority that is trapped in a sector of the labour market where jobs are highly volatile and repeated unemployment endemic. The experiences of this minority may account for a large proportion of the total experience of unemployment. Since the great majority of the studies of the unemployed have used either cross-sectional data sources or panel studies lasting for relatively short periods of time, it has simply not been possible to establish a good picture of these longer-term career experiences of unemployment.

Second, it is still far from clear whether there has been a significant change over time in the vulnerability to unemployment of different types of employee. There has been some suggestion that a distinctive characteristic of the early 1990s was that unemployment increasingly cut its swathe not only through the ranks of manual workers, but also through the clerical workforce and through middle management. If this were the case, it would suggest a decline in traditional class differences in the employment relationship. But the evidence to date is simply insufficient to establish the argument one way or the other. Even if it were the case that a higher proportion of the unemployed had come to be drawn from those in more-skilled occupational positions, it would still need to be seen whether this just reflected the increased share of such occupations in the workforce or a real increase in their relative vulnerability.

Finally, little is known about the longer-term impact of an experience of unemployment. Does unemployment have a relatively transient effect on people's work lives? Or does the loss of a job leave a permanent mark on their careers, condemning them to jobs with poorer conditions of employment than those of similarly qualified people who had managed to maintain continuity of employment? How does the experience of unemployment affect people's commitment to employment and the importance they attach to different features of a job? How does it affect the security that they feel in the jobs they manage to get? While vital for a proper assessment of the wider implications of unemployment, such questions still remain largely uncharted terrain.

THE RESTRUCTURING OF EMPLOYMENT AND THE EXPERIENCE OF WORK

There are grounds, then, for thinking that employers have confronted the changed economic circumstances of the 1980s and 1990s by the widespread introduction of new technologies, by experiments with new forms of employee-relations policies, by relying to a greater extent on non-standard employment contracts, and by substantial reductions in the numbers of their employees. While there is much disagreement about how such changes have affected the nature of work and the employment relationship, few doubt that the implications have been considerable. This suggests that they may also have had a marked effect on employees' subjective experience of work. But, if so, what exactly has been their impact?

This raises the issues both of the subjective quality of working life and of work motivation. First, have such changes been creating a work environment in which people are finding greater involvement and satisfaction in their work or has increased technical efficiency been achieved at the expense of lower job involvement and greater stress? Given what is known about the critical importance of work experiences for self-esteem and non-work life (Kohn and Schooler 1983), this has important implications for the overall quality of people's lives. Second, how have the changes in the work environment affected employees' work motivation: the work effort they are willing to put in, the reliability of their attendance, and the commitment that they feel to the organizations for which they work. It has been widely suggested that the move towards more highly skilled, quality-conscious, work systems puts a particularly high premium on worker commitment (Lincoln and Kalleberg 1990). But have the changes over the last decade had the effect of reinforcing or undercutting such commitment?

In the theoretical literature, the arguments about the factors underlying subjective well-being and high levels of work motivation are very similar indeed. This reflected the fact that it was widely (if controversially) thought that employees' satisfaction with their work was a crucial factor in work motivation. There is, moreover, a remarkable degree of convergence about the aspects of work and the employment relationship that are most influential in these respects. Very broadly, it is possible to distinguish three 'classic' theoretical approaches to the determinants of subjective well-being in work, although each has a wide range of variants. The first underlines the importance of the work task, in particular the extent to which it avoids fragmented, repetitive work and provides employees with scope to use their initiative. The second locates the key to worker satisfaction in the social supportiveness of the work environment, embracing both relations between colleagues and between individuals and their superiors. The third has been concerned with the implications of the degree of participation in work, the extent to which employees were involved in the decisions that affect their everyday working lives.

From the viewpoint of the 1990s, a notable feature of each of these research traditions is their virtually exclusive emphasis on intra-organizational factors. This is perhaps understandable given that they were initially developed in a period of full employment, when job security was regarded as largely unproblematic. However, in the contemporary era, it is difficult to ignore the possibility that labour market experience may also have major implications both for the subjective quality of working life and for work motivation. To the earlier three perspectives, then, it is necessary to add a fourth, focusing on the implications of job insecurity.

The Nature of the Work Task

Perhaps the most influential approach to subjective well-being at work has focused on the characteristics of the job task. This has underlined the importance of the job task itself for people's involvement in their work and their ability to develop themselves through their work. This was the approach that informed a good deal of qualitative research in the Marxian tradition. It also led to the extensive body of 'job characteristics' research, which sought to separate out a number of different task dimensions that were likely to effect subjective well-being (Caplan *et al.* 1975; Hackman and Oldham 1975, 1980; Sims *et al.* 1976). Several of these task characteristics have consistently been found to be associated with job satisfaction: for instance, variety, pace, opportunities for use of initiative, autonomy, feedback, and the extent to which the job is a recognizable whole, a clear and identifiable piece of work (Warr 1987). It is an approach that has led to a strong emphasis on the possibilities for job redesign and task enrichment, and it was an important influence on the Quality of Working Life Movement that emerged in the 1960s and 1970s.

One task characteristic that has become increasingly central in recent years to research on subjective well-being is the degree of discretion or control allowed to the individual. Lack of personal control over the pace of work had been highlighted in the early studies of the negative effects of work on the assembly line (Walker and Guest 1952; Chinoy 1955). Subsequent studies were to repeatedly emphasize the importance of control over methods of work (Patchen 1970). Task discretion also has been an important focus for social psychologists of work. For instance, in their major study *Work and Personality*, Kohn and Schooler (1983: 2) argued for the central importance of 'occupational self-direction' or the 'use of initiative, thought and independent judgement at work'. They concluded that 'occupational self-direction has the most potent and widespread psychological effects of all the occupational conditions we have examined' (ibid.: 81). Self-direction in the work task has been found to be a key variable not only for job satisfaction but also for accounting for vulnerability to psychological distress (Wall and Clegg 1981; Parkes 1982; Karasek and Theorell 1990).

Of the changes that have been considered earlier, it is perhaps the spread of new technologies that could be expected to have the strongest impact on employee work involvement and motivation through changing the nature of job tasks. While theories of automation were in general based on a less sophisticated conception of job characteristics than was developed in research on subjective well-being, their emphasis on the implications of new technologies for skill levels is clearly of major potential importance. However, no a priori conclusions can be drawn about the implications of technological change for employee attitudes. The predictions of these different theories about trends in skill were highly contradictory. Depending upon whether one accepts the upskilling or deskilling argument, it would be necessary to draw quite different conclusions about the likely effects of technical change on employee attitudes. If the upskilling argument were correct, it should lead to heightened job involvement and a closer sense of identification with the employer. Alternatively, if new technologies are deskilling the workforce, it should accentuate the sense of alienation at work and lead to a more conflictual perception of employer–employee relations.

Task characteristics may also be affected by broader managerial philosophies. The adoption of new managerial policies of human resource management may be important in this respect. At one level, a concern for human resources implies investment in training and in the skills of employees. Equally important, to the extent that it involves the decentralization of decision-making to employees, whether as individuals or through forms of team decision-making, it could be expected to encourage the growth of task discretion and autonomy which has been shown to be so important in employee satisfaction and motivation. However, as has been seen, there are still substantial doubts about how far human resource management policies have spread within industry and there are also doubts about how far the emphasis on decentralization of decision-making is rhetoric or reality. Moreover, even if they have been implemented in a relatively full way, there would not be unanimity about their implications for worker satisfaction. For those who view such policies as primarily a means for enhancing managerial control, it could well be the case that they lead to an intensification of work and hence to more negative employee attitudes.

Social Integration in the Workplace

A second well-established tradition of research into subjective well-being might be termed, in its most general form, the theory of social support. This lay at the very foundation of industrial sociology as a subject. It came to prominence with the pioneering research, carried out between 1927 and 1932, at the Hawthorne plant of the Western Electric Company in Chicago (Mayo 1932, 1949; Roethlisberger and Dickson 1939). The central conclusion of this research was that people, even in their work behaviour, were ultimately 'social' not 'economic' actors. Their most fundamental need was to belong to a relatively cohesive microcommunity, and their status within this community was more important in determining their behaviour than any type of economic incentive. In particular, the critical factor for worker satisfaction was the degree of social integration into a workplace community.

With respect to the future development of research, this perspective focused attention on two characteristics of the work environment: the importance of the social cohesiveness of the work group and the significance of the role of first-line supervision. Subsequent research certainly confirmed the importance of the quality of social relationships for employees' experiences and behaviour. The central importance of cohesive work-group organization for satisfaction and work motivation was one of the major conclusions of the research of the British sociotechnical school in the 1940s and 1950s (for instance, Trist *et al.* 1963). The significance of social support has also been well documented through the work of J. S. House (1981).

Theories of automation again differ sharply in their predictions about social integration. At one extreme, automation is viewed as encouraging teamwork which should provide employees with a much stronger sense of community in the workplace (for instance, Blauner 1964), at the other it is seen as leading to social isolation in work (Naville 1963; Chadwick Jones 1969). Indeed, the variety of research findings has been so great in this respect that it is difficult to escape the conclusion that 'team integration' or 'social isolation' at work have little at all to do with the necessities of technology but are better seen as an aspect of work organization which derives from wider managerial philosophies of organization.

In this respect, it is above all the nature of change in management's employee-relations policies that is likely to be crucial. The early research on social integration had, at least in the USA, a fairly direct impact on management thinking. It led to what is commonly termed the 'Human Relations' school, which placed emphasis on the training of supervisors in the skills needed to win employee commitment. However, in the longer term, it became clear that the results of this approach were very limited. Attempts to change the style of first-line supervisors quickly collapsed in organizations that, at a higher level, remained largely hierarchal and even authoritarian in approach (Landsberger 1958; Tannenbaum *et al.* 1974).

The growth of human resource management could be seen, to some degree, as an attempt to resurrect some of the key principles of the Human Relations tradition, but in a way which involved the whole structure of the organization. Its objective was not only to build up the skill base, but also to develop direct ties between employees and the organization that would ensure higher levels of commitment to the objectives of the firm. With the growth of 'human resource management', the role of the supervisor has once more become a prime focus of attention. The active transmission of the organizational ethos through the supervisor is viewed as one of the keys to effective management. At the same time there is a greater emphasis on the role of the supervisor in maintaining the flow of communication between higher management and employees and in providing support for employees.

Yet, while the overt objectives of such policies would certainly lead to the expectation that they would increase support and social integration, they clearly also had the potential of having a quite different effect. The direct relationship between supervisor and employee has long been recognized as a major source of tension, particularly when associated with tight systems of control. It is certainly the case that, with the new policies, the role of the supervisor is conceived in a different way. It was now to involve responsibility for making a retrospective assessment of employee performance, rather than the direct policing of work. But the new emphasis on individual assessment and merit-linked pay could still provide ample fuel for friction, with disagreements about merit assessments leading to the belief that the restored powers of supervisors were being exercised on the basis of favouritism. Whether the new managerial policies achieve their objectives of heightening commitment by generating a more supportive atmosphere and closer links between supervisors and employees or whether they lead to a sense of injustice and more hostile relations on the shop floor is a question that can only be resolved empirically.

Participation

The third broad approach to the subjective quality of working life has focused not on the immediate work task but on the extent to which the employing organization permits employee participation in decision-making. The interest in participation has come from a number of directions. One was from personality theorists that have postulated the growing importance of aspirations for self-realization.[4] A culturally contingent version of this argument was presented by Chris Argyris (1957), who suggested that, at least in Western societies, individuals were socialized into a basic pattern of personality growth in which maturation was associated with a person becoming increasingly independent from others and able to control his/her own behaviour. These generalized expectations for self-control could

---

[4] The three classic texts were Argyris (1957), McGregor (1960), and Likert (1961). For an assessment of their influence on the practice of management, see Lawler (1989).

only lead to frustration and withdrawal in the sphere of work unless organizations were designed in a way that allowed people to participate meaningfully in decisions.[5]

An alternative sociological perspective suggested that there was a pressure towards increased participation deriving from the inconsistency between general citizenship norms, with their assumption of equality, and the persistence of relations based on subordination at work. The classic presentation of this argument was made by T. H. Marshall (1964). For Marshall the institutionalization of legal and political citizenship generated a conception of equality and equal social worth that led naturally to a challenge to the legitimacy of economic inequalities and forms of organization based on traditional hierarchical principles. The principle of citizenship is likely to prove contagious. Marshall saw this affecting work life primarily through the extension of the role of trade unions. 'Thus the acceptance of collective bargaining', he wrote, 'represented the transfer of an important process from the political to the civil sphere of citizenship . . . Trade unionism has, therefore, created a secondary system of industrial citizenship parallel with and supplementary to the system of political citizenship' (Marshall 1964: 103–4).

Whatever the precise theoretical grounds for believing that participation is likely to be important, the research evidence certainly seems to suggest that it does have significant consequences for the subjective quality of work experience (Blumberg 1968; Gill *et al.* 1993; Martin 1994; Frohlich and Pekruhl 1996). There is a striking consistency between the results of studies using quite different research methods: laboratory experiments on small groups (White and Lippitt 1960), field experiments in which organizational change is introduced in different ways among different groups of employees (Coch and French 1948; Morse and Reimer 1956), and larger-scale cross-sectional surveys (Patchen 1970). In contrast, the evidence is much less clear about whether or not increased participation increases work motivation and leads to better job performance (Strauss 1963).

The implications of the recent changes for the extent to which employees can participate in decisions in their organizations is far from clear. It has been seen that the restructuring of employment has seen the decline of the major traditional form of participation—joint regulation through collective bargaining. However, arguably there has been an increase in new forms of direct participation, for instance through involvement in quality circles or through direct meetings with management to discuss developments in the organization. Have the effects for employee involvement of the decline of one form of participation been compensated by the rise of another? It is again difficult to have an a priori view on

---

[5] Other psychological theories of the mechanisms underlying the efficacy of participative practices have also emphasized the importance of group pressures for goal commitment, the increased clarity participation brings between performance and outcome, the greater legitimacy of self-selected goals, and the implications of participation for self-efficacy and the reduction of anxiety (for an overview see Erez 1993).

this. In the first place, there is still very inadequate evidence about the prevalence and efficacy of the new forms of participation. Second, the literature is remarkably uninformative about the relative significance for employee satisfaction of specific modes of participation. Are such very different forms of participation equally effective in giving employees a sense that they have a say in the way their organizations are run? It is again only empirical enquiry that can begin to clarify this.

Job Insecurity

It was noted that one potentially important set of factors received remarkably little treatment in the literature—namely the implications of job stability or insecurity. In part, this reflects the fact that these traditions of research confined their attention very much to the internal nature of the workplace and neglected the interconnections between the dynamics of the workplace and the labour market (Gallie 1988). However, with the return of periods of unemployment in recent decades and the increase in non-standard employment contracts, it is difficult to continue to regard job insecurity as unproblematic. Research evidence on unemployment has shown just how severe its consequences are for psychological well-being (Warr 1987; Whelan *et al.* 1991; Gallie *et al.* 1994*a*). It would be strange indeed if the fear of unemployment had no implications for the way in which employees experienced their work.

However, the significance of the rise of unemployment for job insecurity in the wider workforce is likely to vary a good deal depending on the pattern of unemployment. If unemployment has fallen primarily upon a relatively marginalized minority of the workforce, it may have made very little difference to the way in which the bulk of the employees perceive their own future. If, on the other hand, unemployment has been a wider experience, then it is more plausible that it has come to affect attitudes to the employment relationship more generally. Even if it is the case that job insecurity is fairly pervasive, there still could be rather different views about its precise effects. It might have made people more defensive about existing working arrangements and hostile to technical innovation which is seen as the harbinger of future redundancies. Alternatively, it may have made people more aware of the benefits of having a job, less demanding in their expectations about the quality of work, and more attached to the organizations that continued to employ them.

Work Values, Gender, and Subjective Well-Being

Finally, it should be noted that all of the perspectives discussed above assume that people will generally react in a broadly similar way to particular features of their work or labour market situation. But a number of researchers have suggested that the impact of objective characteristics of the work situation could well be mediated by differences in work values and job preferences both between individuals and between social groups. In the social-psychological literature, this was evident in the development of theories of 'person-environment fit', which emphasized that individuals differed in the factors that were important to them and that it was the lack of valued work characteristics that would lead to psychological demoralization (Caplan *et al.* 1975; French *et al.* 1982). In the sociological literature, the emphasis has been rather on the way that 'orientations to work' could vary across time and between different categories of employee.

At the most general level, there has been periodic concern about the possibility of a decline in the cultural importance of the work ethic. This has been an issue not only in Britain, but also in the USA and in France (Rousselet 1974; Linhart 1981; Barbash *et al.* 1983; Rose 1985). An alternative argument is that there has been a growth of instrumental orientations to work, emphasizing primarily its financial aspects, reflecting a decline in its centrality to people's identity and the increased salience of family and leisure values (Goldthorpe *et al.* 1968, 1969). The presumed changes in the importance attached to work are attributed *inter alia* to the increased opportunities that have become available for self-realization in the sphere of leisure, the marked improvement in housing conditions, and the growth of the 'companionate' marriage.

It has to be said that the type of evidence that would be needed to assess whether such long-term cultural changes were occurring has simply not been available; for the greater part, these theses must be regarded as purely speculative. Moreover these arguments were developed in a historical period when people had just lived through two decades of economic growth, with relatively full employment. Working-class families were experiencing for the first time some degree of economic affluence and it was a period in which there was still a relatively high level of marital stability. It must be open to question whether more recent decades, with the return of periods of high unemployment and rising divorce rates, have provided conditions that were likely to be favourable to the predicted trends.

However, if such developments were occurring, it is clear that they would be important for views about the implications of changes in employer policies for workers' well-being and motivation. More sophisticated motivational policies may produce little net effect if there is a significant decline in the value attached to work due to influences entirely outside the workplace. By the same token, policies that reduce the intrinsic interest of work may not lead to higher levels of discontent if people are coming to place less emphasis on work as a sphere of self-realization.

Quite apart from the issue of whether or not there has been a long-term decline in the centrality of work in people's lives, there may be important differences between social groups in commitment to employment and in what is valued in a job. Much of the initial interest in work orientations in British research centred on the possible differences between social classes and occupational groups (Goldthorpe *et al.* 1968, 1969; Bulmer 1975; Blackburn and Mann 1979;

Brown *et al.* 1983). More recently, the focus has been upon whether or not there are important differences in work values between the sexes (Martin and Roberts 1984; Dex 1988; Hakim 1996*a*). Do men and women respond to work conditions in broadly the same way or are they likely to react very differently, due to the legacy of traditional role conventions in which men were regarded as the breadwinners and women as family makers? If there are important differences in work values, the type of work and employment conditions that would lead to satisfaction and high levels of motivation for women might be quite different from those that were influential for men. Alternatively, if women's identities were less strongly invested in the work situation, their attitudes may have been less affected by change, whether this involved an improvement or a deterioration in employment conditions.

The evidence on gender differences in work attitudes remains remarkably slight. The most substantial study on women's attitudes to employment—the Women and Employment Survey (Martin and Roberts 1984)—provided a rich and nationally representative picture of women's attitudes to work. But in the absence of comparable data for men, it was difficult to judge how far women's attitudes were distinctive. The inadequacy of the data is made more acute by the fact that there are reasonable grounds for suspecting that women's attitudes to employment may have been changing rapidly in the last two decades, given such a marked change in their level of labour market participation. Such evidence as existed for the mid-1980s suggested that women were indeed less committed to employment than men. But was this still the case in the rather different economic and social environment of the 1990s?

THE EMPLOYMENT IN BRITAIN RESEARCH PROGRAMME

It has been seen that, while there has been very considerable speculation about the nature of changes in work over the last decade, well-grounded knowledge about how the employment relationship and people's experience of work have been affected remains very uneven and often quite inadequate. The clearest example of where a systematic effort has been made to examine trends over time is with respect to industrial relations. The Workplace Industrial Relations Surveys provide an invaluable basis for assessing in a systematic way changes in representative institutions and in several other aspects of managerial policy (Daniel and Millward 1983; Millward and Stevens 1986; Millward *et al.* 1992). But there has been no comparable evidence for examining the impact of technical and organizational change on the lives of employees. Qualitative case studies, focusing on workplaces where new technologies or innovative management policies have been adopted, have provided a rich basis for hypotheses about possible trends. But, while providing a fine-grained picture of what such developments consist of in particular instances, it is difficult to know whether the conclusions reached are more generally applicable or reflect conditions that are specific to the particular examples taken.

The Employment in Britain research programme was designed to fill this gap, by providing a more representative picture of the nature of work and the employment relationship in the early 1990s. The overall programme consisted of two large-scale national surveys of people in the labour market. The first involved a survey of people in employment, resulting in an achieved sample of 3,869 people who were either employed or self-employed. The second survey, involving interviews with 1,003 individuals, was of people who were officially registered as unemployed. A response of 72 per cent was achieved for those in work and 71 per cent for the unemployed. Both samples focused on people aged 20 to 60. The technical details of the sampling and weighting procedures are provided in the Technical Appendix. The greater part of the analysis in this volume is based upon the first survey and, more specifically, upon the sub-sample of 3,458 employees (3,469 weighted). However, for certain analyses the combined samples have been used and this is indicated at the relevant point in the text.

The design of the study privileges the perceptions and experiences of employees. This has both advantages and limitations. The major limitation is that there are aspects of organizational functioning and change that may not be easily visible in the workplace. Insofar as they provide a description of institutional practices, such data must be treated with a degree of caution. In compensation, they have the advantage that they provide the most direct and reliable information on the lived experiences of those who have been subject to change. And it is this that is the focal interest of the current study. If there are elaborate managerial policies, with respect to the organization and regulation of work, that employees are unaware of, then there must be doubts about whether they are likely to be very efficacious. If we wish to know whether conditions of employment have improved or deteriorated, the direct knowledge of employees about the changes they have experienced remains by far our surest guide.

While there are good grounds for thinking that the quality of the data provides a rich basis for describing the experience of work and the employment relationship as they were in the early 1990s, our concern with the extent of change over time inevitably poses much more difficult methodological problems. Given the limited evidence available for the past, the assessment of change has to proceed on a narrower and more fragile basis. We have tried three approaches. First, we built into the survey a range of questions that provided comparisons with earlier surveys carried out in the 1980s. In particular, we have included questions that featured in the Class in Modern Britain Survey conducted in 1984 (Marshall *et al.* 1988), and in the Social Change and Economic Life comparative labour market surveys carried out in 1986 and 1987 (Gallie *et al.* 1993). Second, we have collected extensive work histories for each individual, which trace their labour market experiences in detail from the time that they first left full-time education. Finally, we asked people for their own perceptions of how their situations had changed over relatively short periods of time, for instance over the previous two or five years depending on the particular issue. While none of these methods is unproblematic, we have sought to draw a picture which relies on a combination of methods. Where there is strong convergence between different types of data, one can feel more assured that the trends depicted reflect the reality of change in people's experiences.

The aim of our analysis is not only to describe recent changes and difference in the experience of work, but also to explain and interpret them. This involves testing a wide range of predictions drawn from the literature which has been reviewed. While we would not wish to draw strong causal inferences from the evidence, we believe that it is possible in many cases to distinguish between those explanations which are relatively consistent and those which are inconsistent with that evidence. In trying to assess the conflicting arguments about the factors that have influenced the patterns that emerge, considerable care has to be taken to control for other potentially confounding factors. The analyses we present are based upon extensive statistical modelling, designed to assess the robustness of the findings and to check out other possible explanations. In the interests of readability, these analyses are presented only in summary form in the text of the chapters. However, a somewhat fuller account of the forms of data analysis that have been used can be found in the Technical Appendix, which also explains the conventions which have been adopted in presenting tables of statistical findings.

STRUCTURE OF THE BOOK

The structure of the book divides into two main parts. The first part (Chapters 2 to 6) is primarily concerned with the nature of, and patterns of change in, work and the employment relationship. It focuses upon the structural factors that have been thought to be of central importance for employees' experiences: the characteristics of the work task, the way work performance is controlled, the participation of employees in wider organizational decision-making, and job security. The second part (Chapters 7 to 10) seeks to assess the implications of the principal changes that are revealed for the subjective quality of working life and for work motivation.

In Chapter 2 the focus is on the nature of the work task. It examines the trends in skill, assessing the conflicting arguments about the upskilling and deskilling. It then considers the relationship between skill experiences and other features of the job, such as task discretion, the variety of the work, and the opportunities for self-development through work. Finally, it turns to the impact of automated and computerized technologies on the pattern of change in skill and task discretion. Chapter 3 turns to the issue of new types of managerial policy, in particular human resource management or performance-management policies, designed to motivate and to increase the level of social integration of employees. It considers the implications of such policies in the context of the long-standing sociological debate about changing forms of organizational control. It assesses how far individualized performance-management systems have been introduced and takes an initial look at how they affect employees in comparison to other systems of control.

In Chapter 4 the discussion turns to the issue of employee participation in decision-making. It is concerned both with direct participation, examining how far employees feel they can influence decisions about work organization and whether this has changed over time, and with indirect participation through trade union representation or consultative works councils. It examines the relative significance of established forms of union regulation and new forms of 'direct' participation for employees' sense of involvement in decision-making, for attitudes to technical change, and for perceptions of management.

In Chapter 5 the discussion moves away from experiences directly related to the work process to the issue of job insecurity. It begins by examining changes over time in employment stability and in the level of job insecurity across the workforce as a whole. It examines the personal and structural factors that are related to unemployment experience and job insecurity. Finally, it turns to the issue of how the personal experience of unemployment affects people's longer-term career trajectories and the types of job they are able to find. In Chapter 6 the discussion of job insecurity is continued in the context of arguments about the polarization of the workforce between a relatively secure core of skilled and privileged workers, and a highly vulnerable periphery of people confined to low-skilled jobs with poor employment conditions and acute job insecurity. It focuses upon the employment conditions of employees with non-standard employment contracts, in particular part-time workers and workers on different types of temporary contract.

The second part of the book turns from the issue of the nature of structural changes in work and the employment relationship to employees' experience of the quality of work and their work motivation. Chapter 7 starts the discussion by taking up the issue of whether there has been a decline in employment commitment and the work ethic, such that employees are likely to be less concerned about the quality of their conditions of work. It considers the nature of gender differences in work values, and the way these have changed over time, and also examines whether unemployment affects employment commitment and job preferences. The issue of people's subjective quality of life at work is addressed in Chapter 8. It takes three different indicators. The first is the degree of involvement that people have in their work. It seeks to go beyond the conventional interest in job satisfaction, which may merely reflect contentment with an undemanding job, to a concern with a more positive involvement in work. The second is the implications of changes in the work environment for work strain, the fatigue, and mental tension involved in work. Finally, it considers the relationship between work experiences and psychological distress.

The final two chapters take up the issue of work motivation. Chapter 9 examines the influences on people's commitment to the organizations that they work for. In particular, it continues the discussion of the significance of new management policies, looking at whether 'human resource management' or 'performance-management' policies are in practice associated with higher levels of commitment on the part of employees. It also considers whether there are differences between the factors that affect organizational commitment in the private sector on the one hand and in the 'social sector' (for instance, the education, health, and welfare services) on the other. Finally, Chapter 10 seeks to assess the argument that employee motivation has significant implications for labour costs and employee performance. It examines the relationship between organizational commitment and three measures of employee work behaviour that are likely to be of relevance for their productivity: how frequently they are absent from their organizations, how they assess their job performance relative to others, and their intentions about staying in the organization.

2

Skill and the Quality of Work

A wide range of theories have placed changes in the level and nature of skill at the centre of their accounts of the dynamics of wider changes in employment relations. It was seen in the last chapter that three main theses have been advanced. The first argues that there has been a long-term trend of rising skill levels as societies have become more industrialized. The second puts forward the directly contrary view that the long-term trend has been towards an erosion of skill levels, with a concomitant degradation of the experience of work. Finally, it has been suggested that experiences of skill change may have differed very substantially in different segments of the labour market, leading to a growing polarization in employment experiences.

The concern about trends in skill derives in good part from the fact that skill levels are seen as determining much of the employee's experience of the work task and hence the quality of their work life. There are thought to be very close links between skill level, the discretion that employees can exercise in the work task, and the intrinsic quality of work. Writers that have argued that there is a long-term trend towards the upskilling of the workforce have also maintained that employers would increasingly decentralize decision-making to their employees, thereby sharply reducing some of the traditional disadvantages of manual work (Kerr *et al.* 1960; Blauner 1964). Those that have suggested that the central trend is towards deskilling, on the other hand, have underlined the way in which employees would lose control over the work process and be subjected to forms of work that would be increasingly routine, monotonous, and meaningless (Braverman 1974; Crompton and Jones 1984).

How, in practice, have the processes of economic restructuring in Britain since the 1980s affected skill levels and the experience of the work task? The late 1980s and the early 1990s provide an ideal test case for theories of skill change. It was a period in which the types of factors that were viewed as critical in affecting skill were undergoing particularly rapid change. Theorists of upskilling placed a strong emphasis on the role of advanced technology, especially automation, in enhancing skill requirements. From the 1980s, with the advent of microprocessor technology, there was a remarkable increase in the use of automated processes across the employment structure. In terms of these theories, it is clearly a period in which there should have been very substantial upskilling. As higher skill was seen as integrally linked to the decentralization of responsibility, it should also have been a period in which employees came to have far greater discretion over their work.

Theorists of deskilling, in contrast, had been highly critical of the view that technological development had an independent influence on skill or work organization. Instead they placed primary emphasis on the strategies of employers seeking to enhance their control of the labour force. A common criticism of this argument was that it greatly overestimated the capacity of employers to impose their preferred models of work organization; in reality their choices were heavily constrained by the power of organized labour. Yet in the second half of the 1980s, there were grounds for thinking that a significant shift was taking place in the balance of power in the workplace, making it possible for employers to carry through their policies in a relatively unfettered way. There had been a marked decline of the unions' organizational power (Millward *et al.* 1992; Millward 1994). At the same time, the public sector, which had provided particularly strong institutional safeguards for employees, was sharply reduced by successive waves of privatization. Finally, with the second major recession of 1990–2, it had become clear that a high level of job insecurity was not to be temporary. Arguably, the creation of labour markets with endemic insecurity would lead to greater passivity on the part of the workforce and a greater willingness to accept employer dictates. There were grounds for thinking, then, that employers in the later 1980s had much greater freedom of action to implement the types of policies that they preferred than had been the case a decade earlier. If it is true that there is an inner logic to capitalist relations of production that encourages employers to deskill, then it should have revealed itself particularly clearly in the second half of the 1980s and the early 1990s.

It was a period, then, in which one would have expected quite major changes in skill and work organization whichever theoretical position was adopted. Yet, despite the central importance attributed to skill developments, there has been a notable absence of reliable evidence. In part this is because earlier studies failed to give a representative picture from the viewpoint of the employee, often generalizing from limited case-study evidence to the experience of the wider workforce. In part it is because of the lack of reliable evidence across time. Although the arguments were about the nature of trends, such trends have been typically assessed on the basis of evidence collected at only one point in historical time.

The key propositions about the relationship between skill development and the individual's experience of work were also lacking convincing representative evidence. For theorists of upskilling, the improvement in the overall nature of work was seen as virtually inherent in the nature of skill change. Instead of the trend towards more divided and fragmented tasks, there was a reversal of the division of labour in which the conception and execution of work once more became reintegrated. As the intellectual content of work returned, it followed that it would become more varied, meaningful, and interesting. Conversely for theorists of deskilling, the progressive atomization of work remorselessly undercut intrinsic work interest. At the same time, since deskilling strategies derived from a desire by employers to intensify the work process, they would tend to be associated with a general move to curb employee initiative and increase work pressure.

The most relevant research into the relationship between skill and the quality of the work task was by occupational psychologists. This lent considerable support to the view that skill levels are a major determinant of other job characteristics (for an overview, see Warr 1987). However, such studies were mainly small scale and their representativity difficult to assess.

Finally, much of the earlier discussion of changes in skill levels and their implications for the quality of work life was essentially a discussion of developments in the nature of men's work. The archetypal work settings were thought to be those of the craftworker, the assembly-line worker, and the worker in the vast integrated complexes of the chemical and oil industries. In essence, it has been a debate focused on the changing nature of male manual skills in the production industries. As such, it has largely ignored one of the central features of the changing employment structure in recent decades—namely the growth of women's work.

This raises the issue of whether men and women are likely to have had similar experiences of skill change. Given the high level of segregation in the employment structure, much of the growth of female work had been in very different sectors of the labour market from the traditional centres of male work. Women's employment has expanded above all in the service sector industries. It is far from clear that the underlying assumptions of the major theories of technological change can be easily extended to include service sector work, where the opportunities for mechanization are more limited. The very nature of much work in the services is radically different, involving primarily relationships with people and requiring social rather than manual skills. Further, the male and female sectors of employment stand in a very different relationship to the wider international economy. It was especially manufacturing that had received the sharp edge of the intensification of international competition and it is reasonable to suppose that this was the sector on which employer efforts to restructure work were most heavily concentrated. This leaves the possibility that the pattern of skill change may have been very different for men and for women, and that the decade may have seen a growing polarization of skill experiences by sex.

In this chapter, we will first address the issue of the general pattern of change in skill by examining a number of different measures of skill experience. The second part of the chapter turns to the implications of skill change for other aspects of the work task, such as the responsibility involved in work, the intrinsic interest of the work, the nature of working conditions, and the pressures involved in work. We then examine whether there is any evidence that polarization has been occurring along gender lines with respect to either skill or responsibility in work. Finally, we consider two proximate causes of the trends in skill: employers' response to the spread of computer technologies and the growing importance of social skills.

THE RISING DEMAND FOR SKILLS AND QUALIFICATIONS

The general trend from the mid-1980s to the early 1990s was towards an expansion of higher-level jobs, in particular those in management and the professions. It could be argued, however, that this may give a misleading picture of real skill change. Shifts in the occupational structure might reflect in good part changes in the use of titles for describing jobs. Further, it is difficult to evaluate the significance of a growth of higher-level occupational positions without knowing about the trends in skills *within* occupational categories. It might be the case that jobs in the expanding occupational classes were at the same time undergoing a process of deskilling, resulting in little change in the overall distribution of skills.

An evaluation of the different theories of skill change is made more complex by the fact that there is little consensus about the way in which skill should be assessed and, indeed, different perspectives tend to base their arguments upon rather different conceptions of skill (Attewell 1990; Spenner 1990; Vallas 1990; Gallie 1991). Arguments about upskilling have been linked to the view that there has been a long-term shift in the nature of skill, particularly in the manual occupations. The nature of technical change has led to the replacement of traditional craft skills by new forms of conceptual and decision-making skill that primarily require much higher levels of general education. In this respect, the most important measure of skill change is change in the level of educational qualifications that are required to carry out a job. In contrast, theorists of deskilling have continued to view skill primarily in 'craft' terms and have tended to take as an indicator of skill development the duration of the vocational training that people have received.

The approach adopted here has been to take multiple indicators of skill development. We begin by examining the qualifications required for jobs. We then consider trends in the frequency and duration of training and on-the-job experience. Finally, we turn to people's own perceptions of whether or not skills have increased over recent years, leaving them free to define skill in the way that is most relevant to them.

While the major part of our analysis will be based on the 1992 survey, we have some possibility of assessing skill trends not only through retrospective questioning, but also through comparison across time with earlier data. The 1992 survey was designed to include a number of indicators that would allow for comparison with two studies that had been carried out in the mid-1980s. The first of these was the Class in Modern Britain Survey (see Marshall *et al.* 1988), a nationally representative survey carried out in 1984.[1] The second was the Work Histories and Attitudes Survey of the Social Change and Economic

---

[1] It had a total sample of 1,770. The original survey included a wider age range (16–64 for men; 16–59 for women), as well as non-active people. To ensure comparability, our analyses are based on the 945 employed people aged 20–60.

Life Initiative (SCELI), a comparative local labour market study conducted in 1986.[2]

Qualifications Required, Training, and On-the-Job Experience

What has been the trend over time with respect to qualifications? Our measure focuses on employers' demands with respect to new recruits. We asked 'If they were applying today, what qualifications, if any, would someone need to get the type of job you have now?' The emphasis on current requirements rather than on personal qualifications was designed to take account of the fact that people's own qualifications might have been acquired many years ago, when the skill requirements of the work were quite different. The results can be directly compared with those for 1986, drawn from the Social Change and Economic Life Initiative.

The evidence indicates a clear increase over the period in the overall level of qualifications required for jobs. Between 1986 and 1992, jobs requiring no or low-level qualifications have declined (Fig. 2.1), whereas those requiring higher-level qualifications increased. The proportion of jobs where no qualifications were required was 6 percentage points lower in 1992, whereas jobs requiring A level or more had increased by the same amount.

A very similar pattern emerges with respect to training and on-the-job experience (Fig. 2.1). Whereas, in 1986, 52 per cent of employees had received no training for the type of work that they were doing, by 1992 this was the case for only 42 per cent. The requirement for on-the-job experience was measured with a question asking how long it had taken after the employee first started the type of work to do the job well. In 1986, 27 per cent had said that it had required less than a month, whereas by 1992 this was the case for only 22 per cent.

This pattern of higher skill requirements over time can be seen within each broad class category, although it was more marked for some classes than for others.[3] The change in qualification requirements was particularly strong among lower non-manual employees. Whereas, in 1986, 25 per cent of lower

---

[2] This involved a sample of 6,111 people drawn from six localities—Aberdeen, Coventry, Kirkcaldy, Northampton, Rochdale, and Swindon. The analyses presented here are again based on the sub-sample of 3,877 employees aged 20 to 60. Although the SCELI survey was based on localities rather than a representative national sample, the class composition of the aggregate sample turns out to be virtually identical to national estimates for 1986.

[3] The class schema used is that developed by Goldthorpe and his colleagues (Goldthorpe 1980; Erikson and Goldthorpe 1992). The version used is the revised class schema of 1992 (Goldthorpe and Heath 1992). It should be noted that the theoretical interpretation given to the class schema is in terms of differences in employment relationships, rather than skill. Tests carried out by the present authors indicate, however, that it is a more satisfactory proxy of a range of skill measures than the Registrar-General's class schema which purports to be based on skill differences. To simplify the presentation, we have aggregated Goldthorpe Classes 1 and 2 into a joint service class (professional, administrative, and managerial employees) and classes IIIb, VIIa, and VIIb into a class of non-skilled.

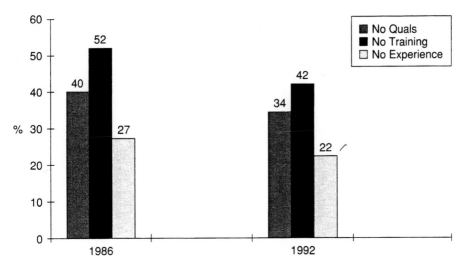

FIG. 2.1 Changes in qualifications, training, and on-the-job experience
requirements, 1986–1992

non-manual employees reported that their job required A levels or a higher
qualification, by 1992 this was the case for 35 per cent. The increase in training
was also particularly clear among lower non-manual employees, together with
technical and supervisory employees. In contrast, the increase in the proportion
needing more than a month's on-the-job experience to be able to do the work
well was greatest among semi- and non-skilled manual workers.

To examine more rigorously whether there had been a narrowing of class differentials with respect to the different skill dimensions, the data for the two periods were pooled and a set of regression analyses were carried out. These analyses
controlled for differences in the age and sex distributions in the two years. The
focus was on the interaction terms between class and year, which showed whether there were specific classes which had improved their position relative to
employees in professional and managerial work, over and above the general
improvement across time. The results pointed in general to a stability rather than
to a significant narrowing of class differentials. There had been a general upwards
movement, but this had affected all classes, leaving the relative class positions
unchanged. The only exception was in the case of lower non-manual employees,
whose qualification levels had risen relative to professionals and managers.

The Experience of Changing Skill Demands

This increase in the skill demands of work is confirmed by people's own
accounts of *changes* in the skills required by their jobs. A majority of employees
(63 per cent) reported that the level of skill they used in their job had increased

TABLE 2.1 *Skill change in the job in last
five years: percentage experiencing an
increase in skill (cell %)*

	1986	1992
Professional/managerial	67	74
Lower non-manual	55	70
Technician/supervisory	56	73
Skilled manual	50	64
Semi- and non-skilled	33	45
All employees	52	63

over the previous five years (Table 2.1). In contrast, only 9 per cent said that the skills they used at work had decreased.

A majority of employees at all job levels experienced an increase in their skills, with the exception of semi- and non-skilled manual workers. The increase was particularly marked among professional and managerial workers (74 per cent), technicians and supervisors (73 per cent), and lower non-manual workers (70 per cent). But it was also the case that 64 per cent of skilled manual workers thought that the skills involved in their work had increased. Moreover, even semi- and non-skilled manual workers were much more likely to have experienced an increase than a decrease in their skills (45 per cent compared with 15 per cent). These changes in skill level were regarded by most people as substantial. Of those who had experienced an increase in the skills they used at work, 85 per cent said that they had increased either a great deal or quite a lot.

How far did such upskilling represent a change in job requirements? Skills may have increased because of changes in the demands of the job or because, with time, people have learned to do their jobs better. While both factors were important, it was changes in the demands of the job that were more commonly cited. Whereas 84 per cent said that their skills had increased because the job required a higher level of skill than before, 64 per cent said they had increased because they had learned to do the same job better.

The view that reports of upskilling represented real changes in job requirements is confirmed if the relationship between skill change and training is examined. Overall, 54 per cent of employees in 1992 had received training in the previous three years. This was a striking increase from the proportion that had received training in the mid-1980s. The Training in Britain Survey (which used an identical measure) showed that only 33 per cent had received training in 1986–7 (Rigg 1989). As with skill change, opportunities for training differed very substantially by job level. More than two-thirds (71 per cent) of managers and professional staff had received training over the three-year period, but this was true of less than half (45 per cent) of skilled manual workers, and of only one-third (34 per cent) of semi- and non-skilled manual workers.

There was a very strong relationship between whether or not people said that the skill requirements of their jobs had increased and training. Only about a third of those who had seen no change in their skills or who had been deskilled had received training (32 per cent and 35 per cent respectively). In contrast, among those who had experienced a small increase in skill, 44 per cent had received training in the previous three years, while the figure rose to 61 per cent among those whose skills had increased quite a lot and to 72 per cent among those whose skills had increased a great deal.

Finally, it is clear that such skill increases were sufficiently real for employers to be prepared to pay for them. Within every occupational class those whose skills had increased received higher gross hourly earnings than those whose skills had remained unchanged. A regression analysis showed that the effect of upskilling on pay remained when age, sex, and occupational class had been controlled. The effect appears to have been virtually identical for men and women.

Finally, the view that the experience of upskilling reflects important changes in the content of jobs is reinforced when changes over time are examined (Table 2.1). Compared to the mid-1980s, the process of upskilling appears to have extended to a substantially wider sector of the workforce. Survey data for 1986 also showed a general tendency for skills to have increased at all job levels. But the overall proportion that experienced an increase in their skills rose from 52 per cent in 1986 to 63 per cent in 1992. This increase had been particularly sharp among technicians and supervisors, lower non-manual workers, and skilled manual workers. Finally, the proportion that had experienced a decrease in their skills was unchanged at 9 per cent. However, a more rigorous statistical test for change in relative class experiences over time, controlling for age and sex, showed that there was no class which had significantly improved its position relative to professional and managerial employees. The more striking feature was the very general nature of the process of upskilling across classes.

While there is good reason to think that reported changes in skill levels reflect to a considerable degree changes in the requirements of jobs, this could still be accounted for in rather different ways. Since respondents had been asked to compare their current work with what they were doing five years before, skill levels might have risen either because a person had been upwardly mobile into a higher-level job or because their existing job had been restructured in a way that increased its skill content. The apparent skill increase within social classes might then be artefactual, reflecting primarily individuals' experience of upward mobility.

A number of analyses have been carried out to assess the extent to which changes in skill levels reflect such mobility processes rather than changes occurring within specific jobs. These focus on those people who had seen little or no mobility, using the detailed work-history data that were collected to compare the jobs that people were in at the time of interview with their jobs five years earlier. The striking feature of the results, however, is that, even among those who had experienced little or no job mobility, a majority had still experienced an increase in their skills. Overall, 62 per cent of those who were in the same type of occupation as five years earlier and 56 per cent of those in exactly the same job said that their skills had increased, while the proportions that had experienced deskilling were only 5 per cent and 4 per cent respectively.

The overall picture is a clear one. The dominant trend was towards an increase of skills. This was the case not only for those who had been upwardly mobile, but also for those who had remained in the same job over the five years. This strongly suggests that a major factor behind the rise in skill levels was the restructuring of work tasks.

SKILL CHANGE AND THE QUALITY OF THE WORK TASK

The Growth of Task Discretion

Writers who have emphasized the trend towards rising skill levels have also tended to argue that this will be accompanied by a significant decentralization of responsibility in work (Blauner 1964; Littek and Heisig 1991; De Tersac 1992). Whereas traditional forms of detailed specification of duties were compatible with the relatively simple tasks associated with a highly developed division of labour, they became, it is suggested, altogether inappropriate with the emergence of more complex tasks requiring more highly skilled personnel. Increasingly, the nature of the work involves anticipating difficulties, and handling unexpected problems in conditions of environmental uncertainty. The detailed regulation of tasks is inherently unsuitable for guiding decisions in situations of uncertainty; rather decision-making power has to be confided to the people closest to the task.

Three approaches have been adopted to assess this argument. To begin with, an index of task discretion was constructed using four questions that had been shown to scale well and that formed a common underlying factor.[4] This included two items asking how true it was that: 'I have a lot of say over what happens in my job' and that 'My job allows me to take part in making decisions that affect my work.' The other two asked people 'How much influence do you personally have in deciding what tasks you have to do?' and 'How much influence do you personally have in deciding how you are to do the task?'

There is strong support for the view that greater job complexity is accompanied by higher levels of task discretion. The index of task discretion ranged from 0.21 among those who have increased their skills to −0.29 among those who had seen no change to −0.55 among those whose skills decreased over the period. This was not simply a reflection of the fact that skill increases were more marked

---

[4] The measure represented a statistically satisfactory scale, with a Cronbach's alpha of 0.64. A principal components analysis showed that the seven items formed a single factor, with an eigenvalue of 2.24, accounting for 32% of the variance. The factor scores have been taken as the measure of intrinsic job interest.

TABLE 2.2 *Change in the responsibility involved in the job in last five years: percentage experiencing increased responsibility (cell %)*

	1986	1992
Professional/managerial	72	79
Lower non-manual	60	66
Technician/supervisory	75	78
Skilled manual	59	62
Semi- and non-skilled	42	50
All employees	60	65

in higher occupational classes. There was a strong relationship between skill change and task discretion *within* each occupational class. Nor could the association be explained away in terms of an association between age and the responsibility given to employees. An increase in skill was related to higher levels of task discretion within each age category. Overall, it seems likely that the increase in skill levels was integrally linked to the increase in responsibility in the job. As tasks become more complex, employers are increasingly obliged to rely on the judgement of individual employees.

A second approach was to ask people directly whether or not the responsibility involved in their job had increased, decreased, or stayed much the same over the last five years. A substantial majority of employees (65 per cent) had experienced an increase in responsibility, while only 26 per cent reported that the responsibility involved in the work had stayed much the same, and 8 per cent that it had decreased. There were considerable variations by class (Table 2.2). Increased responsibility in the job had been most marked among professional and managerial and among technical/supervisory employees, where it was the case for three-quarters of employees. It was least common among semi- and non-skilled manual workers, but, even among these, 50 per cent reported that they had more responsibility than before and only 11 per cent felt that their responsibility had been reduced.

Since increasing responsibility is also likely to be affected by life-cycle factors, it is again important to compare with a similar indicator at an earlier period of time. Data is available for 1986 from the Social Change and Economic Life Initiative. As can be seen in Table 2.2, the tendency for responsibilities to increase had become more marked over time. In the mid-1980s, 60 per cent of people had seen the responsibilities involved in their job increase over the previous five years, compared with 65 per cent in 1992. The most marked changes compared with 1986 were with respect to employees in lower non-manual jobs and in semi- and non-skilled manual work.

The evidence is highly consistent with the view that the increase in skill requirements may have been a major factor underlying the growth of responsibility in

TABLE 2.3 *Task discretion by class, 1986–1992 (cell %)*

	Decides normal daily tasks		Initiates new tasks	
	1984	1992	1984	1992
Professional/managerial	81	75	74	75
Lower non-manual	48	56	48	47
Technician/supervisory	54	69	54	59
Skilled manual	20	33	27	45
Semi- and non-skilled	26	37	29	38
All employees	48	55	48	54

work: 82 per cent of those that had experienced a skill increase felt that the responsibilities of the job had grown greater, compared with only 38 per cent of those whose skills had stayed the same. This relationship was evident within all job levels, although it was particularly strong for lower non-manual workers and weakest for skilled manual workers.

Finally, to provide a stronger assessment of changes in the level of task discretion over time, we introduced two indicators of task discretion that had been used in an earlier national survey carried out in 1984: the Class in Modern Britain Survey. These were put to a random half of the sample. They asked: 'Do you decide the specific tasks that you carry out from day to day or does someone else?' and 'Can you decide on your own to introduce a new task or work assignment that you will do on your job?' In practice these were highly correlated with the main measure of task discretion (0.56 and 0.50 respectively) and when introduced into a factor analysis with the task discretion measure were shown to be part of a single underlying factor.

As can be seen in Table 2.3, both indicators show a clear increase in task discretion over the years. Whereas 48 per cent of employees had discretion over normal daily tasks in 1984, this was the case for 55 per cent in 1992. Similarly, the proportion of employees with discretion to introduce new tasks increased from 48 to 54 per cent.

For both years there were strong class differences in task discretion, with professional and managerial workers distinctly higher than any other category in their degree of self-determination in work, and manual workers relatively rarely having control over their job tasks. However, with respect to change over time, the most marked increases in task discretion are among skilled manual workers and technicians/supervisors. Skilled manual workers were distinctive in that they increased their discretion substantially over both normal and new tasks, whereas technicians and supervisors mainly increased discretion over normal tasks. It is notable that this contrasted with the pattern among professionals and managers who appear to have seen increased constraints on their ability to take everyday decisions.

Overall, the pattern suggests that the second half of the 1980s and the early 1990s saw a marked decentralization of decision-making within organizations. This is consistent with the view that it was a period characterized by significant delayering in which employers were reducing the numbers of middle-level managers, and devolving responsibilities.

Intrinsic Job Interest

Both of the major theories of skill trends posit a close relationship between skill change and the intrinsic interest of the work task. The more pessimistic scenarios of the evolution of work argued that, with deskilling, the ever-increasing division of labour and tightening systems of management control, a sharp deterioration was occurring in the quality of work. As tasks became simpler and more fragmented, work became more repetitive, routine, and monotonous. Further, since the central process of change involved the removal from ordinary employees of the more conceptual and creative parts of the work process, they were increasingly in job tasks where they were unable to utilize the skills they had developed earlier in their careers and they were involved in a type of work that offered few possibilities for self-development through the work itself. In contrast, those that have emphasized the tendency for skills to become more complex and for skill levels to rise have suggested that work will become more varied and intrinsically more interesting. The relationship of skill change to the intrinsic quality of the work task was explored with respect to three key dimensions: the variety of the work, the extent to which people feel that they can utilize their skills, and the opportunities the job provides for self-development.

There were two main measures in the survey relating to the repetitiveness or variety of work. The first asked people whether the variety of their work has increased, stayed the same, or decreased over the last five years. The second asked to what extent the work involved short repetitive tasks, with responses recoded into four categories: three-quarters of the time or more, half the time, a quarter of the time, almost never.

The general relationship between skill change and the variety of work is clearly confirmed. Those who had experienced upskilling were much more likely to have seen the variety in their work increase over the previous five years. While 84 per cent of those in jobs where skills had increased had also experienced an increase in the variety of the work, this was the case for only 39 per cent of those in jobs where the skill level had remained unchanged and for 28 per cent of those whose skills had decreased. This strong relationship between skill increase and variety was evident within each job level. Similarly, there was a highly significant relationship between skill change and the extent to which people were carrying out work that involved short repetitive tasks. Where people were in jobs where skill levels had increased, only 20 per cent of employees were involved in such work for three-quarters or more of their time, whereas among those whose work had been deskilled the proportion rose to 38 per cent.

Upskilling also appeared to be associated with a better match between people's skills and the job. People were asked: 'How much of your past experience, skill, and abilities can you make use of in your present job?' Of those that had increased their skills, 44 per cent felt that they could use almost all their previous experience and skills, while this was the case for 34 per cent of those who were in jobs where the skills had remained unchanged, and 20 per cent of those in jobs where skills had decreased.

Another major aspect of job quality is whether or not the task is one that encourages self-development through the challenges that it provides for new learning. The question used to assess this asked people how strongly they agreed or disagreed with the view that 'My job requires that I keep learning new things.' An increase in skills was strongly associated with the likelihood that people would be in a job that allowed self-development. Overall 33 per cent of those whose skills had increased agreed strongly that they were in a job where they could keep on learning new things, compared with only 13 per cent of those whose skills were unchanged.

Finally, if skill change is genuinely improving the quality of the job task, this should be reflected in the level of involvement that employees feel in the work they are doing. The survey included three measures of job involvement that sought to tap the amount of discretionary effort that people were prepared to put in, the extent to which the work was experienced as boring, and the level of job interest. The items asked how much effort people put into their job beyond what was required, how often time seemed to drag on the job, and how often they thought about their job when they were doing something else. Again, upskilling was strongly related to each of these.

An overall measure of intrinsic job interest was created from the seven items discussed above.[5] A more detailed measure of upskilling was also used for the analysis, taking account not only of the direction of skill change but of people's report of the extent to which their skills had changed. It includes five categories: skill decrease, no change, a little upskilling, quite a lot of upskilling, and a great deal of upskilling. As can be seen in Fig. 2.2, there was a strong linear relationship between skill change experience and the overall score of intrinsic job interest.

A range of other factors were also associated with these aspects of the quality of employment—such as people's age and the occupational class of their job (although not their sex). Yet, the striking fact is that, even allowing for the influence of these, the association between skill change experience and the intrinsic interest of the work still emerges very clearly. Further, a more detailed examination shows that skill change has a consistent and powerful effect in improving job quality *within* each occupational class.

---

[5] The measure represented a statistically satisfactory scale, with a Cronbach's alpha of 0.64. A principal components analysis showed that the seven items formed a single factor, with an eigenvalue of 2.24, accounting for 32% of the variance. The factor scores have been taken as the measure of intrinsic job interest.

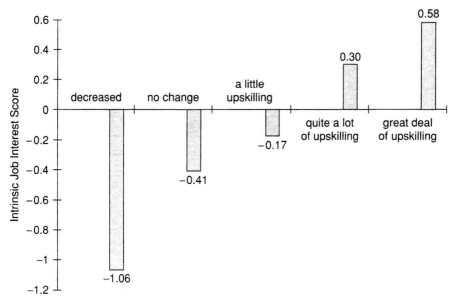

FIG. 2.2 Intrinsic job interest and skill change

Skill Change and the Intensification of Work

For the pessimistic theorists of skill change, the evolution of work involved not only the increased prevalence of repetitive and uninteresting work, but at the same time the intensification of work effort. Indeed, one of the essential reasons behind the supposed desire of employers to simplify work tasks was that it would facilitate greater managerial control and thereby make it easier to enforce higher levels of work effort. In contrast, the theorists that emphasized tendencies for skills to increase had remarkably little to say about the likely implications of such developments for work effort. It has, however, been suggested that upskilling is linked with a tendency to break down the rigidity of traditional skill lines, in particular through a growth of multi-skilling or polyvalence. This greater flexibility in work could be seen to represent a potentially important source of increased work demand.

There were a number of questions in the survey that provide information about the level of work pressure. Two were designed to tap the general level of work pressure, taking account of both physical and mental pressures. People were asked how strongly they agreed or disagreed that 'My work requires that I work very hard' and that 'I work under a great deal of tension.' There were two items on the time pressures in work: whether or not people felt they had enough time to get everything done on the job and whether they often had 'to work extra time, over and above the formal hours of the job, to get through the work or to help

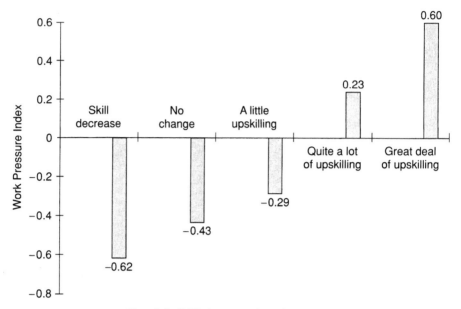

FIG. 2.3 Skill change and work pressure

out'. There was a question asking people whether they were expected to be more flexible in the way they carried out their work than two years earlier. Finally, to get an overall indication of whether or not people felt that there had been a change in the level of effort, they were asked whether, over the last five years, 'the effort you have to put into your job' had increased, stayed the same, or decreased.

It is clear from the results for each of these items that, while the rise in skill levels was associated with an improvement in the quality of work task, it led at the same time to a marked increase in the effort involved in work. A summary picture of the relationship between skill change and work pressure can be obtained by creating an overall index of work pressure.[6] The pattern of the results can be seen in Fig. 2.3. Work pressure is least among those that have been deskilled, followed by those that have experienced no change in their skill level. Those whose jobs have been upskilled a little report greater pressure than those with no change, but the difference is a modest one. It is above all those that have experienced either quite a lot or a great deal of upskilling who report much higher levels of work pressure.

---

[6] The six items discussed above form an acceptable scale with a Cronbach's alpha of 0.70. A principal components analysis revealed a single factor, with an eigenvalue of 2.41, accounting for 40% of the variance. The factor scores have been taken as the values for the index of work pressure. An anova test of the bivariate relationship between the overall work pressure index and the measure of skill change shows these are associated at a high level of statistical significance ($p < 0.000$).

It is possible that this apparently strong relationship between experiences of skill change and work pressure reflects primarily the influence of other factors. For instance, higher-class positions may be both more demanding in terms of workload and more affected by processes of upskilling. A fuller test of the significance of upskilling has been carried out introducing controls for class, age, and sex. Yet even when these factors have been taken into account the influence of skill change is still highly significant. Indeed, its effect appears to be even more pronounced once these factors have been controlled for.

In sum, the trend towards upskilling had rather ambivalent consequences for the quality of working life. In some respects it was clearly very positive. It was associated with an increased devolution of decision-making responsibility for the job and it increased the likelihood that employees would be engaged in intrinsically interesting work tasks that offered variety, the opportunity for people to use their skills fully and to develop their skills through their work. This was consistent with the predictions of the 'optimistic' perspective. But it was also associated with a sharp intensification of the pressures of work, a development that had been central to more pessimistic analyses of the evolution of work processes.

GENDER AND SKILL CHANGE

There can be little doubt that the overall trend has been towards a raising of skill levels, although the implications of this for the quality of work must be judged ambivalent. Yet, have such developments affected men and women in broadly similar ways or have the processes been gender-specific? Those that have argued for a high degree of segmentation in the labour market and for a degree of polarization of skill experiences through the division between core and peripheral sectors of the labour market have also tended to regard the privileged core as more heavily masculine and the secondary or peripheral labour market as more heavily feminine. If this is the case, there should have been very marked differences in skill experiences by sex. Further, if the direction of change has been towards greater polarization of skill experiences, it could be expected that the gender gap would have widened over time.

The Gender Gap in Skill and Responsibility

A comparison of the measures for skill and responsibility for men and women showed that there was a significant gender gap in the early 1990s. Women were more likely to be in jobs where no qualifications were required (39 per cent compared with 30 per cent) and they were less likely to be in jobs requiring at least A-level qualifications (32 per cent compared with 41 per cent). They were also less likely to have experienced an increase in their skills over the previous

TABLE 2.4 *Change in skill and responsibility by class and sex*

	Increase in skill (cell %)		Increase in responsibility (cell %)		Task discretion score (means)	
	Men	Women	Men	Women	Men	Women
Professional/managerial	72	77	80	77	0.44	0.40
Lower non-manual	75	68	77	62	0.13	−0.11
Technician/supervisory	78	56	81	66	0.34	0.10
Skilled manual	65	55	62	57	−0.19	−0.33
Semi- and non-skilled	50	41	55	46	−0.52	−0.35
All employees	66	60	70	61	0.04	−0.04

five years (Table 2.4). Whereas 66 per cent of men reported that the skills they used in their work had increased, this was the case for 60 per cent of women. There are similar gender differences with respect to task discretion and responsibility. On the measure of task discretion, women had a score of −0.04, whereas men had a score of 0.04, a difference that was statistically significant. Similarly, women had benefited less from the growth of responsibility in work, with 61 per cent of women, compared with 70 per cent of men, saying that the responsibility involved in their job had increased.

Women's disadvantage in skill development was evident in all occupational classes other than that of professionals and managers, where they were somewhat more likely to have experienced an increase in skills. The gender gap was particularly marked among technicians/supervisors, manual workers, and lower non-manual employees. Women were also less likely to have seen the responsibility in their job increase in all classes, although the difference between sexes was very slight among those in professional/managerial positions and in skilled manual work. In contrast, there was a 15 percentage point difference among those in technical/supervisory grades, a 14 point difference among lower non-manual workers, and a 9 point difference among those in semi- and non-skilled manual work. In general men have higher levels of task discretion than women even within specific occupational classes; the only class in which women have higher levels of task discretion than men is that of semi- and non-skilled manual workers.

Change Over Time in Gender Differences

While these differences in the experiences of men and women are marked and very consistent, what has been the trend over time? Have the differences between men and women been decreasing or increasing? The evidence suggests that there was a significant decline over the decade in gender disadvantage with respect to skill but not with respect to task discretion.

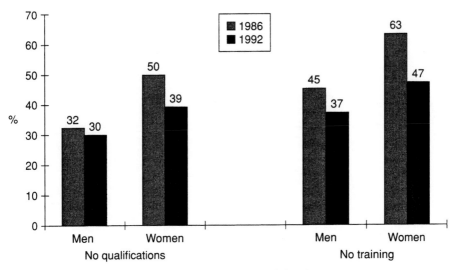

Fig. 2.4 Qualifications and training by sex

To begin with, while there was a rise in the qualifications required for jobs for both sexes, the shift was particularly marked for women (Fig. 2.4). As a result the differential between the proportions of men and women in work without qualification requirements declined from 18 percentage points in 1986 to 9 points in 1992, while the gender gap for jobs requiring degree-level qualifications declined from 7 to 4 percentage points.[7]

Further support for the view that there has been a major change in the relative level of qualifications required for men's and women's jobs is the age pattern. The differential between men and women decreases strongly with age, with the pattern inverted in the youngest age group. The proportion of women in jobs not requiring qualifications is 18 percentage points higher than that for men among those aged 55 or more, 14 points higher for those 45–54, 11 points for those 35–44, and 10 points for those 25 to 34. Among those aged 20 to 25, however, not only does the differential disappear, but it is men that are more likely to be in jobs not requiring qualifications (+7 points). At the other end of the spectrum, the differential between men and women for those in work requiring A level or more, reaches a peak of 16 percentage points among those aged 45 to 54. Among those aged 20 to 25, however, it is women that are more likely to be in work requiring this qualification level (+2 points).

The trends in the gender gap for training and for on-the-job experience follow a similar pattern (Figs. 2.4 and 2.5). The difference in 1986 in the proportions

---

[7] The overall gamma correlation between sex and qualifications required for the job fell from 0.25 in 1986 to 0.16 in 1992.

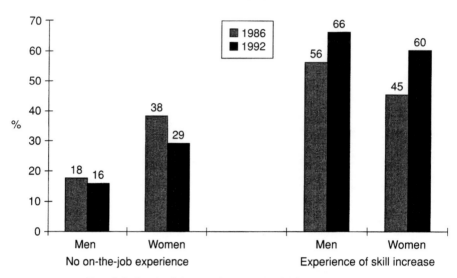

FIG. 2.5 On-the-job experience and skill increase by sex

of men and women who said that they had no training for their type of work was one of 18 percentage points, whereas by 1992 it had been reduced to 10 points. There was a corresponding reduction in the gender differential from 18 to 12 percentage points in the proportions with a year or more training. With respect to on-the job experience, the gender differential declined from 20 to 13 percentage points in the proportions needing less than a month to be able to do the job well, and from 26 to 19 percentage points in those needing a year or more.[8]

As with the evidence for qualifications, training, and on-the-job experience, the data on subjective experiences of skill change indicate that the gender gap is narrowing.[9] In 1986, there was a gap of 11 percentage points in the proportions of men and women that had experienced a skill increase, by 1992, this had diminished to 6 points. The gender gap closed most strikingly at the top and the bottom of the occupational hierarchy: in professional/managerial jobs on the one hand and in semi- and non-skilled manual jobs on the other. Indeed, while, in 1986, women professionals and managers were less likely to have experienced a skill increase than their male equivalents, in 1992 they were more likely to report upskilling. However, in sharp contrast, in lower non-manual work, the relative advantage of men with respect to upskilling had grown greater.

[8] The gamma coefficient for the overall relationship between sex and length of training fell from 0.34 in 1986 to 0.20 in 1992, while that for on-the-job experience fell from 0.43 to 0.32.

[9] The overall gamma correlation between sex and skill change decreased from 0.16 in 1986 to 0.09 in 1992.

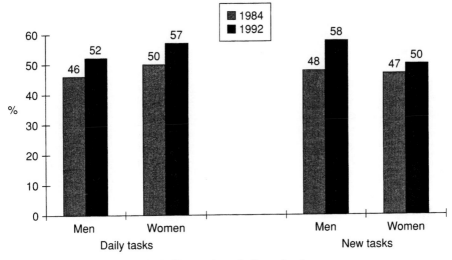

F<small>IG</small>. 2.6 Change in task discretion by sex

A striking feature of the data is that, while there are clear signs of a decline in the gender gap with respect to the various criteria of skill, there is no similar overall pattern for task discretion (Fig. 2.6). There was very little change between 1984 and 1992 in the gender gap with respect to control over usual daily tasks. In both years women were slightly more likely than men to have this type of discretion. However, men appear to have benefited more than women from the general trend towards higher levels of discretion over the introduction of new tasks. The proportion of men exercising this type of control increased from 48 per cent in 1984 to 58 per cent in 1992, compared with an increase from 47 to 50 per cent for women.

These overall figures conceal quite substantial variations between different occupational classes. Among professional and managerial workers, there is a consistent narrowing of the gap between men and women with respect to discretion both over normal tasks and over the introduction of new tasks. In contrast, among lower non-manual employees, the gender gap widened, especially with respect to control over the introduction of new tasks.

ADVANCED TECHNOLOGY, THE GROWTH OF SOCIAL SKILLS, AND SKILL CHANGE

In the literature, the issue of skill trends has been primarily linked to arguments about the implications of new technologies of work, in particular the spread of diverse forms of automation. These were formulated initially with the development of large-scale continuous-process technologies in manufacturing industry in the 1950s and 1960s and were elaborated in the 1980s, when the spread of micro-processor technologies led to a marked renewal of interest in the implications of automation for skill. However, automation may not have been the only change in the character of work with an important bearing on skill. The post-war decades also have been characterized by the growth of the service sector. Much of the work in this sector might be described as 'people-work', involving the direct servicing of people's needs through face-to-face interaction. The skills involved are essentially 'social skills'. In particular, given their concentration in the service sector, a great deal of women's employment is likely to involve work of this type. We will consider then in turn the implications of these two major sources of change in the nature of work tasks.

New Technologies

Our evidence certainly underlines the very rapid change that occurred in work technologies from the mid-1980s. Whereas the 1986 study showed that 39 per cent of employees were working with computerized or automated equipment, by 1992 the figure had risen to over half of the workforce (56 per cent). Its use was most prevalent in the white-collar occupations, affecting the work of 76 per cent of professionals and managers, 80 per cent of lower non-manual employees, and 54 per cent of technicians and supervisors. But even among blue-collar workers, a substantial minority were using technologically advanced equipment (33 per cent of skilled manual workers and 29 per cent of semi-skilled and non-skilled).

Men were more likely than women to be using such equipment in all occupational classes other than lower non-manual work. But, the extent of change since 1986 has been even more marked among women than among men. For men the proportion using advanced technology rose from 46 to 59 per cent, while for women it rose from 33 to 53 per cent. Thus a gender differential of 13 percentage points in the mid-1980s was reduced to one of 6 percentage points by the early 1990s.

What have been the implications of the rapid spread of advanced technologies for skill and responsibility in work? Those who used advanced technology in their work showed markedly higher skill levels than those that did not on all of the measures. For instance, only 19 per cent of those using advanced technology had no qualifications, compared with 55 per cent of those not using it. Conversely, the percentages for those with O level or higher qualifications were 75 and 36 per cent respectively. Those working with advanced equipment were much less likely to be without training (31 per cent compared with 56 per cent) or to have required less than a month on-the-job experience to do the job well (15 per cent compared with 33 per cent). Finally, as can be seen in Table 2.5, the use of advanced technology was strongly associated with having experienced an increase in skill requirements in the last five years (73 per cent compared

TABLE 2.5 *Effects of advanced technology on changes in skill and responsibility (cell %)*

| | Using advanced technology: experiencing increases in | | Not using advanced technology: experiencing increases in | |
	Skill	Responsibility	Skill	Responsibility
Professional/managerial	78	81	63	72
Lower non-manual	72	68	61	60
Technical/supervisory	84	83	59	72
Skilled manual	71	69	61	58
Semi- and non-skilled	60	59	38	46
All employees	73	73	49	55

with 49 per cent). The association of the use of advanced technology with higher skill levels is evident within all occupational classes.

Turning to task discretion, it is again clear that the overall tendency is for the use of advanced technology to be associated with greater responsibility in the job. The index of task discretion was 0.12 for those using computerized or automated equipment, but only −0.16 among those that did not use it. It was also strongly related to recent experiences of increased responsibility (Table 2.5). Whereas 73 per cent of those using advanced technology had seen the responsibility in their work increase in the previous five years, this was the case for only 55 per cent of other employees.

However, there is now a substantial literature which shows that employers can respond in very different ways to broadly similar technological developments and that patterns of work organization are best understood as resulting from the interplay of employer organizational strategies, employee organization, and technical constraints (Gallie 1978; Kalleberg and Leicht 1986; Wall *et al.* 1987; Jurgens *et al.* 1993). An area of considerable speculative contention has been whether, as the result of cultural beliefs or relative power resources, employers tend to adopt rather different patterns of work organization depending upon the sex of their employees (see for instance Steinberg 1990; Wajcman 1991).

To assess this, analyses were run including an interaction term, which indicated whether or not men benefited significantly more than women in terms of skill and responsibility. With respect to skill, this showed that the effects of advanced technology were as significant for women as for men (Table 2.6). There was no difference between men and women in its effects on the qualifications required for the job, the on-the-job experience needed, and the experience of skill development over the previous five years. The exception was training: men were more likely to have received training.

The picture with respect to task discretion is very different. Whether one takes the level of task discretion or recent experiences of change in the responsibility involved in the job, there is a clear gender divide. In both cases, the interaction

TABLE 2.6 *Advanced technology, skill, and responsibility*

| | Without sex interaction | | With sex interaction | | | |
| | Advanced tech | | Advanced tech | | Adv. tech × male | |
	Coeff.	Sig.	Coeff.	Sig.	Coeff.	Sig.
Qualifications required	0.44	***	0.40	***	0.08	
Training	0.22	***	0.10		0.25	**
On-the job experience	0.18	**	0.25	**	−0.14	
Skill increase	0.36	***	0.39	***	−0.06	
Task discretion	0.03		−0.08		0.23	***
Responsibility increase	0.08	**	0.03		0.11	**

Note: The results are drawn from a series of OLS regression models

term shows that men benefited significantly more than women. It might be the case that advanced technology still provided women with more responsibility in the job, albeit to a lesser extent than for men. However, separate analyses for men and women showed that, while it was strongly associated with higher levels of task discretion and increased responsibility for men, there was no such relationship for women.

Overall, it is clear that the development of computerized technology has been powerfully linked with the general movement towards increased skill in work. This has been true for all occupational classes and it has been the case for both men and women. It also has been associated with higher levels of task discretion and increased responsibility in work for men, but this was not the case for women. This suggests that employers have responded to the demands of technical change with rather different organizational policies depending upon the gender characteristics of the workforce, and that such policies play an important role in mediating the relationship between skill and responsibility.

The Growth of Social Skills

The discussion in the literature of the nature of skill change has predominantly taken as its point of departure developments in manufacturing industry. However, the development of the service sector draws attention to a major area in which new skill requirements may have been becoming increasingly important —namely that of social skills.[10] The effectiveness and competitiveness of the service industries depend crucially on the ability to handle people. But there is some indication that there has been a more general increase in the demand for

[10] The problems that this poses for traditional conceptions of skill has led to an increased use of the notion of 'competencies' to capture this wider range of capacities (see Dugue and Maillebouis 1994).

social skills in industry, for instance with respect to communication, teamworking, and handling clients or customers.

As an initial approach people were asked to describe from a list of activities what their work involved. Our list of activities covered nine types of work: caring, dealing with clients, producing with machines, maintaining or servicing machines or vehicles, working on an assembly line, monitoring equipment, handling or analysing information, driving vehicles, and organizing other people. The importance in the modern economy of work directly concerning people emerges very clearly. Taking all those mentioning a particular factor, the data show that more than half of employees (60 per cent) now deal with customers or clients in their jobs, while, in addition, one in five (24 per cent) are now in jobs involving caring for other people, such as the sick or young. Further, more than one-third of employees are called upon to exercise some degree of responsibility for other people in their work: 48 per cent see part of their task as organizing other people; 43 per cent see their jobs as involving some supervisory responsibilities.

These figures, however, do not reveal the relative importance in the job of different types of work. To find out the main type of activity, additional information was collected to find out which types of activity took up more than half the person's time. The great majority of employees (79 per cent) were able to specify a main activity from those listed.

Even with respect to the principal work activity, the importance of what might be termed 'people-work' comes out very clearly. Overall, 14 per cent spent more than half their time caring for other people, 22 per cent dealing with clients and customers, and 10 per cent organizing others. In short, 46 per cent of the overall sample were engaged predominantly in some form of people-work. These figures contrast with the 7 per cent who produced with machines and the 4 per cent involved in maintenance work. Most striking of all, given its frequent position in the literature as the very model of employment, is the very low proportion mentioning assembly-line work. Only 6 per cent of the sample said that part of their working time involved assembly-line work, and only 3 per cent reported that they spent more than half their time doing such work. In contrast, the handling of information had become as common a form of work activity (14 per cent) as the types of work associated with traditional manufacturing production.

There can be little doubt, then, about the central importance of people-work in the modern economy. However, the frequency of this type of work differed sharply by gender. While 60 per cent of women had some type of people-work as their main activity, this was the case for only 33 per cent of men. The biggest difference was with respect to caring work (25 per cent compared with 4 per cent), but women were also much more likely to be dealing with clients and customers (28 per cent compared with 16 per cent). The only type of people-work which was more commonly to be found among men was that of organizing

TABLE 2.7 *Main type of work by sex (column %)*

	Male	Female	All
Caring	4	21	14
Dealing with clients/customers	16	28	22
Organizing people	12	8	10
Producing with machines	9	5	7
Maintenance (machines/vehicles)	8	—	4
Assembly-line work	3	2	3
Analysing/monitoring information	19	12	15
Driving	7	1	4
Other or no main activity	21	21	21
No.	1,803	1,667	3,469

other people: 12 per cent of men spent most of their time on such organizing activities compared with 8 per cent of women.

Involvement in people-work also tended to be particularly high among non-manual employees, but it was by no means exclusively linked to the higher-job levels. It was most common among professional and managerial employees (57 per cent), followed by technicians and supervisors and lower non-manual employees (52 and 49 per cent), and least common among skilled manual workers (13 per cent). But it is notable that a substantial proportion of semi- and non-skilled manual workers (42 per cent) were also engaged in such work. This largely reflects the jobs of those engaged in sales and personal service work, which are classified in the semi- and non-skilled category.

The close link between people-work and the growth of the service sector is immediately apparent from the distribution of types of work by industry. In the welfare service industries, the great majority of employees had people-work as a main activity (83 per cent in social welfare, 80 per cent in medical services, 65 per cent in education). There were also high proportions in retail (61 per cent) and leisure services (59 per cent). In contrast, in metals and chemicals only 17 per cent had a primary involvement in people-work, and the proportions were only a little higher in mechanical engineering/vehicles (20 per cent), electrical and electronics, construction, and food and clothing (22 per cent). Industries such as telecommunications and postal services, banking and finance, hotels and catering, and national and local government were in an intermediate position, with just under half of employees primarily involved in people-work.

Did the process of upskilling primarily involve those in production activities or had it also affected those whose work primarily involved people? If the measure of skill change is examined by broad types of work, it is clear that the highest level of upskilling of all was to be found among those whose work primarily involved dealing with information, where 63 per cent reported that they had experienced either a great deal or quite a lot of upskilling. However, the next

TABLE 2.8 *Type of work and skill change (cell %)*

	Experience of skill change		
	Decrease in skills	Quite a lot of upskilling	A great deal of upskilling
Main type of work:			
People-work	7	35	23
Analysis/monitoring of information	8	33	30
Production/maintenance	9	31	19
Assembly line	22	14	16
Driving	15	27	17
Other/none	11	27	14

highest category was of employees involved in people-work, with 58 per cent reporting substantial upskilling. This compares with 50 per cent among those in traditional manufacturing activities. Most strikingly, strongly confirming the evidence from earlier case-study work, only 29 per cent of those in assembly-line work had significantly upgraded their skills. This type of work was also characterized by the highest level of deskilling (22 per cent).

More detailed analysis reveals considerable variations within the broad categories that have been examined above. For instance, within production, maintenance workers were quite likely to have experienced a substantial upgrading of their skills (61 per cent); it was the machine operators (49 per cent) that were responsible for the relatively low figure in traditional manufacturing activities. Similarly, among those involved in people-work, the high levels of upskilling were particularly evident among those who had responsibility for organizing other people (72 per cent), whereas those involved in caring and dealing with clients were closer to the average (56 per cent and 54 per cent respectively).

Nonetheless, employees in all types of people-work shared in the wider process of upskilling and appear to have benefited more than those in traditional production activities. Indeed, even when sex and class have been controlled for, working with people remained strongly associated with the likelihood of having experienced an increase in skills.

A second, more direct approach, to the significance of social skills was through a series of questions asking employees about the importance of particular types of social skill for their ability to perform their current job competently. The items included: knowing their way around the organization, good social relations with people at work, good contacts with clients or customers, being able to communicate well with other people, and providing leadership.

To facilitate the analysis, a summary scale of 'social skills' was constructed from a factor analysis of the responses. The results confirmed the basic patterns found earlier. Social skills were closely related to the different types of work:

TABLE 2.9 *Social skill and skill change*

| | Social skill score | | |
	Male	Female	All
		Means	
Skills have:			
Increased a great deal	0.07	0.30	0.18
Increased quite a lot	−0.03	0.28	0.11
Increased a little	−0.31	0.14	−0.11
No change	−0.33	−0.01	−0.16
Decreased	−0.28	−0.29	−0.29
Total	−0.14	0.13	−0.01
No.	1,694	1,560	3,253

they were highest among those whose work primarily involved people and lowest among those doing assembly-line work. They were more commonly required in women's work than in men's. They were also far more frequently to be found in the welfare service industries than in manufacturing. However, on this measure, medical services now ranked highest rather than welfare services. A general class gradient emerges, although non-skilled manual workers have slightly higher social skill scores than those in skilled manual work. This again reflects the importance in this category of those involved in sales and personal service work, predominantly women. If the data for men and women are examined separately, male non-skilled workers are less likely to use social skills in their jobs than skilled manual workers, while the reverse is the case for women.

As can be seen in Table 2.9, there is a strong relationship between the measure of social skill and the likelihood that a person had experienced a skill increase in the previous five years. The overall scores for social skills range in a linear way from 0.18 among those who had experienced a great deal of upskilling to −0.29 among those whose skills had decreased. However, it is clear that this is predominantly due to the pattern among women. For men there is a fairly simple division between those whose skills have either increased a great deal or quite a lot and all others. Significant skill increase is associated with greater use of social skills. For women, there is a much clearer gradient across the different categories of skill experience. In particular, the experience of deskilling appears to have been particularly sharp among those whose work involves relatively little in the way of social skills.

A regression analysis makes it possible to assess the relative significance of a range of factors influencing the experience of skill change (Table 2.10). It can be seen that there was no difference between men and women. In contrast, age was clearly important. There is a linear pattern in which the older the age the less likely it is that a person experienced upskilling. The oldest workers (those

TABLE 2.10 *Effects of advanced technology and social skills*
on skill development

	Coeff.	Sig.
(Constant)	3.39	
Sex (Men)	0.05	n.s.
Lower non-manual	−0.15	*
Technicians/supervisors	−0.04	n.s.
Skilled manual	−0.12	n.s.
Semi-/non-skilled manual	−0.67	***
Age 25–34	0.01	n.s.
Age 35–44	−0.10	n.s.
Age 45–54	−0.18	**
Age 55+	−0.46	***
Works with computerized technology	0.38	***
Social skills	0.23	***
Male social skills	−0.11	*

Note: OLS regression; no. = 3,196; R^2 = 0.13

aged 55 or more) were particularly disadvantaged. Job level was also of major importance. In particular, those in non-skilled work have lost out heavily in the upskilling process. It is clear that disadvantage becomes cumulative: those with the lowest skills have the least chance of increasing them.

It was seen earlier that new technologies affected skill change in much the same way for men and women. However, the relationship between social skills and upskilling proved to be significantly stronger for women. In short, it appears clear that the increased requirement for social skills has been a powerful factor underlying the general tendency towards upskilling. This is true even when sex, age, and job level are controlled. Moreover, it is factor that has been particularly important for women's work.

CONCLUSIONS

Our evidence indicates that employers responded to increasingly rapid technological change and to more intense market competition primarily by raising skill levels and by job enrichment rather than by the degradation of work. The most striking feature of our data is the very extensive upskilling of the workforce. At the same time, there has been a significant devolution of responsibilities for more immediate decisions about the work task. The most prevalent employer policy with regard to work organization has been a move towards task discretion.

This process of upskilling and increasing responsibility, however, has had ambivalent implications for employees' experience of the quality of work. In some respects, it was associated with marked improvements. This was particularly the case for the variety of the work, the ability of employees to feel that they were making full use of their skills, and the opportunities for self-development through the everyday work process. But, at the same time, there was clearly a significant negative side to these developments. Upskilling was associated with a substantial intensification of work effort.

Further, the extent of the process of upskilling varied substantially by occupational class. There has been a particularly marked increase in the skills of skilled manual workers, technicians and supervisors, lower non-manual employees, and professionals and managers. However, semi- and non-skilled manual workers continued to have much lower chances of skill development. This points to a process of growing polarization in skill experiences between lower manual workers and other employees.

Our evidence also confirmed the marked differences that remain between men's and women's experiences of opportunities for skill development and self-determination at work. But an important finding was that the trend over time has been for gender differences in skill to diminish, although there was no evidence of any improvement over time in gender differentials with respect to responsibility over work decisions.

The rapid spread of computer-based technologies in the second half of the 1980s was a factor strongly associated with rising skill levels for both men and women. However, once again, there were important differences between the sexes with respect to responsibility in work. While the use of computerized or automated equipment was linked to greater task discretion for men, this was not the case for women. In introducing new technologies, employers appear to have adopted gender-specific organizational policies. A second important factor underlying skill trends has been the increased requirements for social skills in work. This has been particularly significant for women's work. The last decade has seen the growth of more stringent criteria for 'customer' satisfaction across a wide range of service industries and a greater emphasis on personal skills in the organization of work at all levels. It is clear that many of the traditional accounts of skill change are still rooted in an increasingly outdated conception of the work process.

Overall, there clearly has been a rise in skill levels and this in general has been linked to greater responsibility in work, greater intrinsic work interest, and greater work pressure. However, there is no direct translation of changes in skill requirements into changes in wider work roles. The impact of changing skill requirements for the wider work role is mediated by employers' organizational philosophies and their culturally derived assumptions about the capacities of different categories of employee. This is particularly clear in the rather different implications of higher skill levels and automation for men's and women's discretion in work.

3

Discretion and Control

On the evidence of the previous chapter, there were strong trends in the early 1990s for employees to gain increased discretion in their work. Does this mean more than a change in the way work is organized and carried out? Does it mean that employers are now prepared to trust their employees more than they did in the past, to permit them a greater degree of autonomy, and to rely primarily on the individual's own commitment? Does this, in short, herald a transformation of the relationships between the employer and the employee, the manager and the subordinate?

Certainly, the late 1980s and early 1990s witnessed a transformation in the vocabulary of management. Concepts like 'empowerment', 'team working', and 'commitment' came progressively into use. The whole profession of personnel management faced a takeover bid by those who wished to rename it 'HRM'— human resources management, in recognition of the altered role which they saw it taking up. And a variety of surveys of employers, both in Britain and the USA, showed them introducing a range of new practices in the late 1980s or early 1990s, which were at least loosely inspired by 'HRM thinking' (McKersie 1987; Storey *et al.* 1993; Fernie *et al.* 1994; Osterman 1994, 1995; Wood and Albanese 1995).

Yet none of these studies has tested the claim that a transformation of the employment relationship is taking place. That claim, expressed in the terse phrase 'from control to commitment' (Walton 1985*a*), implies not only a growth in practices which foster employees' commitment but also *a progressive withdrawal of managerial control in favour of employee autonomy.*

This is a large claim, for the centrality of power and control has been a recurrent theme of sociologists' accounts of workplace relations (e.g. Mouzelis 1967; Fox 1974; Edwards 1986). The focus on means of control has been continued in recent examinations of human resource management (HRM) practices (e.g. Blyton and Turnbull 1992; Willmott 1993; Wilkinson and Willmott 1995). Through a wide range of case-study research, British sociologists have accumulated evidence of the introduction during the 1980s and 1990s of new or more intensive forms of control over the worker. Examples include monitoring or surveillance based on information technology (Sewell and Wilkinson 1992); the growth of 'flexible' employment contracts and subcontracting (Barnett and Starkey 1994); 'just-in-time' (JIT) systems for production and delivery (Delbridge 1995); the continuing spread of budgetary systems and targets, including in the public sector

(Whittington *et al.* 1994); and the increasing use of individualized financial reward systems (Smith 1992; Wood 1996). Added to this, the workplace developments of the 1980s and 1990s have often been interpreted as a reassertion of the 'right to manage' with an implication that management exercises more authority and more control than previously (for review, see Purcell 1991).

This evidence creates a *prima facie* case that, so far from a retreat from control, there may well have been an extension of control in recent years. On the other hand, case-study evidence in itself does not demonstrate that control has been increasing generally, even if it has been developed in particular circumstances. One of the chief tasks of the chapter is to assess the extent and development of control with nationally representative data, and to establish what are the factors which are driving it. But this task cannot be carried out in isolation from the growth of task discretion. The chapter will also be concerned with a comparison of the effects of discretion and control, and with how, or whether, a growth in control can be reconciled with a growth in task discretion.

PREDICTIONS ABOUT CONTROL

There are three groups of predictions about control which this chapter will test. The first, and simplest, comes from the HRM movement already referred to, and its central claim about a transition from the 'strategy of control' to the 'strategy of commitment'. The American business-school academics whose writings have partly led and partly reflected the HRM movement (Walton 1972, 1985*a*, 1985*b*, 1987; Beer *et al.* 1984; Kochan *et al.* 1994) express this shift in a prescriptive way. Employers are faced with an 'economic necessity' (Walton 1987) of making this shift, for market success depends on a 'superior level of performance' which can only come from employee commitment, not managerial control. The high-commitment strategy delivers higher levels of effort, and also lower levels of worker alienation and conflict (Walton 1972) than those which result from adversarial and coercive control.

Allied to this claim is an assertion that 'commitment cannot flourish in a workplace dominated by the familiar model of control' (Walton 1985*a*). The production of commitment requires a set of policies which are seen as incompatible with high levels of management control. And among these new policies is the cultivation of high levels of task discretion and autonomy.

Recent empirical studies of the spread of HRM or high-commitment systems (e.g. Osterman 1994; Fernie and Metcalf 1995; Wood and Albanese 1995) have rather strangely ignored the claim of HRM as a movement not only *to* commitment, but *away from* control.

There are other foundations for predicting a transformation of the employment relationship. Chief among these are theories based on the changing nature of technology. Robert Blauner (1964) and Shoshana Zuboff (1988) have stressed that changes in technology, in transforming the nature of work, offer opportunities for management to reconstruct their relation with workers. In his discussion of the effects of automated chemical process technology, Blauner points to the possibilities for social integration of the workforce. As managers come to appreciate the responsibility assumed by workers in such advanced technologies (for example, for safe operation of hazardous processes), they may cease to regard them as adversaries. For Zuboff, focusing upon the introduction of information technology, fundamental change comes from managerial recognition of the enhanced cognitive requirements of work under the new technology, which potentially place workers and managers in the same relationship to the productive task.

The Blauner–Zuboff thesis therefore leads to relatively optimistic predictions. With the diffusion of information technology in both manufacturing and services, work at all levels becomes more cognitive in nature, responsibility is dispersed, and the differences between managerial and subordinate roles are reduced. It becomes easier for employers to choose and to implement a system of low control and high trust and discretion.

Given the widespread advance of information technology, as described in the previous chapter, there is an excellent opportunity to test these predictions (formerly supported only by case-study evidence). Employees in workplaces with a substantial use of information technology, according to these theories, should have experienced on average a reduction in control. Similarly, industries with a high intensity of information technology should be leaders in dismantling control and moving towards high-trust systems. Further, these developments (high technology, low control) should be linked to high work effort and low levels of conflict.

A third, and highly contrasting, group of predictions comes from Richard Edwards (1979). Control systems, according to his analysis, reflect the underlying opposition of interests between employers and workers. What is observed in practice is a continuing active development of control systems by employers, with the aims of guarding against worker opportunism, reducing or muffling conflict, and increasing compliance.

Edwards provides a detailed description and interpretation of *three overlapping stages in the development of control systems*. Stage One was 'personal control', in which the dominant factor was the power of the foreman. This individual (nearly always a man) hired, fired, disbursed wages, punished, and allocated work according to his will. A point emphasized by Edwards is that this system of personal control *has never passed away*; it remains widespread in small and medium-sized establishments in the lower-waged, non-unionized segment of the labour market.

Stage Two was the 'technical control' which grew up with the rationalization of production and the introduction of various aspects of 'scientific management', partly following the ideas of Frederick Taylor (see Hirschhorn 1984). Technical control has two chief elements: control by a work process (planned by work study or industrial engineers and often, though not necessarily, linked to machinery), and control by a system of 'payment by results' tied to work standards. Like personal control, technical control has not passed away but remains in widespread use, wherever work can be routinized and standardized.

The third stage, called 'bureaucratic control' by Edwards, was initially developed by large corporations for their growing corps of administrative and technical specialists. Bureaucratic control is a system of individual progression or advancement, governed by organizational norms or codes and by impersonal processes of review, commonly an appraisal system.[1] Progression depends not only on output or performance as such (which may often be difficult to measure), but upon adherence to company norms of conduct, such as co-operativeness or flexibility. In the years since Edwards' book, many employers have further developed systems of this type and they are now often referred to as 'performance-management systems' (Incomes Data Services 1992, 1997).

There are four major predictions from Edwards' theory, as well as numerous minor ones. First, in contradiction to the two other groups of predictions, control over workers is necessary for industrial capitalism and will not be reduced, whether because of market or technological changes. Second, changes in technology or the mode of production change only the form of control, not the degree of control. Third, because control is required in order to maintain the power of employers, systems of control will be applied particularly to reduce opposition to that power: for example, where unions continue to exert an influence. Fourth, since Edwards observed that employers had on the whole obtained the results they wanted from control systems, he implies that their progressive development into more sophisticated forms will tend to reduce conflict.

Edwards' view of task discretion is also very different from that of Walton, Blauner, or Zuboff. For the latter, discretion is expected progressively to replace control, and the consequences are predicted to be benign, with more commitment and higher productivity following. For Edwards, on the other hand, the advance of discretion in itself implies a weakening of work intensity. But management will counter this through new forms of control (bureaucratic control, or whatever lies beyond that) and it will be the control, not the discretion, which assures productivity.

ASPECTS OF SUPERVISION

In describing recent developments, we will use Edwards' three stages of control as a convenient framework. The first questions to consider, then, are how widespread and how strong personal supervision remains. With the growth of task

---

[1] Earlier in its development, this system was famously described by Max Weber (1947).

TABLE 3.1 *Supervising and being supervised, by class (cell %)*

	Professional/ managerial	Lower non-manual	Technician/ supervisory	Skilled manual	Semi- and non-skilled
Has supervisor	94	96	96	94	95
Has supervisory responsibilities	68	34	70	35	21

discretion, together with other developments such as flat organization structures and team-working, it might well be supposed that supervision is on the decline, and this has often been supposed in the past (for review, see Armstrong 1986).

In fact, as Table 3.1 shows, the great majority of employees continue to have a supervisor, and the rarity of any exceptions is striking (94 per cent of employees had a person they recognized as their 'immediate superior or supervisor'). One cannot, then, interpret increased task discretion as implying a disappearance of the supervisory relationship. Nor is there any sign that the growing managerial and professional class, which leads the shift towards task discretion, is absolved from being supervised. Even in this group, the proportion seeing themselves as supervised scarcely differed from semi- and non-skilled employees.

Remarkably, more than four in ten employees (43 per cent) stated that they themselves held supervisory responsibilities.[2] This, of course, does not mean that they had the word 'supervisor' in their job title: the class of supervisors-as-such is a small one (some 5 per cent in our sample). Rather, the result demonstrates that the responsibilities of personal supervision are widely diffused. Among the large managerial and professional class, two-thirds have such responsibilities, but they are also held by one-third of routine non-manual workers and skilled manual workers, and by one in five of semi- and non-skilled manual workers (Table 3.1).

Two surveys offer a rough comparison over time. Rose and colleagues (1987), reporting results from the 1984 survey of class, found that 32 per cent positively answered the question 'Do you supervise or have management responsibility for the work of other people?' The Social Change and Economic Life Initiative (SCELI) survey of 1986 asked 'Are you directly responsible for supervising other employees?' The proportion responding positively, 34 per cent, was almost the same as the 1984 result. Differences in question wording should make one cautious in interpreting the apparent change by the 1990s. But the picture is hardly one of supervision wasting away.

It could be, however, that while supervision continues, its intensity is weakened and its role circumscribed, as has often been predicted before (for example, by

---

[2] The question ran as follows: 'As an *official* part of your main job, do you supervise the work of other employees or tell other employees what work to do?'

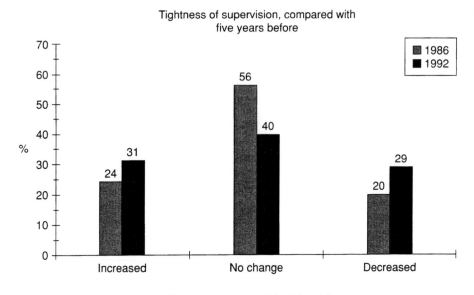

Tightness of supervision, compared with
five years before

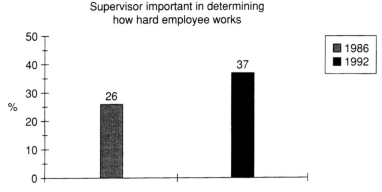

Supervisor important in determining
how hard employee works

FIG. 3.1 Changes in supervisory control, 1986–1992
Sources: 1986, SCELI Survey; 1992, Employment in Britain

both Blauner and Edwards, and by Braverman 1974). The survey included an item on the 'tightness of supervision over your job', which was compared with the position five years previously. Nearly one-third (31 per cent) did indeed report a decrease in the tightness of supervision. But nearly as many reported an *increase* in the tightness of supervision (29 per cent) as reported a decrease.

This result can be compared with those from 1986 (the SCELI survey), which used exactly the same question. In 1986, the degree of change was less *in both directions*, as shown in Fig. 3.1. The implication is that, over recent years, there has been some polarization of control. Another comparison between the

TABLE 3.2 *Strength of supervision, by class (cell %)*

	Professional/ managerial	Lower non-manual	Technician/ supervisory	Skilled manual	Semi- and non-skilled
(a) Tightness of supervision in last 5 years:					
increased	24	31	41	34	32
decreased	42	29	25	29	21
(b) Supervisor has great deal or fair amount of influence on:					
Deciding how you do tasks	41	53	55	56	64
Deciding tasks	65	68	74	83	78
Deciding how hard you work	61	70	75	72	73
Deciding standards of quality	70	75	75	78	80
(c) Very/fairly easy for supervisor to know:					
How much you do	54	68	71	76	76
Quality of work	67	79	75	79	79
(d) Supervisor attaches great deal of importance to:					
Arriving on time	54	52	73	56	67
Not being absent	61	54	68	63	66
Hard work	74	71	75	69	69
High-quality work	82	79	81	82	77
(e) Supervisory favouritism: very true/true	37	35	37	50	38

two surveys comes from a question about whether supervision is 'important in determining how hard you work in your job'.[3] Here the trend was unequivocally up, not down, with *26 per cent giving a positive response in 1986 and 37 per cent in 1992* (Fig. 3.1).

If the tendency is for supervision to become polarized, who are the winners and losers? The evidence (see the first panel of Table 3.2) shows plainly that the decrease in tight supervision has been concentrated on the managerial and professional ranks, while manual workers have experienced a substantial tightening of supervision.

A check on this conclusion is to consider in a more detailed way the supervisor's direct influence over task allocation and methods, and over the pace and quality of work performance. Only half of the employees (52 per cent) said that their supervisor exerted a 'great deal' or a 'fair amount' of influence over how they carried out their tasks, and this is reasonably consistent with wide and growing task discretion. But a considerably larger proportion, 70 per cent, stated that

[3] The question was also posed for a number of other potential influences on work effort, which will be considered later.

supervisors influenced what tasks should be done, so their role of task allocation remains widespread. Further, 65 per cent attributed a great deal or a fair
amount of influence to supervisors over how hard they worked, and 74 per cent
over the quality of their work. For each item, there was a clear though fairly
gentle class gradient in supervisory control from management and professionals
(least), via routine non-manual workers, to manual workers (most). In these respects, then, the traditional notions of supervision still appear to be widely applied.

Again, while it may be supposed that the variability and complexity of
'knowledge work' makes supervisory monitoring difficult, this was not borne
out by the survey. Two-thirds of respondents said it was very easy or easy for
their supervisors to know how much work they were doing, and this rose to nearly
three-quarters (74 per cent) for the quality of their work. The class gradient was
somewhat steeper here, yet two-thirds of managers and professionals felt that
the quality of their work was easily judged.

A potentially important role for supervisors is in transmitting the standards
or norms of behaviour expected by the organization. Most employees saw their
supervisors as communicating the importance of timekeeping (59 per cent attach-
ing a 'great deal' of importance to this), attendance (61 per cent), hard work
(71 per cent), and care over quality (79 per cent). Differences between classes
were clear in the case of timekeeping and attendance, where more semi- and
non-skilled workers perceived a strong supervisory norm. Class differences were
minimal when it came to the importance attached to work effort and quality.

Finally, the survey asked how true it was that the respondent's supervisor 'treats
some employees better than others'. This taps into notions of favouritism and
personal power, and in this way is connected to the style of personal control
described by Edwards. Nearly one in six employees (16 per cent) stated that
this was 'very true', and a further 23 per cent stated that this was 'true'. So
nearly four in ten seemed, to some degree, concerned about a lack of fairness
in supervision. There was no class gradient here, but skilled manual workers
were somewhat more likely than others to be aware of favouritism.

If favouritism is characteristic of personal supervision, it might be concentrated, as assumed by Edwards, in the small or medium-sized 'sweat shops'
of traditional industries. However, the survey data revealed no association of
supervisory favouritism with either size of establishment or industry. It was not
only very common but very widely and evenly spread.

To conclude, supervision remains the predominant form of control, and most
employees are conscious of its influence.[4] Its importance for the control of work
effort has been growing, not declining. Class differences in many aspects of supervision are marked, with managers and professional workers experiencing a

---

[4] Rose *et al.* (1987) also concluded that supervisors continued to have wide and important functions, but they based this upon the reports given by supervisors about what they did, rather than
employees' reports of how they were supervised.

loosening of supervision but manual workers a tightening. And the personal power of the supervisor, as manifested in favouritism among subordinates, remains widespread and by no means confined to backward or secondary areas of employment.

TECHNICAL CONTROL: MACHINES AND INCENTIVES

If supervision flourishes, what is the position with Edwards' 'second stage' of control, known variously as 'technical control', 'scientific management', 'Taylorism', or 'Fordism'? The chief elements of such a system, from an employee viewpoint, are the discipline of working at a pace which is externally dictated (often by a machine), and the dependence of financial rewards on measured performance.

Since so many discussions of control have focused on assembly-line systems, it is interesting to start with these. The survey asked whether the respondent worked at all on an assembly line, and also whether more than half the time was spent working on it. Only 6 per cent of employees worked on an assembly line, and 3 per cent spent the majority of their time there. Two further questions directly addressed the extent of control exerted by machine pacing, including assembly lines. The first asked whether or not 'a machine or assembly line' was 'important in determining how hard you work in your job'. The second question asked, in a similar way, whether a machine or assembly line was important in determining 'the quality standards to which you work in your job'. The results for these questions underline how exceptional it is for individuals to be controlled by a machine-paced or assembly-line system. *Only 6 per cent of employees said a machine or assembly line was important for their work effort, and only 4 per cent for their work quality.* The 1986 SCELI survey asked only about work effort (not quality); 7 per cent at that time reported that a machine or assembly line was important for their work effort.

Although pure machine-pacing or assembly-line work is rare, there may be other situations in which people are tightly constrained or driven by the system. Two questions to assess this asked how often the respondent's work involved (i) 'working at very high speed', and (ii) 'carrying out short repetitive tasks'.[5] As shown in the first panel of results in Table 3.3, nearly one in four of respondents (24 per cent) were working mostly[6] at very high speed, and exactly the same proportion were mostly engaged in short repetitive tasks. So this type of constrained work-pattern seems much more widespread than machine-pacing or assembly-line work in itself. As might be expected, the proportions were higher among manual workers, especially in regard to repetitive tasks, but the proportions were substantial for all groups.

[5] The questions were derived from a survey on conditions of work in Europe, sponsored by the European Foundation for the Improvement of Living and Working Conditions.

[6] Answers to the question were on a six-point scale, ranging from 'all the time' to 'never', but for simplicity the top half of the scale—ranging from 'all the time' to 'three-quarters of the time'—is grouped into a single upper value.

TABLE 3.3 *Individual control and incentives, by class (cell %)*

	Professional/ managerial	Lower non-manual	Technician/ supervisory	Skilled manual	Semi- and non-skilled
(a) *Pacing of work*:					
Working at very high speed mostly	18	25	19	33	28
Short repetitive tasks mostly	9	23	23	35	38
(b) *Individual incentives*:					
Incentive based on work pace	13	19	6	28	4
Effort set by incentives	16	20	14	28	20
Quality set by incentives	9	10	6	10	11
(c) *'Technical control' index*:					
2+ aspects	17	29	23	46	35
Only 1 aspect	26	26	31	25	27

For the other aspect of technical control, pay incentives, two types of questions are available: (i) whether or not the individual received an incentive[7] based on his or her own work pace, and (ii) whether pay incentives were important for determining 'how hard you work in your job' or 'the quality standards to which you work'. The results appear in the second panel of Table 3.3.

An incentive bonus based on the individual's own work pace was received by one in six (16 per cent) of employees: this excludes 'merit awards', which will be discussed in the next section. A more broadly worded question asked whether 'pay incentives' were influential over how hard individuals worked. One in five of respondents felt that pay incentives were important for effort. The identical question was asked in the 1986 SCELI survey, and the comparison indicates that the importance of individualized payment systems has been increasing. In 1986, only 14 per cent stated that pay incentives were important in determining their pay effort (Fig. 3.2).

To obtain an overall measure of 'technical control', the five items shown in Table 3.3 can be combined with the three items discussed earlier concerning assembly-line and machine control. If a person is defined as working under 'technical control' when at least two of the items apply to them, then *just under three in ten of all employees* (29 per cent) would be so classified. This result is summarized in the last panel of Table 3.3.

It seems from the table that employers have focused their use of technical control especially on skilled manual workers. However, this apparent finding does not allow for other factors which may influence employers' choice of control system, factors which may coincidentally be associated with the proportions of skilled or semi-skilled workers present. Later in the chapter it will be

---

[7] The precise wording was 'Do you receive any incentive payment, bonus, or commission based on . . . ?'

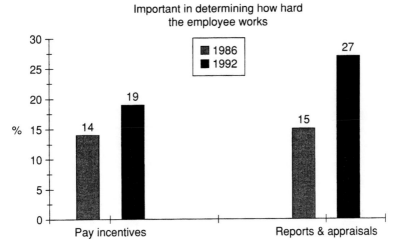

FIG. 3.2 Importance of individual control systems, 1986–1992
Sources: 1986, SCELI Survey; 1992, Employment in Britain

shown that, when these other factors are taken into account, manual workers as a whole—skilled and semi-skilled alike—are equally the focus of technical control systems.

BUREAUCRATIC CONTROL SYSTEMS

While technical control applies particularly to manual workers, the white-collar classes are more likely to work under 'bureaucratic control' systems. In current management parlance, these are widely known as 'performance-management systems'.[8] In such systems, formalized procedures of target-setting and/or personal appraisal, are linked to rewards such as merit increases and promotion or regrading reviews. These ideas are naturally linked with the white-collar tradition of career progression, but may be spreading to manual work through a gradual process of harmonization of treatment (Millward 1994).

The availability of internal routes of career progression is a key assumption of bureaucratic control, as presented by Edwards. The survey asked 'If you were trying to get a better job, generally speaking which would offer you the best opportunities—staying with your current employer or changing employer?' Just over four in ten employees (41 per cent) stated that staying with the current employer offered the best prospects, a proportion which did not vary greatly across job levels (see the first panel of Table 3.4). If internal progression is the

[8] The term 'performance management' originated in the USA and, in Britain, a more limited concept of 'reward manangement' is often substituted (Incomes Data Services 1992; Smith 1992).

TABLE 3.4 *Elements of 'bureaucratic control', by class (cell %)*

	Professional/ managerial	Lower non-manual	Technician/ supervisory	Skilled manual	Semi- and non-skilled
(*a*) Expects internal progression	44	46	55	44	34
(*b*) Has appraisal Appraisals affect:					
pay	26	29	14	13	8
prospects	46	42	43	22	16
training	50	42	41	21	16
pay rises on merit	48	44	32	29	24
(*c*) Targets affect:					
effort	60	50	54	50	36
quality	36	28	32	28	20
(*d*) 'Bureaucratic control' index:					
3+ items	57	53	45	34	23

foundation of performance management, then in many organizations that foundation seems weak or lacking. Furthermore, comparison with 1986 (the SCELI survey), when the identical question was posed, indicates that there has been a marked decline of faith in internal progression. In the more buoyant 1980s, the internal path was thought the better option by 55 per cent of employees.

A little more than one-half of employees (53 per cent) stated that they had a periodic *formal* assessment of performance. About one in five saw the appraisals as influencing pay, one in three saw an influence on prospects, and the same proportion perceived an effect on training and development (which would presumably influence prospects also). All these aspects of performance management were much more prevalent among white-collar than among manual workers (see the second panel of Table 3.4).

The survey also asked whether 'reports and appraisals' were seen by the respondent as 'important in determining how hard you work in your job'. The wording was exactly as in the SCELI 1986 survey. As shown in Fig. 3.2, there was a sharp increase on this measure between 1986 and 1992: from 15 to 27 per cent. There was also a shift in the importance of reports and appraisals relative to pay incentives. Whereas in 1986 the proportion thinking reports and appraisals important for effort was about the same as saying this about pay incentives, by 1992 reports and appraisals were important for more employees (27 per cent against 19 per cent).

An element found in many performance-management systems is to vary individual pay increases on the basis of a rating of personal performance, referred to as merit pay. The term, however, is not necessarily widely understood; the survey therefore asked whether pay rises, at the respondent's place of work, were given to 'those workers who work hard and perform well'. Here 36 per cent answered in the affirmative, nearly twice as many as reported incentives based on individual work-rate. White-collar workers were more likely to report merit pay than were manual workers, but the differences were less sharp than in the case of appraisal systems.

The final element in many performance-management systems is the setting of objectives or targets. The respondent was asked whether 'targets you are set' were among the things which were 'important in determining how hard you work in your job', and also 'the quality standards to which you work'. Targets were, in fact, widely perceived to be important, especially in relation to work effort. They were particularly important for managers and professionals; they were also important, and to an equal degree, for routine non-manual and skilled manual workers. Even among semi- and non-skilled manual workers, who were least influenced in this way, targets were more commonly seen as important than appraisal reviews (see the third panel of Table 3.4).

How widespread are bureaucratic control systems as a whole? A substantial degree of bureaucratic control is likely to exist where three or more of the items are reported. On this basis, as shown in the final panel of Table 3.4, more than half of the managerial, professional, and routine non-manual groups work under bureaucratic control, and even among skilled manual workers (nearly half of whom work under technical control systems) the proportion under bureaucratic control is one-third.

The most striking points in this section of the evidence concern the growth of appraisal systems. Although employees became less confident in internal career paths between 1986 and 1992, the control exercised by personal appraisals appeared to increase substantially over the same period, in both absolute terms and relative to pay incentives (which were also growing). There is no corresponding comparison concerning individual or team-based target-setting, but this too was certainly being widely used by the 1990s. It seems that many employers have pressed ahead with the chief instruments of bureaucratic control even in a period when chances of progression were in decline.

CONTROL VS. DISCRETION?

The evidence shows that, in the late 1980s and early 1990s, not only was task discretion increasing, but so too was control, whether represented by personal supervision or by the formal systems of technical control and bureaucratic control. An obvious interpretation, if one assumes an opposition between discretion and control, is in terms of polarization. This is unlikely to have been simply on the basis of class, because the evidence shows all classes experiencing high or growing control in one or more respects. Another possibility is polarization among employers, with some adopting high control and low discretion while

Discretion and Control

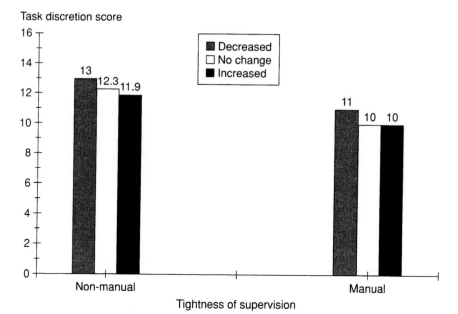

FIG. 3.3 Task discretion by change in tightness of supervision,
by non-manual and manual

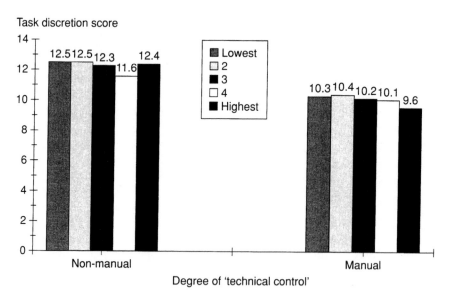

FIG. 3.4 Task discretion by degree of technical control,
by non-manual and manual

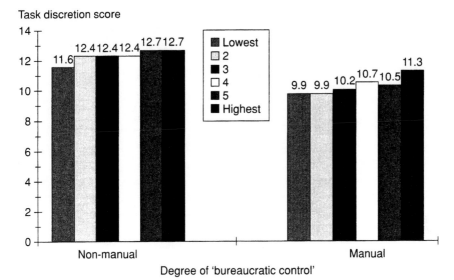

FIG. 3.5 Task discretion by degree of bureaucratic control,
by non-manual and manual

others adopted low control and high discretion. If that were so, then discretion
and control would be inversely related at the level of the individual employee.

This interpretation obtains some support in the case of the changing tightness of supervision (Fig. 3.3). For both non-manual and manual employees, a
decreasing tightness of supervision was associated with higher average levels
of task discretion, by comparison with those having unchanging or increasing
levels of supervision. The relationship shown in the figure is in statistical terms
a highly significant one.

This simple relationship broke down, however, when formal systems of control were considered. Technical control was not systematically related to task
discretion (Fig. 3.4). Among manual employees, there was a marked fall in discretion only at the most intensive level of technical control system, affecting
just 12 per cent of this group. Among non-manual employees, the relationship
was U-shaped, with higher task discretion for those at both extremes of technical control, and lower discretion in the middle. A statistical test for a linear
relationship was negative in both groups.

Finally, in the case of bureaucratic control (Fig. 3.5), there was a fairly clear *positive* association with task discretion. The greater the development of bureaucratic
control, the higher was the average level of task discretion, and this relationship was statistically significant for both the non-manual and manual classes.

So the idea of a simple trade-off between discretion and control was not sus-
tainable. Employers' practices about task discretion do not appear to map in

any straightforward way onto their policies of control. Depending on the type of control used, higher control might be associated with lower discretion, higher discretion, or something in the middle. To arrive at an understanding of why employers choose various forms and intensities of control, requires a more extended analysis, and that is the aim of the next section of the chapter.

ORGANIZATIONAL CHOICE OF CONTROL SYSTEMS

The theories reviewed at the start of the chapter imply a number of clear predictions about the circumstances which lead employers towards particular kinds of control system, or towards a general reduction of control. The most direct of these, perhaps, is the influence of *advanced technology*. Writers like Blauner and Zuboff suggest that automation and computers should swing the pendulum towards more discretion and less control, whether of the personal type or through formalized systems. Control should be at a lower level both at establishments with relatively high levels of advanced technology, and in industries with a high investment in technology. For Edwards, on the other hand, the intensification of technology should lead to the decline of personal supervisory control but the substitution of more formal systems of control, especially bureaucratic control.

Predictions about the relationships between control and unionization or union power are less simple. The issue is most fully addressed by Edwards. He sees tight personal supervision as a feature of the non-unionized secondary sector of employment, so high supervisory control should be associated with low union presence or power. But unionization evokes other, more formal systems of control, so the association between formal control and union presence or power should be positive. American writers on human resource management also tend to see a strong union presence as a barrier to developing a high-commitment strategy (Foulkes 1980), so 'high trust and low control' tends to be linked to newer non-union establishments.

To assess these predictions, we carried out a series of multivariate analyses, for each type of control in turn. The analyses estimated the net effects of technology, unionization, size of establishment, and industry, on the degree of control as experienced by the individual employee. The gender and social class of the employee were also controlled in the analyses.

Technology was represented by the proportion of workers using computers or information technology, divided into three bands. Unionization was represented by five categories. The first category was where unions were not recognized at all, and the other four were derived from a question about how much influence the union (where recognized) had over the organization of work. The size of the establishment was also in five bands, by number of employees, and there were fifteen industry dummies.

TABLE 3.5 *Effects on increases in the tightness of supervision*

	Coeff.	Sig.
Gender	1.00	n.s.
Occupation: (comparing with semi- or non-skilled)		
Professional/managerial	0.61	***
Technician/supervisory	1.57	*
Lower non-manual	0.93	n.s.
Skilled manual	1.13	n.s.
Size: (comparing with size 10–24 employees)		
fewer than 10	0.51	***
25–99	0.88	n.s.
100–499	0.82	n.s.
500+	0.99	n.s.
Advanced technology (compared with low level)		
Medium level	1.00	n.s.
High level	1.09	n.s.
Union power (compared with no union)		
Low union power	1.39	**
Some union power	1.70	***
Medium union power	1.97	***
High union power	3.16	***

Note: Logistic regression model, no. = 3,235; Multiplicative effects on odds. Results for industries are not shown.

The Tightness of Supervision

Supervision has many different facets, which it is impossible to reduce to a single scale. We decided to focus the analysis here on the question about the change in tightness of supervision over the previous five years. Those experiencing an increase in tight supervision were compared with those who had experienced either no change, or a decrease in the tightness of supervision.[9] The results of the analysis are shown in Table 3.5. The industry effects are not included in the table, but some of these will be introduced in the text.

The prediction that information technology leads to a decline in the tightness of supervisory control received little support. The extent of information technology at the workplace was unrelated to an increase in supervision. A parallel analysis, not shown here, indicated that it was also unrelated to a decrease in supervision.

The results concerning the changing tightness of supervision, and union influence, were unequivocal. The probability of an increasing tightness of supervision

[9] In a more complete analysis, both decreases in close supervision and increases in close supervision were simultaneously compared with 'no change'. This makes full use of the information available but produces rather a complex set of results. The simpler analysis yielded a similar pattern of findings even though it used the information less efficiently.

in recent years was much higher in unionized than non-unionized establishments. It also rose progressively with the perceived level of union influence in the workplace. There was every indication, then, that a strong union presence was a motive for employers to increase the tightness of supervision.

Employees in the smallest establishments (those with less than ten employees) were much *less* likely to experience increasing closeness of supervision than those in other organizations. There was no clear size difference apart from this. The influences of unionization and size were broadly consistent. Of the smallest workplaces, which were least likely to be experiencing a tightening of supervision, 75 per cent were non-unionized, a far higher proportion than elsewhere. (The percentages without unions in the other size groups were: 10–24 employees, 55 per cent; 25–99 employees, 45 per cent; 100–499 employees, 32 per cent; 500-plus employees, 15 per cent.) The fact that large workplaces were as likely as any others to be increasing the tightness of supervision (and more likely than the smallest workplaces), tells strongly against the idea that personal supervision is the favoured control system only of informal or archaic organizations. A tightening of supervision appears to have been selected as a policy by a very wide range of employers.

Technical Control

A composite measure of technical control, as experienced by the individual employee, was formed from the questions described earlier in the chapter concerning control by a machine or assembly line, working at a high speed, working on short repetitive tasks, obtaining individual incentive earnings, and having work effort or quality controlled by these incentives. The total score on this measure ranged from 0 to 6; however, as few people scored at the high end, a score of 6 was amalgamated with 5.[10]

The results of the analysis of technical control are shown in Table 3.6. Employers were more likely to apply technical control to manual workers as a whole than to non-manual. Any apparent difference, in the earlier descriptive tables, between skilled manual and semi- or non-skilled manual workers vanished with the more extensive set of factors now taken into account. Among the non-manual workers, there was a clear difference between managers and professionals on one hand, and supervisors, technicians, and routine non-manual jobs on the other. Employers used technical control to a lower degree on the managers and professionals. It was also the industries with high proportions of manual workers (extractive, manufacturing, and construction) which tended to apply technical control more. The exceptional case was financial and business services: despite being composed almost entirely of white-collar workers, it was an above-average user of technical control.

---

[10] The six items used had a reliability (Cronbach alpha) of 0.53.

TABLE 3.6 *Effects on the use of 'technical control'*

	Coeff.	Sig.
Gender	0.08	n.s.
Occupation: (comparing with semi- or non-skilled)		
Professional/managerial	−0.58	***
Technician/supervisory	−0.37	***
Lower non-manual	−0.28	***
Skilled manual	0.06	n.s.
Size: (comparing with size 10–24 employees)		
fewer than 10	−0.17	**
25–99	0.14	*
100–499	0.06	n.s.
500+	0.23	***
Advanced technology (compared with low level)		
Medium level	0.07	n.s.
High level	0.12	**
Union power (compared with no union)		
Low union power	0.12	*
Some union power	0.16	**
Medium union power	0.12	n.s.
High union power	0.60	***

Note: Ordered probit analysis, no. = 3,165. Results for industries are not shown.

Where most of the work was done with the aid of computers or automation, the workers experienced on average a *higher* degree of technical control. Here, the findings directly contradicted the predictions of writers such as Blauner or Zuboff, who expected advanced technology to generate higher-trust, less-controlled work systems. The above-average use of technical control in the financial and business services industry, noted in the previous paragraph, fitted with the technology effect, since the industry is a particularly intensive user of information technology.

Technical control was used more in unionized than in non-unionized settings, even after taking account of all the other factors, such as industry and size, which are linked to unionization. However, the degree of technical control did not change systematically with the degree of union influence. Only when union influence was at its highest point (covering a mere 3 per cent of the sample), was the use of technical control also at a particularly high level. The main distinction, then, was simply between the union and non-union establishments. A reasonable interpretation is that the presence of unions disposes firms to the use of technical control. It is possible, additionally, that where technical control has been established, workers are more likely to support and maintain union representation. In either case, the findings in this respect are broadly consistent with the predictions of Edwards.

TABLE 3.7 *Effects on the use of 'bureaucratic control'*

	Coeff.	Sig.
Gender	−0.15	***
Occupation: (comparing with semi- or non-skilled)		
Professional/managerial	0.67	***
Technician/supervisory	0.51	***
Lower non-manual	0.42	***
Skilled manual	0.17	***
Size: (comparing with size 10–24 employees)		
fewer than 10	−0.14	*
25–99	0.18	**
100–499	0.26	***
500+	0.40	***
Advanced technology (compared with low level)		
Medium level	0.26	***
High level	0.48	***
Union power (compared with no union)		
Low union power	0.05	n.s.
Some union power	0.12	**
Medium union power	0.22	***
High union power	0.34	***

Note: Ordered probit analysis, no. = 3,165. Results for industries not shown.

Bureaucratic Control

Table 3.7 summarizes the results of the analysis of bureaucratic control, carried out in the same way as for technical control. The items contributing to the scoring of bureaucratic control were those described earlier in the chapter, and covered internal progression, appraisal systems, merit pay, and performance targets.[11]

There are both striking differences and clear similarities between the analyses concerning technical and bureaucratic control. The relationships of bureaucratic control to social class were *the reverse* of those for technical control. Managers and professionals had the highest exposure to bureaucratic control; next came supervisors, technicians, and routine white-collar workers; then, with a considerable gap, skilled manual workers, while semi- and non-skilled manual workers had the lowest exposure of all.

As in the case of technical control, however, there was a positive relationship between advanced technology (computers or automation) in the workplace, and the use of bureaucratic control. Here the relationship was in fact still clearer. While the use of technical control systems increased only when advanced technology was at a high level, the use of bureaucratic control systems increased

[11] The seven items used had a reliability (Cronbach alpha) of 0.65.

even at a moderate level of advanced technology and still more at the higher level. The financial and business services industry, with its intensive use of information technology, and above-average use of technical control, also had the highest use of bureaucratic control, even after controlling for the social class of employees. Once again, and still more strongly, the predictions of writers such as Blauner or Zuboff were rejected.

Bureaucratic control was more likely to be present in unionized than in nonunionized establishments, and the use of bureaucratic control became more intensive when union influence was also at higher levels. Here again the findings were consistent with the predictions of Edwards, who described bureaucratic control specifically as a policy to counteract the spread of unionism into white-collar and professional groups. The indication is, then, that the presence of unions, especially white-collar unions, disposed employers to use bureaucratic control systems. It is also possible, in some cases at least, that where bureaucratic control systems were introduced, workers became more disposed to support unions.

CONTROL AND WORK PRESSURE

The previous section explored some major structural influences on the adoption of control systems by employers, but did not directly address the prime motives for using them. Presumably, employers use control systems because they believe that they will achieve certain results with them, and it is to some of these expected results that the analysis is now directed.

The most obvious purpose of control is to extract hard work from employees. But the theories reviewed at the start of this chapter have very different views about the effectiveness of control in achieving that purpose. For Edwards, assuming that relations at work are based on a simple opposition of interests, workers would not produce at a profitable level for the employers if left to themselves. From this viewpoint, control systems are necessary (for employers) to coerce workers into levels of effort which are profitable. For Walton and other advocates of human resource management, in contrast, there is no necessary opposition of interests between employers and employees, because individuals have personal goals of self-actualization, which—when nurtured through high-commitment strategies including increased autonomy—can evoke much higher work effort than under direct control systems. The aim of the next analysis is to test these contrasting positions. How far is work effort influenced by task discretion, and how far by systems of control?

A thorny issue is how to get sensible information about 'hard work' from individuals' testimony about themselves. The approach adopted here is to consider the *work pressures*, or work intensity, which individuals experience. The assumption is that when people drive themselves, or are driven to, high levels of effort, they will tend to experience a range of pressures in their work.

The measure of work pressures is a composite score derived from ratings of agreement with the following four statements:[12]

'My job requires that I work very hard.'
'I never seem to have enough time to get everything done on my job.'
'I often have to work extra time, over and above the formal hours of my job, to get through the work or to help out.'
'I often move from one sort of work to another.'

The analysis seeks to estimate the net effects of task discretion and control systems on this measure of work pressure, while taking account of numerous other potential influences. Some of these are the obvious controls for social class, or other personal and organizational characteristics which have in the previous analyses been shown to be linked to the use of control systems or to differences in task discretion. Two further groups of variables were also introduced into this analysis to make the controls more complete. One group provided a coverage of organizational practices, which might be associated either with control systems or with task discretion. These included various aspects of supervision,[13] the role of the workgroup,[14] and participation in 'quality circles'.[15] The second group concerned the recent experience of organizational change.[16] A high rate of change might generate increased work pressures, whatever the control system or level of task discretion.

The main findings are summarized in Table 3.8. Because so many other factors were included in the analysis, it would be cumbersome to show the complete results: those shown focus solely on task discretion and on control systems. The

[12] The four items used had a reliability (Cronbach alpha) of 0.60.

[13] The aspects of supervision covered were whether the supervisor helps workers to learn to do their jobs better; monitoring of performance (a combination of the two items in panel (c) of Table 3.2); and the importance attached by the supervisor to various kinds of behaviour (a combination of the four items in panel (d) of Table 3.2).

[14] Some 46% of employees with supervisors also described themselves as members of a work group. Further items established what influence the group had over various aspects of work organization, and whether the group's responsibility for work organization had increased, remained static, or decreased over the past five years. These questions distinguished four levels of 'team-working': work mainly on own (no team-working), work in group but it has little or no responsibility (low team-working), work in group with some responsibility (medium team-working), and work in a group with a lot of responsibility (high team-working). Some 15% worked in a group with high team-working.

[15] The term 'quality circles' comes from Japan where such groups were originally established with the aim of improving quality. The wording of the question was as follows: 'Some organizations have groups of employees who meet regularly to think of improvements that could be made within the organization. These are sometimes called quality circles. Are you involved in a quality circle or similar group?'

[16] Four types of change were considered: a reduction in the workforce affecting the individual, introduction of new computer systems, introduction of new equipment other than computers, and change in the work organization. In each case, the respondent was asked whether the change had taken place, or not, during the past two years.

TABLE 3.8 *Effects of discretion and control on work pressure*

	Coeff.	Sig.
(a) *Non-manual employees*		
Task discretion	0.18	***
Increased supervision	0.47	***
Reduced supervision	0.54	***
Bureaucratic control	0.11	***
Technical control	0.15	**
(b) *Manual employees*		
Task discretion	0.18	***
Increased supervision	−0.15	n.s.
Reduced supervision	−0.37	*
Bureaucratic control	0.05	n.s.
Technical control	0.18	***

Notes: OLS regression; no. for non-manual = 1,753, for manual = 1,369. R^2 for non-manual = 0.25, for manual = 0.22. Results for control variables not shown.

measures used for these variables are just as in previous sections of this chapter, and in the case of task discretion, as in Chapter 2.

The effects varied in important ways between non-manual and manual employees, so the analysis was carried out separately for the two groups. Within the non-manual analysis, account was taken of the average difference in work pressure between those in routine white-collar jobs on one hand, and those in managerial, professional, and supervisory jobs on the other. Similarly, within the manual analysis, account was taken of the average difference in work pressure between those in skilled jobs and those in semi- or non-skilled jobs.

The results provide clear support for the view that people whose jobs leave them with more discretion tend to commit themselves to a higher intensity of work. The idea that employees will use task discretion to avoid or reduce work pressure is rejected, and it is rejected as much for manual workers as for non-manual.

At the same time, however, there is clear support for the view that tight supervisory control, and more formal control systems, get people to work under higher work pressure. Technical control systems provided the clearest evidence, being linked to higher work pressure for both non-manual and manual groups. Bureaucratic control was linked to higher work pressure only in the case of non-manual employees, its effect being non-significant for manual workers.

Changes in the tightness of supervision also appeared to influence work pressure, but provided the main complications of the analysis. For non-manual employees, 27 per cent had experienced an increase in tight supervision while

36 per cent had experienced a reduction. Both an *increase* in the tightness of supervision and a *reduction* in the tightness of supervision were linked to higher work pressures, by comparison with no change in the level of supervision. A plausible interpretation is that where supervision was reduced for non-manual employees, personal responsibility was increased, and this was the underlying factor explaining above-average work pressure. In that case, the reduction of supervisory control is effectively an alternative indicator of growing task discretion. The interpretation was supported by a supplementary analysis: where tightness of supervision decreased, responsibility increased in 84 per cent of cases, by comparison with 78 per cent of cases where supervision increased.

While the results for non-manual employees suggest that employers were applying two contrasting policies of supervisory control, each of which was producing or sustaining a high level of work intensity, similar supervisory policies appeared to have little impact on manual workers. These had been experiencing an increase in tight supervision (32 per cent) more commonly than a reduction (22 per cent). However, an increase did *not* result in higher work pressures. Where supervisory control had become less tight, on the other hand, work pressure was significantly *lower*. Consistent with this, an increase in responsibility was experienced by only 60 per cent of manual workers with decreasing tightness of supervision. This was not only a much lower proportion than in the case of non-manual employees who experienced decreasing supervisory control, but lower even than for manual workers with increasing tightness of supervision, 64 per cent of whom reported an increase of responsibility.

Overall, then, there is rather a clear pattern distinguishing non-manual from manual employees, in terms of what forms of control produced or sustained work pressures. In the case of non-manual employees, newer forms of control were effective, including the use of 'performance-management' systems with targets, appraisals, and merit pay; and reduced supervisory control coupled with more personal responsibility. In the case of manual employees, 'performance management' seemed relatively ineffective if compared with more traditional technical control (based on machine-pacing and incentives); close supervision was on the increase, and where it was relaxed, this was not linked sufficiently to broader responsibility and led to a weakening of work pressure.

It could be argued that these differences between non-manual and manual employees reflect traditional attitudes towards work among the social classes. One has to question, however, whether this difference in 'class attitudes' is more than a reflection of the policies adopted by employers towards the different classes. Here the result concerning task discretion is a telling one. For manual and non-manual employees alike, higher levels of task discretion were associated with higher work pressure.

Despite the complications, the overall picture is clearly one in which both high task discretion and high control produce more work pressure. The opposition between the two strategies, in terms of this result at least, is false.

DISCRETION, CONTROL, AND INDUSTRIAL CONFLICT

Employers, however, do not only adopt control systems to enforce hard work. They choose particular forms of supervision and control so as to suppress or avoid conflict with employees. Such at least is the theory of control put forward by Edwards, who sees control as regulating and diverting a continuing opposition of interests in the workplace. In contrast, writers such as Blauner or Walton, who envisage a transformation of the employment relationship, would predict that conflict will be reduced directly through higher levels of autonomy in work. These contrasting predictions offer a further crucial test of the relative strengths of these theories.

The focus of the ensuing analysis is upon organized, collective industrial action, but our interest is in what influences the individual to take part in and support such action. The first step is to analyse the circumstances which are connected to the occurrence of industrial action. The second stage is to consider what circumstances influence the individual to take part in such action, in those cases where it has occurred.

To identify the occurrence of industrial action, we rely on the report of individual employees. They were asked whether one or more from a list of six forms of industrial action was taken at the workplace during the preceding two years.[17] There is an issue about how reliable such reports might be. To check this, the incidence of industrial action reported here can be compared with the incidence reported in the Workplace Industrial Relations Survey (WIRS) of 1990, the major national source of information based on reports from employers (Millward *et al.* 1992).[18] The comparison has to take account that (i) WIRS covered only establishments with twenty-five or more employees, and (ii) WIRS asked about the previous year rather than the previous two years. Altogether, WIRS reported industrial action of some kind at 12 per cent of establishments in the previous year. Of the present respondents, 17 per cent reported industrial action in the previous two years, but this rose to 21 per cent for respondents in establishments with twenty-five or more employees. Since some establishments will experience industrial action in successive years, the reports of the employees do not seem much adrift.

Industrial Conflict at Establishment Level

The first stage of analysis aims to establish influences on industrial action at the establishment level. As explanatory factors, one must use variables which are also interpretable at the level of the establishment. The measure of task

[17] The six types of industrial action listed were: a strike; a work-to-rule; an overtime ban; a short stoppage; picketing; and a public demonstration.

[18] Although WIRS did not use exactly the same list of instances of industrial action, most industrial action consists of strikes, overtime bans, or working to rule, which were common to both lists.

discretion cannot be used, for it describes only the individual's position and not that of employees in general. We can, however, use the measures of advanced technology in the workplace, which are predicted to promote autonomy and reduce conflict. The greater the adoption of advanced technology, therefore, the lower should be the probability of conflict.

Strictly speaking the available measures of technical control and bureaucratic control, like those relating to discretion, refer only to the individual: the survey did not obtain information about how extensively the workplace had adopted these control systems. Nonetheless, such systems would usually only be introduced if applied to a substantial group of employees. So they can reasonably be interpreted as workplace variables, but they will if anything understate the presence of the systems, which could be in use for groups of employees other than the respondent's. Because of these uncertainties, one should not place too much reliance on the establishment-level analysis, which is needed primarily as a preliminary step for the analysis at individual level.[19]

The analysis took account of background organizational characteristics which could be expected to increase or reduce the average chances of industrial action taking place. These included size, industry, unionization and level of union influence, and the occurrence of various kinds of workplace change.

The results of the analysis are summarized in Table 3.9. The background influences on conflict at workplace level operated much as expected, which increases confidence in the analysis. There was a clear size trend, conflict becoming more prevalent with increasing size, and by far the largest factor was the simple presence or absence of unions. Differences in the perceived influence of unions, however, did not make any clear difference to the probability of industrial action. This may be because, as unions grow more powerful, employers adopt a more conciliatory approach.

The analysis once more failed to support the contentions of technological optimists. There was no indication that advanced technology reduced the occurrence of industrial action. If anything, where there was a high concentration of advanced technology, there was some indication of an *increased* tendency to conflict, although it fell short of statistical significance.[20] There was some support, on the other hand, for the prediction that formal control systems, and especially bureaucratic control, would be used to reduce or limit conflict. A more developed bureaucratic control system was indeed linked to a significant reduction in the likelihood of conflict at workplace level. The findings for technical control systems were less clear. Technical control showed some tendency to be connected

[19] Participation in industrial action can only be assessed for those who are employed at an establishment where industrial action has taken place, and these employees may be different from the remainder in ways which are not measured in the survey. The linked analysis at establishment level permits a statistical correction against this possible source of bias. See Maddala (1983) for further explanation.

[20] The t-statistic was 1.82 with $p = 0.07$.

TABLE 3.9 *Effects of discretion and control on workplace industrial action*

	Coeff.	Sig.
Advanced technology (compared with low level)		
medium level	0.03	n.s.
high level	0.14	n.s.
Technical control	0.05	n.s.
Bureaucratic control	−0.04	*
Changes at the workplace:		
reduction in numbers	0.15	*
new computers/automation	0.11	n.s.
other new equipment	0.10	n.s.
work organization	0.15	*
Union power (compared with no union)		
low union power	0.71	***
some union power	0.61	***
medium union power	0.78	***
high union power	0.57	***

Note: Bivariate probit estimates, no. = 3,098; jointly estimated with model for Table 3.10. Results for size and industry not shown.

to *increased* conflict, but this was not statistically significant. The shortcomings of the control system variables for an establishment-level analysis must be borne in mind here.

Individual Participation in Workplace Conflict

The second part of the analysis was confined to those who reported workplace conflict; these were asked if they had ever taken part in any of the conflict they reported. At this individual level, the analysis could be more complete and more direct. It considered the influence of task discretion on the individual's activity in conflict, the effects of increasing or decreasing tightness of supervision, and the influence of formal control systems which applied to her or him.

As before, the analysis included controls for a wide range of other variables. As well as gender, class, and age, these included work pressure in the job, involvement in team-working, the extent to which skills had increased, whether the employment contract was permanent or temporary, and tenure (time in the job). Finally, several of the variables used for the analysis of workplace industrial action were also included here: the size of the establishment, the presence and power of unions, and the occurrence of the various kinds of change. The rationale for including these again was that each might affect the outlook of individuals towards conflict. For example, one might expect to find more alienated attitudes among the employees of large organizations and this could lead to more militancy.

TABLE 3.10 *Effects of discretion and control on individual
participation in workplace industrial action*

	Coeff.	Sig.
Task discretion	0.01	n.s.
Bureaucratic control	−0.11	***
Technical control	0.01	n.s.
Change in supervision: (compared to none)		
increase	−0.06	n.s.
decrease	−0.05	n.s.

Note: Bivariate probit estimates, with selection on workplace action
occurring (see Table 3.9), no. = 536. Results for control variables
are not shown.

Table 3.10 summarizes the key results of the analysis of individual participation in conflict, excluding the numerous control variables. In this case, the results are particularly simple. The higher the level of bureaucratic control they experienced, the less likely were individuals to take part in industrial action. This effect was estimated with a high degree of reliability. The other aspects of control, including change in the tightness of supervision, were unrelated to individual participation in conflict, as was the degree of task discretion.

Thus, organizations which develop bureaucratic control systems can on average expect a lower incidence of workplace industrial action *and* lower levels of individual support for such action when it takes place. As predicted by Edwards, bureaucratic control systems constitute a powerful means of avoiding conflict and achieving compliance. On the other hand an increase in task discretion offers no barrier to individual participation in conflict, and this again seems more consistent with Edwards' predictions than those of Blauner or Walton.

To summarize this section, the single salient finding has been the effectiveness of bureaucratic control systems in diminishing conflict or increasing compliance. This stands out all the more in view of the lack of influence of the other aspects of control over the worker, and similarly the lack of influence of task discretion.

SUMMARY AND CONCLUSIONS

Task discretion has been increasing in Britain, but this does not mean that employees have been disengaged from systems of control. On the contrary, this survey's picture is one of extensive and expanding control systems.

Almost everyone is supervised, and a remarkable four in ten supervise others to some degree. In the managerial and professional class supervision is being rolled back in favour of more personal discretion and responsibility, but more manual workers are experiencing tight supervision than before. Formal systems of control are also in widespread use. Technical control, based on work-pacing and pay incentives, is applied to about three in ten employees, and bureaucratic control (or 'performance management'), based on target-setting, appraisal, and merit pay, to about four in ten employees, with a particular concentration in the managerial and professional class where discretion is also greatest.

This general picture does not lend support to those who have predicted a transformation of employment relationships under advanced technologies or under the influence of a new high-commitment strategy of human resource management. It is much more in accord with those who, like Edwards, stress the continuing necessity for employers to control their employees and, indeed, to develop new forms of control as the mode of production changes. Edwards' theory if anything *understates* the resilience of 'older' types of control. For example, supervisory favouritism was reported by 40 per cent of employees, spread evenly over every kind of organization. And large organizations were as likely to be tightening personal control by supervisors as were small or medium-sized organizations.

Advanced technology has been portrayed, by writers such as Blauner and Zuboff, as the key to an era of less coercion and more trust, co-operation and equality at work. In fact, however, advanced technology did nothing to check increases in tight supervision, and strongly favoured the application of both technical and bureaucratic control systems.

Another circumstance which strongly favoured both tighter supervision and the use of formal control systems, was the presence and power of unions in the organization. A reasonable inference, then, is that intensive control systems were preferred by organizations where managerial power was contested. This is consistent with Edwards' theory, which interprets control as (in part) a means of strengthening employer power or of weakening and dividing employee opposition.

The chapter tested conflicting predictions about how discretion and control affect work pressure or intensity. The prediction from one side was that discretion would undermine work pressure, which could only be sustained through managerial control. The prediction, from the other side, was that control could only produce moderate levels of work intensity, and that high levels would depend upon a relaxation of control and an increase in discretion—in short, a movement to a high-trust type of relationship.

The evidence to some extent supported both sides, but to some extent supported neither. The strength of control systems was strongly associated with high work intensity—bureaucratic control for non-manual employees, technical control for both non-manual and manual. Changes in personal supervision were also linked to high work intensity for non-manual though not for manual employees. Thus, the view that supervision and control systems enhance effort, rather than reduce or inhibit effort, was supported. On the other hand, task discretion also proved to be strongly associated with high levels of work intensity, rather than low.

There were also competing predictions about how discretion and control affect workplace conflict. The findings here were particularly simple. Task discretion did nothing to reduce individuals' participation in workplace action. It was bureaucratic control which produced compliance with management, just as the Edwards theory of control would suggest. This finding needs to be linked to the earlier point, that bureaucratic control was installed particularly where unions were present and influential. Of course, bureaucratic control is not simply and solely a means of producing compliance, but it would be hard to argue that that is not *one* of its important roles.

In conclusion, the supposed opposition between discretion and control does not appear to apply in practice. In recent years, employers have frequently pursued both more task discretion and more control, and on the whole workers have responded to both in much the way that employers would have wanted. There is therefore no reason to suppose that task discretion, in itself, points to a transformation of the employment relationship.

4

Participation and Representation

It was seen in Chapter 2 that there had been a striking rise in the skill levels of the workforce and that this was associated with an increase in the discretion that people could exercise in the work process. But were such developments accompanied by an increase in the level of participation in wider organizational decision-making? In the literature, there are highly diverse scenarios of the likely trends. Some have depicted the growth of participation as a necessary consequence of changes in skill and forms of technology; others have argued that there has been a reassertion of traditional managerial prerogatives, in which the influence of employees has been heavily cut back.

The argument for a necessary evolution towards more participative organizational structures takes a number of different forms. Perhaps the most general thesis was developed by Marshall, who suggested that there was an inner logic to the development of notions of citizenship. Where people had come to be treated as equal citizens with respect to civil and political rights, then it was likely that they would expect to be treated as equal citizens with respect to social and industrial rights. Over the longer span, then, there would be strong pressures for the extension of citizenship rights into the sphere of work. Others have stressed the implications of rising levels of skill. Employees with higher levels of education and skill might be expected to have greater expectations of being treated as full citizens of the organization in which they work. They will be more aware of the importance of their work for the organization and they will have the specialist knowledge that will give them confidence in putting forward their views.

From a rather different perspective, a number of writers have emphasized the advantages of participation for management. A considerable amount of research has suggested that participation is beneficial for employee motivation and the acceptance of organizational change (Touraine *et al.* 1965; Blumberg 1968; Gill *et al.* 1993). Denial of participation to higher-skilled workers would be particularly likely to lead to demotivation and to a decline in the quality of work. Given that such employees have greater responsibilities and discretion over the way they carry out their work tasks, the costs to the organization of low employee commitment could be very high indeed.[1] At the same time, management may get greater positive benefit from the involvement of more skilled employees

[1] This point was made very clearly by the US Commission on the Future of Worker-Management Relations (1994) in its report to the Secretaries of Labor and Commerce.

in wider organizational decision-making, since they possess valuable levels of knowledge about organizational activities. This may be reinforced by a growth in the interdependence of activities within the organization, resulting from the use of more sophisticated technologies. Wider organizational participation could help to make employees familiar with a broader range of activities and enhance their capacity to interact constructively with others.

There were some indications that from the mid-1980s an increasing number of employers were beginning to take seriously the need to enhance employee participation, partly influenced by the reputed success of Japanese work practices. For instance, there was an increased interest in the introduction of 'quality circles' where employees could meet regularly to express their views about ways of improving work practices (Hibbett 1991). These could be seen as representing a reversal of the trend towards the ever-increasing division of labour embodied in Taylorism, by giving employees a collective arena in which they could address wider issues about the organization of work. However, the prevalence of such practices remained unclear and there have been doubts about whether it is possible to transplant a system which developed in a culture where there was a strong sense of group loyalty and a much more consensual approach to the employment relationship. Some case-study research has raised a question mark about how real employee involvement has been and suggest that such practices may be regarded by employees primarily as a way in which management can intensify work (Martin 1994: 214–30; Geary 1995).

Moreover, in the context of the 1980s and early 1990s, there were grounds for doubt about whether any significant extension of participation was likely to have taken place. It was a decade in which public policy had placed a strong emphasis on the need to reinforce managerial prerogative and to cut back what was seen as the excessive influence that workplace representatives had acquired in British industry in the 1960s and 1970s. This led directly to a wide range of legislative measures designed to curb union power, by progressively abolishing the closed shop and making it more difficult to engage in strike action (see especially Undy *et al.* 1996). At the same time, as was seen in Chapter 3, it was a period which saw considerable innovations in managerial control strategies, with a growing emphasis on the individualization of relationships through the setting of individual performance targets, personal appraisals, and the linking of pay to performance. Arguably, an organizational culture based upon such direct relationships between management and individual employees would fit uneasily with the encouragement of forms of collective representation.

It also could be argued that the changing character of the national economy was detrimental to a participative ethos. The growth of competitive pressure and the rapidity of the restructuring of industry since the early 1980s were likely to have put a premium on management's ability to take hard decisions rapidly, unfettered by the delays that might accompany an attempt to win consensus through involving employees in the decision-making process. Further, there were signs of a trend towards an increased fragmentation of the workforce. Many service sector activities involved a high degree of geographical dispersal of the workforce, and, even in traditional manufacturing, the move towards the subcontracting of activities was undercutting the homogeneity of the workforce. If one adds to this the growth of non-standard working hours—with the development of part-time work, weekend, and shift work—there were grounds for thinking that the very process of organizing effective systems of workforce participation might have been becoming progressively more difficult.

In considering these different scenarios, it is important to recognize that participation has come to be conceptualized in very diverse ways. For some, it is a matter of direct individual involvement in decision-making, for others it is primarily a question of the growth of collective organizations that can defend employee interests at a higher level. There are also marked differences in the degree of employee control that is implied. The notion of participation covers procedures that at one end of the spectrum involve a willingness to consult the views of employees, with no commitment to acting upon them, to, at the other, processes of joint regulation where employee representatives have a right of veto.

Very broadly we can distinguish between four main modes of participation: direct participation, in which employees are personally involved in decisions that go beyond their immediate work task; communicative involvement, where employee involvement is sought through the development of better (two-way) communications; indirect consultative participation through some type of works council or joint consultative committee; and union representation, where the terms and conditions of employment are jointly regulated by the employer and representatives of the employees.[2] The extent to which these different forms of participation are mutually dependent or contradictory has been a matter of considerable debate. In principle, however, trends may well have been rather different with respect to each.

We will begin by looking at the extent of development of each mode of participation and the stuctural factors that influence this. We will then consider whether employers deliberately encouraged direct and consultative forms of participation with a view to undermining trade union representation. Finally, we will consider the implications of different types of participation for employees' responses to technical and organizational change and, more broadly, to management.

---

[2] While there is a general recognition of the value of conceptualizing these various channels of potential employee influence in terms of 'participation', there is still little in the way of an agreed set of distinctions. However, our concept of 'direct participation' is very similar to that of the European Foundation's EPOC study, where it is defined as 'Opportunities which management provide, or initiatives to which they lend their support, at workplace level for consultation with and/or delegation of responsibilities and authority for decision-making to their subordinates either as individuals or as groups of employees relating to the immediate work task, work organization and/or working conditions' (Geary and Sisson 1994: 4). Unlike the EPOC study, we exclude job enrichment from the concept of direct participation. Their concept of 'task participation' is broadly similar to the notion of 'task discretion' used here (Frohlich and Pekruhl 1996: 4).

Direct Participation

Direct participation is most likely to occur with respect to relatively local decisions about work organization, since these are the types of decision where management is particularly likely to benefit from employees' technical knowledge (Brannen 1983: chs. 4 and 5). Our indicator is based on a question asking people whether they personally would have any say in a decision made at their place of work that changed the way they did their job.

The first point to note is that only a small minority of employees in Britain (32 per cent) felt that they had any significant degree of influence over changes in work organization. Only 11 per cent thought they had a great deal of influence, and a further 21 per cent quite a lot of influence. In contrast, 50 per cent thought that they had no influence at all. The pattern was very similar for men and women, with 33 per cent of men reporting they had a great deal or quite a lot of influence compared with 30 per cent of women. In short, only about one-third of employees in British industry felt that they had any substantial degree of influence over decisions that could result in changes to their work.

Moreover, it is clear that this widespread lack of influence was perceived negatively by employees. Overall, 49 per cent thought that they should have more say over such decisions. While this was more the case for men than for women, the difference was relatively small (51 per cent compared with 47 per cent). Dissatisfaction with the level of participation was very closely linked to people's beliefs about how much say they currently had over decisions. Among those who felt that they already had a great deal of influence, only 18 per cent wanted more say. In contrast, 52 per cent of those who had little influence, and 65 per cent of those who had no influence, wanted an increase in participation.

Has direct participation, or at least satisfaction with it, changed over time? There is comparable data from a series of the British Attitudes Surveys, which suggests that the proportion of employees with a say in such decisions has actually declined over time (Hedges 1994: 49). Whereas in 1985 62 per cent felt that they could exercise some degree of influence, by 1992 the figure had fallen to 50 per cent. The more detailed sequence of the data suggests that the decline in participation occurred in the later 1980s. There has been a corresponding sharp decline in the level of satisfaction with existing opportunities for participation— from 63 per cent in 1985 to 51 per cent in 1992. Overall, there is no evidence of a trend in recent years towards higher levels of employee satisfaction with their level of involvement in decisions; rather this appears to be becoming an area of increasing employee discontent.

What factors, then, help to explain variations in the opportunities for direct participation? We shall consider in turn the impact of higher levels of skill, the presence of advanced technologies, managerial control policies, and structural characteristics such as the size and complexity of organizations.

TABLE 4.1 *Skill change and participation in decisions about work organization (cell %)*

	All	Men	Women
Occupational class			
Professional/managerial	45	49	40
Lower non-manual	28	29	28
Technician/supervisory	32	33	28
Skilled manual	25	25	27
Semi- and non-skilled	20	17	23
Skill experience			
Increased a great deal	45	47	42
Increased quite a lot	34	34	32
Increased a little	25	24	26
No change	27	31	23
Decreased	17	13	20

Skill. Does the evidence support the argument that higher skill levels tend to be associated with increased organizational participation? Taking first occupational class, as the broadest proxy of skill position, it is clear that there is a sharp break between professional and managerial employees, who have relatively high levels of say over changes in work organization, and all other employees (Table 4.1). Whereas 45 per cent of professional and managerial employees had either a great deal or quite a lot of say, the proportion fell to 28 per cent of lower non-manual, 32 per cent of technical and supervisory, 25 per cent of skilled manual, and 20 per cent of semi- and non-skilled manual workers.

If one turns from skill level to the extent to which people had experienced a change in their skills in recent years, there is again strong support for the view that employers may respond to a higher-skilled workforce by increasing organizational participation (Table 4.1). The main division is between those who have experienced substantial skill change and all others. Among those whose skills had increased a great deal, 45 per cent could participate in decisions about work organization, and the same was true for approximately a third (34 per cent) of those who had experienced quite a lot of change. In contrast, only 17 per cent of those whose skills had decreased felt they could participate, while 61 per cent said they had no influence at all.

In short, there seems considerable support for the view that the rise in skill levels is a powerful dynamic encouraging higher levels of direct participation.

Advanced Technology. Some have argued that there will be a growing pressure on organizations to become more participative as a result of the 'requirements' of advanced technologies. New computer-based technologies are seen as a major factor driving up skill levels, which in turn increases pressure for participation. Further, by permitting improvements in communications, they make possible a reduction of the managerial hierarchy and a move towards organizations with flatter structures. As traditional 'mechanical' organizational structures

TABLE 4.2 *Participation in decisions about work organization and organizational characteristics (row %)*

	Level of direct participation				No.
	No influence	A little	Quite a lot	A great deal	
New technology					
Uses new technology	45	19	24	12	1,926
Does not use it	56	17	17	9	1,494
Size of Establishment					
<10	39	17	25	19	531
10–24	49	19	21	11	486
25–99	49	19	21	11	786
100–499	57	19	22	7	790
500+	50	20	22	8	708

are discarded, employers are likely to adopt more open, horizontal systems of an 'organic' type, which encourage higher levels of employee participation.

As can be seen in Table 4.2, the use of computerized or automated equipment did indeed increase the chances that people would be involved in decisions about work changes. Overall, 36 per cent of those using advanced equipment thought they had a considerable say over such decisions, compared with 26 per cent of those that did not. However, this was primarily due to the difference that advanced technology made to men's experiences. This parallels the finding in Chapter 2 that, while advanced technology led to increased task discretion for men, this was not the case for women.

Why was advanced technology associated with higher levels of participation? It was seen in Chapter 2 that there was a close link between advanced technologies and skill. Establishments using new technology were likely to have substantially more skilled personnel, and, as has been shown, skill position was strongly linked to the likelihood of participation. When class and the qualifications required for the job are controlled for, the initially strong effect of advanced technology is heavily reduced, although it still remains significant in the full model (Table 4.4). The higher skill requirements of advanced technologies then are an important factor accounting for the greater participation in these work settings. But there is also some support for the view that such technologies have a wider effect in creating a work environment more favourable to participation, possibly by creating a greater interdependence in the work process.

Organizational size. Whereas rising skill levels and the adoption of more advanced technologies have been seen as favourable to increased participation, a powerful obstacle is frequently thought to be the large scale of organizations in advanced economies. As organizations increase in size, their rule structure becomes more elaborate and the decision-making process more complex. At the same time, there is an increasing specialization of functions. To ensure overall coherence of activities, there is a strong incentive for the decision-making process to become more centralized. As a result it may become more difficult to sustain a system in which individual employees are given any significant degree of direct involvement in decision-making.

Our data confirm that that there is a clear relationship between size of establishment and direct participation (Table 4.2). Employees in very small establishments (ten or fewer people) were the most likely to feel they could influence decision-making. While 44 per cent of those in the smallest-sized establishments thought they had either a great deal or quite a lot of influence, this fell to 32 per cent in establishments with between ten and twenty-four employees. However, there was relatively little difference by size in organizations that were larger than this. In the largest establishments (500+), the proportion feeling that they had significant influence was 30 per cent, only a little less than in the 10–24 category. The size effect persists even when a wide range of controls has been introduced (Table 4.4).

Taken on its own, size of establishment is perhaps a misleading guide to the degree of centralization of decision-making. The majority of employees (73 per cent) were in establishments that were subunits of a bigger organization. Arguably, even small establishments that are part of larger organizations are likely to be subject to central controls that may severely limit discretion at local level. This comes out rather clearly if size effects are compared for those who are or are not in independent establishments. Where employees were in establishments that were subunits of a wider entity, size makes no difference at all to the opportunities for direct participation.

Management systems of control. As was discussed in Chapter 3, a notable feature of the preceding decade was the effort by certain categories of employers to restructure the nature of the control systems in organizations. The more traditional forms of control system were either based on direct supervision or involved what has been termed 'technical' control, that is control through machine-pacing and pay incentives. The new emphasis has been on the need to move to more individualized systems based on target-setting, appraisal, and merit pay (variously termed 'performance management' or 'bureaucratic control'). To the extent that there has been a shift to more individualized forms of bureaucratic control, what effects could this be expected to have had on opportunities for employee participation?

There could be two rather different hypotheses about the implications of performance management or bureaucratic systems of control for employee influence. The positive view would be that they provide a better-structured and more readily available channel for employees to make their opinions known to supervisors. Such systems tend to be constructed around regular formal meetings between supervisors and supervisees, in which objectives are set and performance is assessed. Such discussions might lend themselves readily to consideration of the features of current work organization that needed to be changed, thereby giving employees a greater sense of participation. The negative view would be that such control systems, by individualizing the relationship between employee and manager, tend to weaken the collective strength of the workforce. By dissolving the cohesion of the workgroup, bureaucratic forms of control will release management from the need to concede to employees' view and allow it to exercise a relatively unhampered form of managerial prerogative.

As described in Chapter 3, the measure of bureaucratic control involved a number of different indicators. These included whether the effort and quality standards of work were influenced by targets that had been set, whether pay rises were given on the basis of merit, whether there was an appraisal system affecting the individual's rewards and prospects, and, finally, perceptions of chances for upward mobility in the organization.

Our evidence consistently points to the fact that the implementation of policies of bureaucratic control was associated with higher levels of employee influence. Taken individually, each of the items used to construct the scale was associated with higher employee influence, although target-setting had rather weaker effects than the chances of individual mobility, appraisal procedures, and merit pay. Similarly, the overall scale for bureaucratic control had a strong positive association with participation. In the case of bureaucratic control, it is particularly important to examine the impact of other variables, since (as was seen in Chapter 3) it was a system that was likely to be used for more highly-skilled employees who could be expected to have higher levels of participation on other grounds. However, even when factors such as class, qualifications, advanced technology, and establishment size were controlled for, the positive effect of bureaucratic control procedures for employee participation still came through at a high level of statistical significance (Table 4.4).

Quality circles. The developments discussed above were primarily concerned with constructing more individualized relationships between supervisors and employees. However, there was another direction in which deliberate efforts were made by management to counteract the demotivating qualities of large-scale organization. This involved trying to involve employees collectively in quality improvement. Small groups were set up at shop-floor level to give advice, and in some cases take decisions, about work procedures. Perhaps the most frequently cited practice was the creation of quality circles. The effectiveness of this type of experiment is rather poorly charted. Some evidence suggests that quality circles are received by employees with indifference or collapse rapidly as people become aware that they have little real influence. Given that the higher echelons of organizations usually remain based on hierarchical and centralized principles, such experiments, it is suggested, can appear as charades or, worse, as devices for manipulating employee opinion. Other research, however, has pointed to cases where they did appear to improve employees' sense of involvement in decision-making, with significant benefits for their motivation at work.

Given the range of experiments of this type, the survey question was deliberately couched in a broad way: 'Some organizations have groups of employees who meet regularly to think of improvements that could be made within the organization. These are sometimes called quality circles. Are you involved in a quality circle or similar group?' It is clear that such experiments involved only a relatively small minority of employees. Overall 20 per cent of employees reported that they belonged to a quality circle or equivalent, with membership slightly higher among men (22 per cent) than among women (18 per cent). Involvement was more common among non-manual than among manual employees: 30 per cent of professional and managerial employees, 18 per cent of lower non-manual employees, and 24 per cent of technical and supervisory employees belonged to a quality circle, whereas this was the case for only 12 per cent of skilled manual workers and 13 per cent of semi- and non-skilled manual workers.

The argument that one of the principal motives for setting up such groups was to counteract the alienating effects of large-scale organizations is supported by the fact that their frequency varied significantly by size of establishment. They were more than twice as common among employees in establishments with over 100 employees as among those in small establishments (ten or fewer). They were also more likely to be found in public service organizations than in private.

But were they seen by their members as providing any degree of real influence? The evidence indicates that their effects were generally quite positive. Of those who belonged to a quality circle 77 per cent thought they had either a great deal or quite a lot of influence, with 28 per cent using the strongest form of endorsement. Moreover, they could be effective in both small and large-scale establishments. If those in the smallest establishments were the most likely to attribute them a great deal of influence, it is notable that 77 per cent of those in both the largest and the smallest establishments thought they had either quite a lot or a great deal of influence.

The conclusion that they increased employees' sense of participation is confirmed by comparing the extent to which members and non-members felt they had a say over changes in work organization. Among those who belonged to a quality circle, 47 per cent thought they had a great deal or quite a lot of influence over changes in work organization, whereas among those that did not this was the case for only 28 per cent. The same basic pattern emerges for both men and women, although, as can be seen in Table 4.3, quality circles made a bigger difference to men's sense of participation than to women's.

Moreover, the adoption of such policies does much to undercut the effect of organizational size. As can be seen in Table 4.3, larger establishments with quality circles provided levels of direct participation that were very similar to those of small establishments without them. Indeed, the very largest establishments (500+) with quality circles had even higher levels of participation, perhaps reflecting the greater sophistication of such experiments in organizations with better-equipped personnel departments.

TABLE 4.3 *Quality circles, organizational size, and participation in decisions about work organization (cell %)*

	In quality circle: with great deal or quite a lot of influence	Not in quality circle: with great deal or quite a lot of influence
All	47	28
Men	53	28
Women	39	28
Size of Establishment		
<10	58	42
10–24	43	30
25–99	51	27
100–499	41	19
500+	48	24

Communicative Involvement

A second mode of participation involves the development of formal procedures for the diffusion of information about organizational activities. In contrast to the forms of direct participation discussed earlier, which could be quite localized in form and centred on the workgroup, this tended to be an organization-wide procedure. The aim was not to give employees a say about their immediate work environment, but rather to give them a stronger sense of participating in the life of the organization as a whole by providing information on company or organizational developments. This was the form of participation that has been shown to have grown most rapidly from the mid-1980s (Millward 1994; Wood and Albanese 1995).

The nature of such procedures could vary a great deal. At one end of the scale they might consist of little more than the diffusion of information on notice boards or in newsletters. At the other end, they might involve not only face-to-face meetings with employees, but provision for employees to make their own views known about wider organizational issues. A sequence of three questions was asked to establish the types of procedure used. They asked whether management issued information—for instance notices or newsheets—to give news of what was happening in the organization; whether management organized meetings where employees were informed about what was happening in the organization, and, finally, whether management held meetings where employees could express their views about what was happening in the organization.

Each of these forms of information provision was quite widespread, although it was more common for employers merely to distribute information (76 per cent) than to hold meetings in which employees could express their opinions (63 per cent). Given that the same organization might use multiple channels, our more detailed analysis uses an index of communicative involvement that has been created by summing the answers to the three questions.

As with direct participation, communicative involvement was strongly linked to skill level. The higher their skill, the more likely it was that people would have good access to information. More than three-quarters of professional and managerial employees reported that there were meetings about wider organizational matters in which they could express their views, whereas this was the case for only about half of manual workers, whether skilled or non-skilled. Similarly, there were wide class differences: in the overall measure of communicative involvement. The disadvantaged position of manual workers remained clear-cut even when a wide range of other factors had been taken into account (Table 4.4).

TABLE 4.4 *Factors related to direct participation and communicative involvement*

	Direct participation		Communicative involvement	
	Coeff.	Sig.	Coeff.	Sig.
Class (ref. Prof./manag.)				
Lower non-manual	−0.34	***	−0.12	*
Tech./supervisory	−0.24	**	0.00	n.s.
Skilled manual	−0.51	***	−0.21	**
Semi- and non-skilled	−0.47	***	−0.19	***
Skill increase	0.08	***	0.08	***
Size (ref. < 10)				
10–24	−0.39	***	0.35	***
25–99	−0.37	***	0.50	***
100–499	−0.60	***	0.48	***
500+	−0.47	***	0.61	***
Intensity of advanced technology (ref. low)				
Medium	−0.07	n.s	0.12	*
High	−0.08	n.s.	0.13	*
Personally uses advanced technology	0.11	*	0.04	n.s.
Public sector	−0.17	***	0.14	***
Part of wider org.	−0.31	***	0.53	***
Technical control	−0.03	n.s.	0.00	n.s.
Bureaucratic control	0.03	**	0.10	***
Quality circle	0.39	***	0.48	***
Non-standard contracts				
Part-time (ref. full-time)	−0.13	*	0.02	n.s.
Contract (ref. permanent)	−0.14	n.s.	−0.05	n.s.
Temporary (ref. permanent)	−0.39	***	−0.27	**
Constant	2.35		0.52	

Note: OLS regressions. The full models included age. The adjusted R^2 for direct participation was 0.16 (no. = 2,844), for communicative involvement 0.31 (no. = 2,868).

The use of new technologies is also linked to higher levels of communicative involvement. It is not so much whether or not the employee personally works with a computer or automated equipment that matters, but rather the general environment of the workplace. Once a quarter or more of the workforce is working with advanced equipment, there is a marked improvement in the quality of communications in the organization. Doubtless new technologies make the very process of communication much easier for management. But it should be remembered that the measure focuses specifically on information about wider developments in the organization, so it would still require a specific policy commitment by management.

In sharp contrast to direct participation, communicative involvement was far more frequent in larger organizations. The development of such formal channels may have been one of the ways in which management tried to compensate for the anonymity of large centralized workplaces. The differences are striking. Less than half of employees working in the smallest-sized workplaces reported any type of formal provision of information whether through documents or meetings. In contrast, in the largest establishments (500+), 92 per cent of employees received some type of printed information, 84 per cent said that there were meetings to inform them about what was happening, and 71 per cent said they were able to express their own opinions in such meetings. Presumably part of this difference may reflect the fact that information circulates more easily on an informal basis in small establishments, so there is less need for formal meetings. But it is interesting to note that the index of communicative involvement rises quite consistently through the different categories of establishment size, indicating that there are other mechanisms at work as well.

One probable explanation of this is that the successful implementation of a policy of communicative involvement depends upon the existence of a relatively well-resourced administrative department for handling employee issues. The existence of such departments is likely to favour a more strategic approach to the involvement of employees. At a practical level, the production and diffusion of regular information and the organization of meetings is easier where there are well-equipped personnel or employee relations departments. Such specialized departments are more likely to be found in large establishments, given the greater complexity of their overall organization of recruitment, labour allocation, and promotion. The Workplace Industrial Relations Survey showed that, as size of establishment increased, organizations were more likely to have specialist personnel staff and those with responsibility for such matters had higher levels of educational qualification. Moreover, in larger establishments more time was specifically dedicated to personnel matters, personnel management was more likely to be represented at board level and to have felt that its influence in the organization had been increasing over time (Millward and Stevens 1986: 20–44).

A further indication of the importance of the administrative sophistication of organizations is evident from the other aspects of employee management that were strongly linked to the degree of communicative involvement. Where management was committed to 'bureaucratic' control systems involving target-setting, appraisals, merit pay, and the development of internal labour markets, they were significantly more likely to have good communication systems. The same was true for organizations that had experimented with quality circles and for organizations that had adopted advanced technologies.

Policies of informational involvement, then, characterized large organizations, and were notably more frequent in those that had developed more sophisticated techniques for controlling employee performance. This points to their being part of a more or less conscious strategy on the part of management to counter the alienating effects of large-scale organization.

Indirect Consultative Representation

The two forms of participation examined so far involved a degree of direct contact with employees. However, a third approach was to provide indirect opportunities for consultation with the workforce through a body composed of elected representatives. The most familiar example of this is the works council, although, in practice, there has been a wide range of different types of consultative arrangement, with varying terminologies (Brannen 1983: chs. 2 and 3).

To assess whether some type of arrangement along these lines existed, people were asked 'Is there a works council or similar consultative committee in your establishment?' The first point to note is that this was by far the least common of the different modes of participation. Only 22 per cent of employees reported that there was a works council or equivalent in their establishment. This is consistent with evidence from employers, which indicates that fewer than a quarter of workplaces had consultative committees. Indeed, the proportion declined in the 1980s from 24 per cent in 1984 to 18 per cent in 1990 (Millward 1994: 77).

Moreover, where such a body existed it was rarely seen by employees as having significant influence. A series of questions were asked to establish the influence of the works council over five policy areas: the introduction of new types of equipment, the way work was organized, changes in working practices, the level of pay, and work hours. As can be seen in Table 4.5, its powers were seen to be greatest with regard to changes in working practices, but even in

TABLE 4.5 *Influence of works councils over organizational decisions (row %)*

	None	Not much	Fair amount	Great deal	No.
Introd. new equipment	21	38	34	8	657
Way work is organized	17	44	32	8	658
Changes in working practices	12	40	37	11	659
Level of pay	32	37	24	8	663
Work hours	30	35	27	7	660

this case 52 per cent thought that it had little or no influence. The proportion sceptical about the influence of their works council increased to 59 per cent for decisions about new equipment, 61 per cent for the way work was organized, 65 per cent for work hours, and 69 per cent for pay.

In many respects the factors that were associated with the presence of a works council were similar to those affecting whether or not organizations had developed formal informational policies. In particular, they were primarily characteristic of larger establishments. For instance, only 10 per cent of those in workplaces with ten to twenty-four employees and 7 per cent of those where there were less than ten people had any type of formalized consultative committee. In contrast, the proportion rose to 31 per cent of those in establishments of between 100 and 499 people and to 38 per cent in the largest establishments (500+). Even when a wide range of controls were introduced in a regression analysis, the likelihood of having a works council increased in a nearly linear way with size of establishment (Table 4.8). The effect of being in the public sector was also particularly marked. Whereas 30 per cent of those in public corporations and 29 per cent of those in public services reported the existence of formal consultative committees, the proportion fell to only 12 per cent in routine private services.

Employers that had generally adopted more sophisticated employee relations policies were particularly likely to have established works councils. For instance, there was a highly significant association between the implementation of forms of performance control based on targets, appraisals, and merit pay on the one hand and the existence of consultative committees on the other. Similarly, employees in organizations that had set up quality circles were also more likely to have formalized indirect representation. Potentially this might reflect the fact that bureaucratic control systems and quality circles were more frequently to be found in larger establishments. However, even when size was taken into account, both factors remained highly significant (Table 4.8).

There were, however, two important differences between the factors associated with works councils and those that affected communicative involvement. In the first place, there was no stable relationship with skill. While a lower proportion of semi- and non-skilled manual workers reported the presence of a works council, this effect was not significant once other factors were controlled. Second, there was no clear relation between the use of advanced technology and this type of representation. The main determinants of whether or not there were works councils appeared to lie in the problems posed by the complexity of large-scale organizations and in the general orientation of management towards more sophisticated procedures of employee control and motivation.

Trade Union Representation

The fourth and, in Britain, the most established form of participation is the influence that employees can exert through trade union representation. To the extent that

TABLE 4.6 *Influence of unions over organizational decisions (row %)*

	None	Not much	Great deal/fair amount	No.
New equipment	33	43	24	1,821
Work organization	33	43	25	1,865
Work practices	20	36	44	1,857
Pay	11	26	63	1,867
Hours of work	17	29	54	1,866

this embodies a right to negotiate, it is likely to provide a stronger form of employee participation (albeit of an indirect type) than the type of consultative procedures discussed earlier.

The measure taken is that of personal membership in a trade union. While unions may be present despite the fact that the individual is not a member, the assumption is that membership remains essential for a person to receive the full benefits of representation. One of the difficulties in assessing union membership strength is the diversity of names for organizations performing broadly similar representative functions. A particularly tricky borderline is between staff associations and conventional trade unions. Staff associations originally developed among non-manual employees as a form of representation that preserved a status distinction with respect to manual-worker organizations. In practice their functions can be rather varied. They may or may not exercise negotiating rights. The question used in the survey was devised so as not to exclude such bodies where they were effectively exercising trade union functions. It took the form: 'Are you currently a member of a trade union or similar organization?' Overall, 37 per cent of employees said that they were, with a somewhat higher proportion among men (44 per cent) than among women (37 per cent).

In marked contrast to the perception of works councils, unions were thought to exercise a significant influence with respect to certain areas of policy-making. People were asked how much influence the unions in their establishment had over five types of decision: the introduction of new types of equipment, the way work was organized, changes in working practices, the level of pay, and work hours. Perceptions of union influence varied a good deal depending on the specific issue. As can be seen in Table 4.6, it was thought to be greatest with respect to pay and to work hours and least with respect to the introduction of new equipment and patterns of work organization. While 63 per cent believed that the unions had either a great deal or a fair amount of influence over pay, only 25 per cent thought this was true for the way in which work is organized. However, on all issues, the unions were attributed substantially greater power than works councils.

Unlike the forms of participation discussed earlier, participation through union representation has declined heavily. Over the 1980s the unions experienced a sharp drop in their membership levels. Comparing with the level of membership recorded by the 1986 SCELI survey, which used exactly the same question format, union density had fallen from 45 per cent to 40 per cent. The data published by Department of Employment, derived from the Labour Force Survey, shows a decline to 37 per cent in 1991 (Beatson and Butcher 1993: 676–89). However, these figures conceal an important difference by sex. Union density appears to have held up much better among women than among men. Whereas 51 per cent of men were members in 1986, this was the case for 44 per cent in 1992. In contrast, among women, union membership rates remained virtually unchanged (38 per cent in 1986, 37 per cent in 1992).

There is already an extensive literature on the sources of this decline in membership (Millward 1994; Gallie *et al.* 1996). While it can be attributed partly to government legislative measures designed to weaken the unions, it is clear too that the unions were heavily affected by the economic restructuring of the 1980s, which cut employment particularly sharply in their traditional strongholds in manufacturing industry. But, while the broader context affecting union membership strength has been well charted, much less is known about the possible effects of the changes occurring in employment policies within firms. What has been the impact of the extensive changes in skill and technology on union membership? How far do the new forms of managerial control and new practices for worker involvement affect their position? How does the growth of a workforce employed on non-standard contracts affect union allegiance? With a cross-sectional study, causal imputation clearly needs to be handled with considerable caution. However, it is possible to see whether the data are consistent with the hypothesis that new managerial employment policies have tended to cut the ground away from the unions.

Skill level and skill change. Taking first the issue of changes in skill, is it likely that the shift to an occupational structure more heavily based on professional and managerial skills is less favourable to the unions? The trade union movement was the creation of the manual working classes, and it might be expected that strong class differentials have persisted. Given the fact that the share of the manual working class in the overall working population has declined dramatically since the early 1980s, occupational change might be seen as remorselessly leading to the demise of union strength.

However, if the distribution of membership by class is examined in Table 4.7 the striking feature is that union membership has become, if anything, even more common among professional and managerial (47 per cent) than among either skilled manual (42 per cent) or semi- and non-skilled manual employees (37 per cent). Further if the changes in union density since 1986 are examined by class, the smallest decline has been among the professional and managerial employees (1 per cent). This compares with an 11 per cent decline among skilled manual workers and 5 per cent for non-skilled manual workers.

TABLE 4.7 *Trade union membership by class, 1986–1992 (cell %)*

	All		Men		Women	
	1986	1992	1986	1992	1986	1992
Prof./managerial	48	47	46	43	53	52
Lower non-manual	39	35	52	42	35	33
Tech./supervisory	49	43	55	46	31	30
Skilled manual	53	42	55	43	38	33
Semi- and non-skilled	42	37	55	47	30	28

Why should union membership have held up better among professional and managerial employees than among manual workers? It appears to have had little to do with the process of skill change *per se*. Those who had experienced a very substantial increase in their skills were no more likely to be union members than those who had not. This was true for employees as a whole and for the specific case of the professional and managerial employees. Rather it reflected the fact that professional and managerial employees tended to be employed in a working environment that was more favourable to union recruitment and membership. Most crucially, they were more likely to be employed in larger organizations and they were more likely to be in the public sector. When size of establishment is controlled for, there is no longer a difference between professional and managerial employees and skilled manual workers. When account is taken of the public–private divide, it is the skilled manual workers that show a stronger tendency to be unionized, although professional and managerial employees remain as likely to be members as semi- and non-skilled manual workers. As can be seen in Table 4.8, this pattern persists even when a much wider range of controls has been introduced.

In short, skill change *per se* appears to be largely neutral in its implications for trade union membership. There is little difference in levels of unionization by class and such differences mainly reflect features of the working context other than skill level. There is no relation between the experience of skill change and the likelihood of being a trade union member.

Advanced technology. If skill *per se* is of little importance, what about the shift towards more advanced types of technology? There have been sharply contrasting arguments about the implications of automation for union membership. It has been suggested that, as a result of its effect in increasing skills, encouraging responsibility in the work task, and improving employment conditions, advanced technology would undercut much of the basis for union loyalties and lead instead to a high level of employee identification with the company (Blauner 1964). In contrast, others (for instance, Mallet 1969) were of the view that it would lead to stronger collective organization among employees, partly

TABLE 4.8 *Effects of employment characteristics on presence of works council and union membership*

	Works council		Union membership	
	Coeff.	Sig.	Coeff.	Sig.
Class (ref. Prof./manag.)				
Lower non-manual	1.01	n.s.	0.93	n.s.
Tech./supervisory	0.95	n.s.	0.73	n.s.
Skilled manual	1.52	*	1.51	**
Semi- and non-skilled	0.88	n.s.	1.11	n.s.
Skill change (ref. decrease)				
No change	0.88	n.s.	1.40	*
A little upskilling	1.19	n.s.	1.50	*
Quite a lot of upskilling	1.15	n.s.	1.22	n.s.
A great deal of upskilling	1.34	n.s.	1.30	n.s.
Size (ref. < 10)				
10–24	1.12	n.s.	1.99	***
25–99	2.08	***	2.59	***
100–499	3.41	***	3.16	***
500+	4.74	***	5.33	***
Intensity of advanced technology (ref. low)				
Medium	1.04	n.s.	1.04	n.s.
High	0.99	n.s.	1.08	n.s.
Uses advanced technology	0.85	n.s.	0.96	n.s.
Public sector	0.06	***	3.58	***
Part of wider org.	2.03	***	2.49	***
Technical control	1.00	n.s.	1.09	*
Bureaucratic control	1.13	***	0.92	***
Quality circle	1.71	***	1.02	n.s.
Non-standard contracts				
Fem. Part-time (ref. male ft.)	0.62	**	0.42	***
Contract (ref. permanent)	1.37	n.s.	0.81	n.s.
Temporary (ref. permanent)	0.77	n.s.	0.42	***

Note: Logistic regression analysis; multiplicative effects on odds. No. = 2,610 for works council; 2,836 for union membership.

because higher skill levels were associated with a greater awareness of the implications of organizational decisions and partly because such employees, with their firm-specific skills, were less easy to replace and had greater job security.

In practice, those that used advanced technology in their work were a little more likely to be unionized (43 per cent compared with 38 per cent). Similarly those who were in organizations where more than a quarter of employees at their job level were using computerized or automated equipment were more likely to be union members. But, as with skill level, these effects appear to result from the association between advanced technology and other aspects of the work situation. When other factors are controlled for (Table 4.8), technology is no longer significant. On closer analysis, the decisive factor is organizational size. Larger organizations are substantially more likely to make use of advanced technologies; once size is controlled for the effect of technology itself disappears.

New management employment policies. Rather more important are the new managerial policies for controlling work performance. Earlier a contrast was drawn between more traditional technical forms of control, relying on machine-pacing and output-based financial incentives, and bureaucratic forms of performance management that placed an emphasis on formal appraisal, merit-grading, and the provision of opportunities for promotion. Whereas the former tended to emphasize the common position of employees, the latter was designed to individualize the relationship between management and employees. Some have seen such policies as deliberately designed to undermine trade union influence. But even without adopting a conspiratorial view of managerial intentions, there was clearly a potential tension between the individualizing tendency of such policies and the sense of common fate and collective interest which has been seen to be the foundation of union solidarities.

Again it is essential to assess the impact of these policies taking account of the effects of organizational size and ownership sector. Both of these forms of control were more frequent in larger organizations. Similarly, bureaucratic control was more common in the public sector and technical control in the private sector. Taking just the direct relationships, there was no significant association between technical control and union membership, while bureaucratic control was linked to higher membership levels. But, once other factors are controlled for, the picture looks very different. Technical control is associated with a higher, and bureaucratic control with a lower, likelihood of being unionized. The argument that policies to individualize the relationship between management and the workforce tend to undercut the unions seems then to get a measure of support. Looking at the component parts of bureaucratic control, the two features that were particularly strongly linked to lower union membership were merit pay and performance-assessment procedures that affected earnings.

The other aspect of management policies that weakened the unions was the increased employment of staff on non-standard contracts. Female part-time workers were substantially less likely to be union members. Whereas 44 per cent of male and 42 per cent of female full-time workers were in unions, the proportion fell to only 30 per cent among female part-timers. In a similar way, short-term temporary workers (with contracts less than a year) were much less likely to be unionized (24 per cent) than either those on fixed-term contracts of one to three years (40 per cent) or permanent employees (42 per cent).

Ownership sector. Finally, it is clear that a fundamental factor affecting union strength was whether or not the organization was in the public or the private sector. This has been consistent across the decade. In the 1986 survey, 69 per cent of employees in the public sector were in unions, compared with only 31 per cent in the private sector. In 1992, although union density had declined in both sectors, the same general pattern emerges, with 63 per cent unionized in the public sector and only 28 per cent in the private.

Moreover, ownership sector influences the effect of other factors on union membership. To begin with, it mediates in an important way the commonly noted relationship between size of establishment and unionism. With the exception of the smallest establishments (with fewer than ten employees), size made very little difference at all in the public sector. For instance, 61 per cent of employees in workplaces with between ten and twenty-four employees were unionized, whereas among those in the very largest establishments (500+) the proportion was 68 per cent. In contrast, in the private sector, size of establishment had a strong effect. The proportion who were union members rose in a linear way across the difference size categories, from 8 per cent in the smallest establishments to 53 per cent in the largest.

Further, ownership sector makes a difference to the impact of the new types of managerial policies for controlling performance. It was seen earlier that technical control, which relied primarily on machine-pacing and output incentives, tended to increase union strength, while bureaucratic performance management reduced it. However, if the analyses are run separately for the public and private sectors, it turns out that the nature of the managerial control system only has an impact in the private sector.

Finally, while the employment of people on non-standard contracts is associated with lower levels of trade unionism in both sectors, the effect is much less marked in the public sector. This is not immediately apparent in terms of the simple percentage differences. However, once other factors such as age, job level, size of establishment, and technology are controlled for, the difference stands out clearly. The relative odds of unionization of temporary workers compared with those of permanent workers fell from 50 per cent in the public sector to 29 per cent in the private sector. Similarly, the relative odds of union membership of female part-timers compared with those of male full-timers fell from 49 per cent in the public sector to 32 per cent in the private.

In short, whether or not a person works in the public sector has a major impact on the likelihood they will be unionized. Moreover, several of the other factors that make an important contribution to general union levels—such as size of establishment, management employment policies, and the use of non-standard contracts—seem mainly to have their effect in the private sector.

EMPLOYER STRATEGIES AND FORMS OF PARTICIPATION

The main focus of employer policy in the last decade has been on the development of direct and communicative forms of participation, while trade union

TABLE 4.9 *Change in employers' attitudes to trade union membership, 1986–1992 (cell %)*

	Employees reporting that employers encouraged membership		Employees reporting that employers accepted membership	
	1986	1992	1986	1992
All employees	21	14	41	33
Men	22	13	42	36
Women	20	14	39	29
Employees in trade unions	37	24	55	52
Non-unionized employees	7	5	27	18

representation has been in sharp decline. Does this suggest that the development of new forms of participation represented a deliberate policy by employers to undermine traditional forms of participation through union representation? The assumption underlying such a view is that there has been an important shift in employer attitudes to representation over the decade. In the past, British management was widely thought to have come to an accommodation with union power, preferring the relatively ordered character of joint regulation of the terms of employment to the unpredictability of an unmediated relationship with the shop-floor. However, the experience of considerably enhanced union power in the 1970s, together with the emergence of radical conservative policies in government, have encouraged a reassertion of employer prerogatives (Beaumont 1987).

We do not have the type of direct data that would be needed to establish whether there was such a change in employer policies. However, it is possible to examine employees' perceptions of the attitudes of their employers to trade union membership. People were asked whether their employer encouraged trade union membership, accepted it, discouraged it, or whether it wasn't really an issue at their workplace. The responses can be compared to those from the 1986 SCELI survey.[3]

As can be seen in Table 4.9, the evidence does suggest that there had been a hardening of employer attitudes to the unions since the mid-1980s. Whereas in 1986 a majority of employees (62 per cent) thought that their employers were either encouraging union membership or at least willing to accept it, by 1992 this was true of less than half (47 per cent). The picture is very similar whether

[3] While the response options were identical for the two items indicating employer favourability to the unions, there was a difference in the wording for one of the responses that suggested that conditions were unfavourable to the unions. Instead of 'it wasn't really an issue at the workplace', the 1986 wording of the fourth option was that the employer was 'unaffected because the employees weren't interested in them (i.e. the unions)'. This may affect the distribution of responses between the two items indicating an unfavourable environment for the unions. The analysis will focus on the items indicating favourability, which are directly comparable.

one takes the reports of male or of female employees. The change did not simply reflect the growth of non-unionized workplaces. Even people that were in unions were substantially less likely to report employer encouragement (37 per cent compared with 24 per cent). All occupational classes were affected, although the decline in employer favourability to the unions was most marked for professional and managerial, technical/supervisory, and skilled manual employees.

There are certainly signs that employers have become less favourable to unions, but did this mean that they encouraged other forms of participation as a means of undermining union strength?

There is no evidence that this was the case. Where employers were seen as discouraging union membership, employees were no more likely to report direct participation over issues of work organization than where employers were favourable to unionism. Whereas 26 per cent of employees with anti-union employers had either a great deal or quite a lot of say over such issues, this was the case for 30 per cent of those with employers that encouraged membership and 28 per cent of those that accepted it. A similar pattern emerges with respect to communicative involvement. It was where employers either encouraged or accepted union membership that employees were most likely to have access to meetings in which they could express their views about wider organizational matters (72 per cent in each case compared with only 56 per cent where employers discouraged membership). Finally, works councils were also more likely to be found where employers were favourable to the unions. Among employees in workplaces where membership was encouraged 37 per cent had a works council compared to 33 per cent where it was accepted and 22 per cent where it was discouraged.

In short, there is no evidence that the implementation of non-union forms of employee participation was a deliberate policy adopted by employers to discourage union representation. Employers that sought to discourage union membership were also likely to discourage all other forms of participative involvement.[4]

Nonetheless, it might still be the case that the introduction of such practices had the effect (whether willed or not) of weakening commitment to trade unionism. This can be examined by looking at the relationship between participative policies and people's reports of how their attitudes to the unions have changed in recent years. A question in the survey asked: 'Thinking back over

[4] This is consistent with the evidence from the employer-based Workplace Industrial Relations Survey, which showed that (with the exception of meetings with senior management) the greatest frequency of, and the principal growth in, forms of internal company communication occurred in establishments where unions were recognized (Millward 1994: 86, 93). Case-study research also confirms the positive effect of the presence of unions and casts doubt on the extent to which non-union employers developed coherent alternative strategies for involving employees (Marchington *et al.* 1992: 40, 49; McLoughlin and Gourlay 1994). Finally, European-wide research has found that there is now relatively little union distrust of forms of direct participation, which is regarded as complementary to and, possibly, helpful to the influence of union representatives (Regalia 1995; Frohlich and Pekruhl 1996: ch. 6).

the last five years, has it become more important or less important for you to be a trade union member, or has there been little or no change in its importance?' Despite the major erosion of trade union membership, it is interesting to see that there is no evidence of an overall decline in people's attachment to the principle of trade unionism. Although 15 per cent of employees had become less favourable, 19 per cent had come to feel that it had become more important to become a trade union member. This confirms the primarily structural nature of the forces affecting membership and undermines arguments emphasizing the role of deeper cultural shifts (for instance, the growth of more individualistic attitudes towards employment).

However, there is some indication that the provision of alternative forms of representation did weaken the sense that unions were essential. Employees that were in organizations that provided direct participation were less likely to have become more favourable to the unions (14 per cent compared with 21 per cent). That this is relatively independent from employers' intentions is confirmed by the fact that the relationship remained highly significant even when employers' attitudes to trade union membership were taken into account.

In short, there is little evidence that employers deliberately set out to introduce specific forms of employee participation in order to undercut union influence. There is some support for the view that direct participation has the indirect effect of reducing employees' sense of the necessity for union membership.

THE IMPLICATIONS OF PARTICIPATION

What are the implications of the different types of participation for employees' attitudes to technical and organizational change? A wide array of case studies, using diverse methodologies, have pointed to its beneficial effect in reducing the anxiety accompanying change and thereby leading employees to adopt a more positive attitude to new forms of work organization (Blumberg 1968). The limitation of these studies, however, is precisely that they were case studies.[5] While their design frequently meant that they were able to make relatively strong statements about the direction of causality, showing that higher levels of motivation followed rather than preceded specific procedures for introducing change, it remained unclear how far the results could be generalized to the wider workforce. There are good reasons for thinking that the organizations that are most willing to open their doors for in-depth case-study research are far from representing a random cross-section. They are likely to be organizations where, as a

---

[5] An exception is the European Foundation's large-scale survey of participatory practices in European companies, based on interviews with managers and employee representatives (Gill *et al.* 1993). This study, however, also has major problems of representativity since it excluded smaller enterprises and those in which there was not some type of formal employee representation. Nonetheless, the general findings of this study on the effects of participation for both social relations and decision-making in the organization are highly consistent with our own results.

result of the characteristics of both management and employees, there is a relatively consensual culture and an established pattern of good employment relations. A central question, then, is whether the favourable impact of participation on attitudes to change can be shown to hold across very different types of work situation. Further, given the variety of forms of participation that have been practised, are some forms more effective than others?

If participation leads to a more favourable response to technical and organizational change, it might also be expected to have longer-run consequences for employees' attitudes to management. The successful resolution of the potential tensions surrounding the reorganization of work practices is likely to contribute to an enhanced sense of common purpose and to a belief that management makes a genuine effort to take account of the interests of its employees. Over time, this could affect the extent to which management is perceived as legitimate by its workforce.

Participation and Attitudes to Technical and Organizational Change

To assess the extent to which people had experienced some type of technical or organizational change over recent years, people were asked: 'In the last two years have any of the following changes occurred in your organization that have affected your job?' The list comprised:

- a reduction in the number of people employed,
- the introduction of computerized or automated equipment (excluding word-processing),
- the introduction of word-processing,
- the introduction of other new equipment, and
- a change in the way work is organized.

The commonest forms of change that people had experienced were changes in work organization (51 per cent) and workforce reductions (46 per cent). However, recent experience of technical change was also widespread. Over a third of employees (38 per cent) had been affected by the introduction of automated equipment, 20 per cent by the introduction of word processors, and 36 per cent by the arrival of other types of new equipment. Mens' jobs had been more affected by technical change than womens'. For instance, 43 per cent of men, but only 32 per cent of women, mentioned the introduction of automated equipment. However, there was much less difference with respect to the experience of work reorganization, which had been very common for both men and women (52 per cent and 50 per cent respectively).

How did employees react to this very widespread process of change in the working environment? Did they view it as undermining work practices and relations to which they had had a strong attachment, for benefits that largely accrued to management? Or did they feel that employees themselves had shared in

TABLE 4.10 *Effects of participation on attitudes to technical and organizational change (with control variables)*

	Direct participation		Information		Works councils		Unions	
	Coeff.	Sig.	Coeff.	Sig.	Coeff.	Sig.	Coeff.	Sig.
Positive towards recent								
Workforce reductions	0.11	***	0.04	n.s.	−0.08	n.s.	0.02	n.s.
Computerized or								
automated equipment	0.07	**	−0.02	n.s.	−0.02	n.s.	−0.05	n.s.
Word processors	0.10	**	0.00	n.s.	−0.07	n.s.	−0.05	n.s.
Other technical change	0.05	*	0.02	n.s.	−0.11	*	−0.12	*
Work reorganization	0.18	***	0.06	*	0.06	n.s.	−0.12	*

Notes: The results are drawn from a series of OLS regressions. Control variables were age, sex, establishment size, occupational class, experience of skill change, and public/private sector.

the benefits and had improved their position as a result of the changes? Those affected by a change were asked how much they thought that it had benefited the employees. In practice, the nature of the response depended very much on the type of change in question. A substantial majority (70 per cent) were of the view that the employees had not benefited from reductions in the workforce. In sharp contrast, the introduction of new equipment was very widely seen as having brought advantages. For instance, 75 per cent of employees whose jobs had been affected by the introduction of automated equipment or word-processing thought that employees had benefited a great deal or a fair amount, while 79 per cent said the same with respect to the introduction of other types of new equipment. The most significant split in opinion was with respect to changes in work organization. While, overall, this was more frequently seen as having brought advantages, a substantial minority (40 per cent) took a rather more negative view.

Was it the case that employees that were in establishments that had introduced participative practices were more favourable to technical and organizational change than employees subject to more unilateral management authority? It is clear that the effect on employee attitudes to change varied very substantially depending upon the particular type of participation (Table 4.10). The least effective were works councils and communicative involvement. Works councils made no difference to the way employees viewed most types of change, although they were associated with a more negative view about the introduction of more traditional technologies. Similarly, the adoption of good channels of information about organizational developments had only one significant effect: it made employees more favourable to changes in work organization.

Union representation and direct participation had more extensive effects, but in quite opposite directions. Union membership was associated with greater scepticism that employees benefited either from traditional forms of technical change or from changes in work organization. In sharp contrast, where employees could directly participate in decisions about work organization, they were more positive about *all* forms of change. At least with respect to direct participation, our nationally representative data provide striking confirmation of earlier case-study research that the greater the involvement of employees in decisions about work organization, the more favourably they will react to the changes that affect them.

Participation and Attitudes to the Employer

If participation increases satisfaction with the outcomes of decision-making in an area as potentially difficult as that of technical and organizational change, it is possible that it has a wider impact on the way management is viewed. Two issues are of particular interest. The first is how it affects perceptions of the effectiveness of the organization and the second is its implications for views about the quality of relations between employers and employees.

The implications of participation for effectiveness is perhaps the most controversial issue. It may heighten efficiency by bringing a wider range of relevant knowledge to bear on decisions. Further, by producing a higher level of consensus about decisions, it may lead to more co-operative work patterns and to a greater willingness to put in discretionary effort. However, against this, participative systems may reduce effectiveness by making it less easy to reach decisions rapidly and by encouraging a form of decision-making that gives priority to the protection of vested interests.

The notion of effectiveness covers not just cost minimization, but the capacity for coherent longer-term planning and the ability to harness well the resources available to the organization. It must be emphasized that the data derives entirely from employees and may therefore give a rather partial view of effectiveness: the view from the shopfloor. Nevertheless, it would be surprising if this did not reflect an important component of effectiveness as more broadly conceived. The questions that have been used as measures of effectiveness focus specifically on the perception of the wider organization, rather than on the person's assessment of their own work performance. (The influence of participation on personal effort will be considered in a broader discussion of the determinants of individual work performance in Chapter 10.) People were explicitly asked to think about the whole organization rather than just the part in which they worked, and to assess their organization on a range of criteria using a five-point scale. Four items are of particular relevance here:

- How well are changes planned in this organization?
- How much coordination is there between departments in this organization?
- How keen is management on new ideas and improvements in this organization?
- How efficiently is work carried out in this organization?

TABLE 4.11 *Effects of participation on perception of employer efficiency*
(with control variables)

	Direct particip.		Information		Works councils		Unions	
	Coeff.	Sig.	Coeff.	Sig.	Coeff.	Sig.	Coef.	Sig.
Planning	0.18	***	0.16	***	−0.03	n.s.	−0.15	**
Coordination	0.18	***	0.13	***	−0.05	n.s.	−0.11	*
Innovation	0.12	***	0.28	***	0.10	*	0.09	*
Efficiency	0.08	***	0.08	***	−0.05	n.s.	−0.11	**

Notes: The results are drawn from a series of OLS regressions. Control variables were age, sex, establishment size, occupational class, experience of skill change, and public/private sector.

Both direct participation and communicative involvement were associated with a stronger belief that the organization was effective on all of the measures. For instance, where people were in organizations that allowed employees a great deal of say in decisions about work organization, 64 per cent were positive about the quality of coordination in the organization and 56 per cent about the quality of planning. In contrast, where there was no provision for employee involvement, the proportions fell to 39 per cent and 30 per cent respectively.

In Table 4.11, the relationship between participation and the different measures of effectiveness are presented, taking account of the effects of age, sex, establishment size, skill composition, and ownership sector (public/private). The effects are estimated simultaneously for the different types of participation. It is notable that, even when allowance is made for all of these factors, both direct participation and the quality of information provision come through as highly significant on each measure. Indeed, these effects persist even with more elaborate models that take account of the utilization of advanced technology, the nature of organizational control systems, and the contractual status of employees.

The pattern that emerges for the presence of a works council in the organization is very different. Whether one takes the simple relationships, or the effects once other structural factors have been controlled for, the conclusion is the same. There is no evidence that works councils are linked to better planning or coordination. The one area in which works councils may enhance effectiveness relates to the openness of management to consider new ideas and improvements. However, the fact that this stands quite apart from other considerations of effectiveness suggests that works councils are only likely to have positive implications for performance when linked to other forms of participation.

Finally, participation through union representation was associated with a generally more negative view of the effectiveness of work arrangements. As union representation was so closely linked to organizational size, it is particularly important to consider the effect of control variables. However, as can be seen in

Table 4.11, the basic pattern of the data remains unchanged, although the significance levels are lower than in the simple relationships. It is only when account is also taken of organizational control systems, technology, and the contractual status of employees, that the relationship becomes non-significant with respect to coordination and efficiency, although unionism still remains associated with poorer planning. Clearly the precise nature of this link between unionism and perceived effectiveness is debatable. The process of joint regulation may slow down decision-making and lead to compromise solutions that have a significant impact on effectiveness. Alternatively, unionism as an oppositional form of representation may highlight organizational inefficiencies and colour perceptions of management competence.

In short, the impact of participation on organizational effectiveness seems to depend heavily on the specific mode of participation under consideration. There seems little doubt that, at least in the eyes of employees, participation in the form of direct involvement in decisions about work organization and in the provision of good communications is linked to higher levels of organizational effectiveness. However, there is no comparable evidence for works councils, while union representation is associated with perceptions of lower effectiveness at least with respect to the planning of change.

The second potential effect of participation for the wider organization relates to employees' sense of social integration. There are a number of ways in which it might heighten social integration. The process of involvement in decision-making may encourage closer working relations between employees and lead to a stronger sense of social cohesion and social support. There is a considerable body of theory that suggests that this in turn will lead to more positive attitudes to the organization more generally. Second, the provision of opportunities for participation may influence directly interpretations of the characteristics of management, encouraging an image of management as genuinely concerned for the welfare of its employees.

The measures of social belonging and of the orientation of the organization to employee welfare followed the same procedures as the measures of effectiveness. The wording of the items was as follows:

- How much of a friendly, family atmosphere is there in this organization?
- How much does this organization care about employees' well-being?

Again both direct participation and the quality of communications in the organization behaved in fundamentally the same way, although they had strong effects independently of each other. In both cases, they were linked to a sense of the organization as a friendlier place to work in and they appeared to encourage an image of the employer as concerned about employee welfare. The percentage differences were substantial. Among those who had a great deal of say in questions about work reorganization, 67 per cent thought that there was a friendly atmosphere at work and 72 per cent considered it an organization that cared

TABLE 4.12 *Effects of participation on the perception of employer–employee relations (with control variables)*

	Direct participation		Information		Works councils		Unions	
	Coeff.	Sig.	Coeff.	Sig.	Coeff.	Sig.	Coeff.	Sig.
Friendliness	0.25	***	0.10	***	−0.16	**	−0.27	***
Caring	0.28	***	0.16	***	0.04	n.s.	−0.25	***
Communications	0.32	***	0.31	***	0.06	n.s.	−0.20	***
General relations	0.19	***	0.11	***	0.01	n.s.	−0.22	***

Notes: The results are drawn from a series of OLS regressions. Control variables were age, sex, establishment size, occupational class, experience of skill change, and public/private sector.

about the well-being of its employees. In contrast, among those that felt they had no say over such matters, these proportions fell to 40 per cent and 39 per cent respectively. Factors such as the size of the establishment may clearly be very important in determining the anonymity or warmth of social relations. But, as can be seen in Table 4.12, the effects of these forms of participation remained highly significant when establishment size was controlled for, along with age, sex, skill composition, and ownership sector. Indeed, this remained the case even with more complex models taking account of forms of organizational control, technology, and contractual statuses.

In contrast, there was no evidence that the existence of a works council or of union representation were associated with more favourable attitudes in these respects. The presence of a works council was somewhat surprisingly related to the perception of social relations as cooler rather than warmer. While this might have simply reflected the fact that works councils were more common in larger organizations, the effect persists even with the full range of controls for structural characteristics. Further works councils made no difference at all to whether or not the organization was perceived as caring. Trade union representation was associated with negative attitudes to the social atmosphere in the organization and to the importance given to employee welfare. It is possible that oppositional forms of participation lead to a degeneration in the social climate in the organization. Alternatively, a poor social climate may be a factor that helps to encourage union organization.

Given their positive influence on employees' sense of social belonging and their perception of the importance that the organization attached to employee well-being, it might be expected that direct participation and more developed communication systems would also lead to a generally favourable attitude to the quality of management–employee relations. This is quite clearly supported by the data.

People were asked: 'In general, how would you describe relations in your organization between management and employees?', with responses on a five-point scale running from very good to very bad. Among those who had a high level of involvement in decisions about work reorganization, 46 per cent described relations as very good, whereas among those that had no say the proportion fell to only 16 per cent. The effect of more developed communications was also significant but rather less dramatic: 27 per cent of those who reported that the organization held information meetings in which employees could express their views thought relations were very good, compared to 20 per cent where this type of meeting did not exist. As can be seen in Table 4.12, these effects persisted even given a wide range of controls. Consistently, given the earlier findings, the presence of a works council made no difference at all to the perceived quality of relations once structural factors had been controlled, while trade union representation was associated with a more conflictual sense of management–employee relations.

CONCLUSION

The picture that has emerged with respect to the opportunities for employees to participate in wider decisions in their workplace is a good deal less optimistic than that for developments in skill and in the quality of work tasks. The two most significant forms of participation that emerged from the analysis were direct participation and trade union representation. Direct participation in issues of work organization affected only a third of the workforce and there is some evidence that it declined from the mid-1980s. Trade union membership also declined heavily over the period and by the early 1990s less than 40 per cent of employees were members of unions. Overall the capacity of employees to affect their employment conditions appears to have diminished over time, although they may have become better informed about organizational activities.

An important part of the explanation of this is likely to be a shift in employer culture towards employee representation. There was evidence of a significant decline in employer favourability to union membership compared to the mid-1980s, which affected employees in all occupational classes. It might have been thought that employers that discouraged unions would have been actively seeking to develop other forms of participation as a way of winning the loyalty and motivation of their employees. But this was not in fact the case. Employers that discouraged unions also appeared to be the least likely to have set up forms of direct participation or even to have developed good communication systems in their organizations.

However, the evidence suggests that there were other aspects of change that may have more positive long-term implications at least for non-union forms of participation. Upskilling and the use of new technologies were associated with more frequent involvement in decision-making about work organization, providing support for the arguments that employers have an interest in increasing the involvement of a more skilled workforce in a steadily more complex and integrated work environment. It was also notable that, where employers had gone furthest in adopting relatively sophisticated performance-management systems, employees were more likely to be directly involved in wider decisions and they were more likely to have formalized and two-way communication systems.

It is clear that the sheer scale of organizations was an important factor affecting the prevalence of different types of participatory system. Direct participation was more common in small establishments, while the use of formalized communication systems and indirect representation either through works councils or unions were much more characteristic of large-scale establishments. Nonetheless our evidence suggests that there is nothing deterministic about the effects of size. Where employers in large-scale organizations had taken the initiative to develop quality circles, they were able to give their employees a significant sense of direct participation over their work environment.

The rather weak level of development of participatory arrangements that has prevailed in British industry appears to have had serious implications for the way employees viewed their organizations. It is notable that, where there were opportunities for direct participation, employees were more favourable to technical and organizational change, they were more likely to report that their organizations were efficiently run, they found the organization friendlier and more caring, and finally they reported far better relations between employees and management. This is entirely consistent with the reported benefits of participative arrangements in countries such as Japan and even the USA (Lincoln and Kalleberg 1990). In a period, then, of exceptionally rapid technological change and competitive pressure, British employers very widely failed to carry through the institutional reforms in their organizations that would have enhanced cooperation in employment relationships and led to a higher level of social integration of their employees.

5

The Growth of Job Insecurity

One of the most salient changes in the labour market over the last two decades was the return of mass unemployment. It was a period marked by two major recessions in the early 1980s and the early 1990s in which unemployment returned to a level comparable with the interwar years. The broad pattern of change is familiar from official statistics. In this chapter, we want to consider how the rise of unemployment has affected people's experience of employment. Did the rise in unemployment mean that there was an overall decrease in the stability of people's experience of employment? What factors accounted for people's vulnerability to unemployment? What was the effect of unemployment on individuals' subsequent work careers?

Taking a historical perspective, we first consider how employment stability has changed over time, looking at the way that career trajectories have altered since the 1950s. Unemployment is only one of the factors that helps to determine the overall stability of employment careers. Work careers also can be disrupted by people leaving the labour market whether out of choice or constraint. The trends in these different sources of instability may reinforce or counterbalance each other. In order to assess the overall stability or instability of people's experience of employment, it is essential to be able to take into account the various potential sources of instability and the way these interconnect in the experience of individuals. This requires a relatively rare type of data—detailed individual work histories—that makes it possible to reconstruct in detail the changing pattern of people's experience over time. The data are examined separately for men and for women and then we seek to assess whether there has been a polarization over time in men's and women's experiences of employment stability.

Our second objective was to look at the factors that affect people's job security, examining first vulnerability to unemployment in the work histories, and second people's current perception of the security of their employment. The analysis of vulnerability to unemployment focuses on the issue of whether it is primarily the characteristics of the individuals themselves that increase their risk of unemployment or whether it is the nature of the structural conditions in which they find themselves. The subsequent analysis of the job insecurity of the currently employed examines more closely the implications for job security of the organizational contexts in which people worked, for instance the effect of working in particular economic sectors or of being in an organization that had adopted advanced technology.

Finally, we examine the impact of experiences of unemployment on the quality of people's jobs in their subsequent work careers. Did unemployment have consequences for the types of jobs that people were able to get later and hence for their longer-term work careers or was it a relatively transitory experience, with people recovering their occupational position in a relatively short period of time? This analysis also requires data that enables us to go back into people's work histories, so that we can compare the quality of the current jobs of those with and without previous unemployment experience.

HISTORICAL CHANGE IN CAREER STABILITY

In reconstructing the changing nature of career trajectories across time, we are able to make use of the rich work-history data in the surveys. We have used not only the employee sample, which has formed the basis of the analysis in earlier chapters, but the *combined samples* of 3,855 people in employment (employed and self-employed) and of 1,000 unemployed people. In earlier periods, a large proportion of the currently unemployed were in employment and our historical picture of change might be severely distorted if their experiences were not taken into account. The employed and unemployed work-history datasets were merged and weighted to give as accurate a picture as possible of the general labour market population.

The work-history data we collected contains all labour market events of at least one month recalled by the 4,855 respondents since they first left full-time education. It includes not only periods of employment (with detailed information on the types of jobs held), but also people's experiences of unemployment and the periods when they were out of the labour market altogether (for instance to bring up children).

There are potential pitfalls with retrospective work-history data. Exact dates may sometimes be hard to recall, so some approximation has to be allowed. More seriously, it seems plausible that people may have a tendency to forget short events, especially those in the distant past. If short events in the work histories tend to include non-work events such as unemployment or sickness, then unemployment experience will be underestimated. In practice, a recent assessment of the types of event that are likely to be forgotten (Campanelli and Thomas 1994) indicates that this occurred most frequently with respect to job changes where the person had remained with the same employer (e.g. promotions etc.). This was followed by changes that involved jobs of a highly casual nature or part-time jobs which were taken while looking after family dependants. Unemployment spells were actually third on the list and tended to be forgotten only if they were very short.

Charting Career Stability

The method that has been adopted for examining work careers is that of sequence analysis.[1] Sequence analysis can be used to show the trends in patterns of employment stability and to trace the frequency of the occurrence of non-work events in people's lives. For every respondent a ten-year period is selected (for instance, when they were aged 20 to 29), which is referred to here as a 'life-stage'. The activities recorded during this time are either scanned and categorized as a whole into certain predetermined groups or searched for the occurrence of a particular activity (such as unemployment). If respondents are of different ages at interview, the patterns found for each life-stage will correspond to different years in historical time. The proportions of people with specific patterns can thus be plotted against their birth cohort to obtain a chronological picture of changing employment patterns.

In this analysis three life-stages have been chosen: the 20–9-year-old period, the 30–9-year-old period, and the 40–9-year-old period. These will be referred to as the twenties', thirties', and forties' life-stages respectively. As our oldest respondents are 60 years old, this gives us details of employment patterns as far back as the 1950s for the twenties', 1961 for the thirties', and 1971 for the forties' life-stages.

The nature of each individual's employment history has been examined for each of these life-stages, with the objective of categorizing the period as one of 'employment stability' or of 'employment instability'. Our general definition of stability is a continuous period of work for the ten-year period. However, clearly there are very different potential sources of disruption of employment, which may have rather different later career implications. We have then also constructed less stringent definitions of stability which regard the career period as stable even if there are breaks resulting from maternity leave or participation in education or government schemes.

In the following analysis a number of definitions of employment stability are used, each allowing for a wider range of events to contribute to the classification of a life-stage as stable. The first only considers a continuous experience of full-time employment as stability; the second includes both full-time or part-time employment, the third allows for full-time, part-time, or self-employment, the fourth adds in maternity leave, and the fifth adds in spells spent in education or on government schemes.

Trends in Stability over Time

Taking one life-stage at a time, we have broken down by birth cohort the proportions with stable and unstable periods. The pattern for the successive cohorts

[1] The approach is loosely based upon the recent work of Berger *et al.* 1993 which looks at changes in the employment patterns of people during particular periods in their life (when they were in their twenties, thirties, etc.) combined with a historical approach in which these changes can be plotted over time.

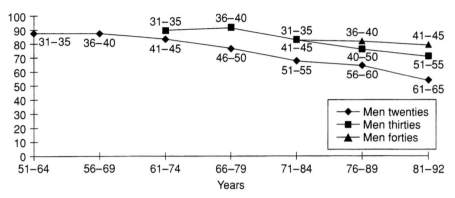

FIG. 5.1 Stability of men's employment by life-stage

Note: Figures on points indicate birth cohorts

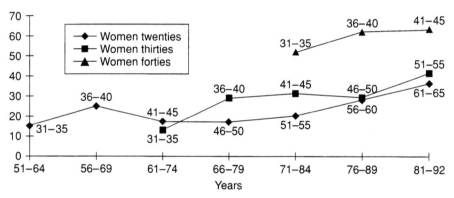

FIG. 5.2 Stability of women's employment (including maternity leave) by life-stage

Note: Figures on points indicate birth cohorts

during their twenties is given in Figs. 5.1 and 5.2. The birth cohort gives a proxy for real time. Hence the first cohort (born between 1931 and 1935) were in their twenties during the period 1951 and 1964. The oldest person in this cohort (born in 1931) would have first entered their twenties in 1951, while the youngest (born in 1935) would have been 29 years old not later than 1964. Hence, as we move to the right in these charts, time is increasing. In this way, we are able to construct a picture from the 1950s through to the 1990s for each life-stage. These real times overlap, but this probably acts as a smoothing influence allowing easier interpretation.

The picture that results is striking. We can see that for men in their twenties and thirties (whichever definition we use) stability appears to be declining. For men in their forties the decline is not very sharp using the broader definitions of stability (incorporating self-employment and periods in education or on government schemes) and, indeed, is slightly improving using the narrower definitions confined to full-time or both full-time and part-time work. The most notable feature appears to be the decline in stability over time for young men.

But, for women, the converse appears to be the case. Whichever measure we use, there is a general upward shift in stability as we move towards the 1990s. The sharpest rise appears to be among women in their twenties and thirties and the least sharp for women in their forties, although older women are much more stable in general. One interesting point to note is that for the twenties life-stage there is a dramatic rise in the stability of the 1936–40 cohort. This corresponds roughly to the 1960s which did indeed see the sharpest rise in the female participation rate in employment. Unfortunately we cannot go back this far with the thirties and forties life-stages to see how this period affected their stability. The overall impression then is that women are enjoying more employment stability, while men are finding it more difficult to maintain stable employment patterns.

Factors Contributing to the Stability of Women's Careers

What has contributed to the increased stability of women's careers? Three factors stand out as particularly important: the decline of housework, the increased uptake of maternity leave, and the growth of part-time work.

A major factor contributing to the increase in stability has been the fact that women have become less likely to leave the labour market to spend periods as housewives. In all periods, housework plays a major role in breaking up stable employment sequences for women, but the reduction in housework spells over time emerges clearly using this type of analysis. There is a general decline in the occurrence of housework in the histories for all three life-stages.

One potentially important factor affecting this is maternity leave provision. If we compare in Fig. 5.3 the measure which only includes employment events with the measure that includes maternity leave in the definition of stability, it is clear that maternity leave has a greater impact on stability as we move towards the 1990s. There was less impact for women in their thirties and virtually no impact for women in their forties.

This, however, does not in itself tell us about the extent of uptake of maternity leave, only about its contribution to continuous employment. A more direct approach is to distinguish all periods in which there was a spell of maternity leave (Fig. 5.4). This shows a general increase over time in the use of maternity leave among young women. It rises from 3 per cent of women in their twenties in the earliest period to 19 per cent in the last period. However, it is clear that the number of women who take maternity leave and who do not subsequently return to work is also quite marked. Sequences in which maternity leave is combined solely with employment events are increasing, but they still lag well behind the overall number of sequences with a spell of maternity leave.

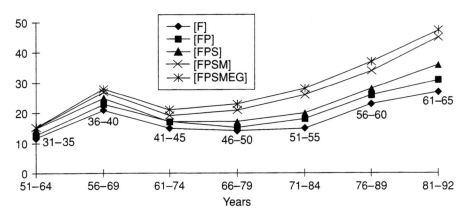

FIG. 5.3 Stability of women's employment in their twenties using different measures of stability

Note: Figures on points indicate birth cohorts
F = full-time; P = part-time; S = self-employed; M = maternity leave;
E = education; G = government scheme.

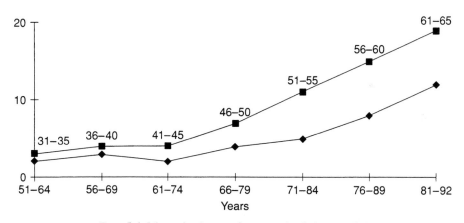

FIG. 5.4 Maternity leave of women in their twenties

Note: Figures on points indicate birth cohorts
The bottom line is maternity leave in combination only with employment, the upper line is all occurences of maternity leave.

Finally, it is clear that part-time work has played an important role in contributing to the stability of employment patterns for women. We can see the implications of part-time work for the employment stability of women in their twenties looking at Fig. 5.3. The first two lines on each cohort show the difference between the measures based on constant full-time employment and on either full- or part-time work. This effect becomes very much more marked at later life-stages. For women in their thirties and forties it typically doubles the number who are stable.

An analysis of all periods with a part-time work event shows a general tendency for a rise in part-time work. For women in their twenties there is a general rise over the successive birth cohorts from 25 per cent to over 40 per cent. For women in their thirties part-time work is generally at a much higher level, and the trend over time is again upwards. But for women in their forties, a high level of part-time work is accompanied by a downward trend over time.

Unemployment and Employment Stability

The most significant factor contributing to greater instability in careers was unemployment experience. In Table 5.1, it can be seen that the proportion of people who had experienced a spell of unemployment rose through each successive time-period and for both sexes. As would be expected, the sharpest rise occurred in the periods that cover the major recessions of the early 1980s and early 1990s.

In the later periods, the sample expands as more young people are eligible for inclusion. This could have a distorting effect on the figures as the young tend to be more vulnerable to unemployment. However, as can be seen in Table 5.2, the basic picture remains the same even if the analysis is confined to the changing experiences of those who were in the labour market in the mid-1970s.

Further, the trend cannot be discounted simply in terms of a greater likelihood that people will recall very short periods of unemployment more frequently in recent periods. We tested this by restricting the analysis to people who had had spells of unemployment of three months or longer (Table 5.3).

TABLE 5.1 *Percentage of people with a spell of unemployment by period (cell %)*

	1973–7	1978–82	1983–7	1988–92
Men	7.4	12.0	15.0	20.2
Women	3.5	6.9	7.5	8.5
All	5.6	9.6	11.6	15.0
No.	2,361	2,997	3,699	4,473

TABLE 5.2 *Percentage of people with a spell of unemployment by period (those in the labour market in the mid-1970s only) (cell %)*

	1973–7	1978–82	1983–7	1988–92
Men	7.4	9.6	11.3	17.7
Women	3.5	5.3	5.3	7.0
All	5.6	7.6	7.6	12.7
No.	2,361	2,361	2,361	2,361

TABLE 5.3 *Percentage of people with a spell of unemployment by period (those in the labour market in the mid-1970s with spells of three months or more) (cell %)*

	1973–7	1978–82	1983–7	1988–92
Men	6.6	8.5	10.4	15.6
Women	3.1	5.2	4.9	5.6
All	5.0	7.0	7.3	11.0
No.	2,361	2,361	2,361	2,361

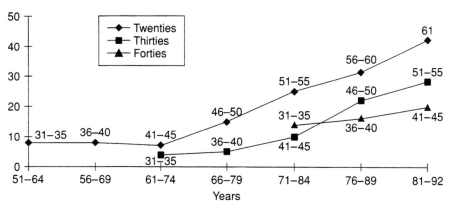

FIG. 5.5 Incidence of unemployment for men by life-stage

Note: Figures on points indicate birth cohorts

The fact that the rise over the years is less sharp than in Table 5.2 is consistent with the view that there may be a tendency for shorter periods of unemployment to be recalled more frequently when they are more recent. While the figures for spells over three months for the period 1973–7 are very close to those for all spells of unemployment, there is a greater divergence in these figures for the period 1988–92. This would fit with the view that in earlier periods people tend mainly to recall longer spells, whereas for recent periods they also include shorter spells. Yet, the overall effect is fairly small and, even when it is taken into account, the same overall pattern emerges. The proportion of the labour force at risk of a spell of unemployment has been rising strongly. The main difference is that this effect is now only clear-cut for men; among women the pattern has been stable since the period 1978–82.

The graphs mapping the changing vulnerability of men and women to unemployment over the decades show clearly how profoundly the labour market has altered and how perilous it has become, especially for the young, in the last two decades. Figs. 5.5 and 5.6 show how the incidence of unemployment has been

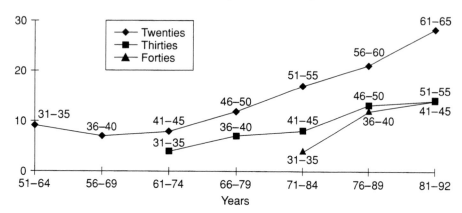

FIG. 5.6 Incidence of unemployment for women by life-stage

Note: Figures on points indicate birth cohorts

increasing over time for men and women at different life-stages. Taking first the graph for men, it is clear that an experience of unemployment in the first three time-periods, which span from the early 1950s to the early 1970s, was relatively rare and there was very little change over time. This is consistent with the official statistics which show the rise of unemployment beginning after the first oil crisis in the early 1970s.

It is notable that, when it came, the rise was particularly sharp for young men in their twenties. Indeed, it is striking to note that, for the most recent birth cohort (the 1961–5 cohort), over 40 per cent of men experienced at least one spell of unemployment. However, the increased vulnerability to unemployment was by no means confined to the young. There is also a marked upward trend for those in their thirties. While youth unemployment is sometimes regarded as a partial outcome of the tendency to try out different jobs in order to explore the labour market and as possibly less serious in its consequences because of the lack of family dependants, this is clearly not the case for people in their thirties. The rise of unemployment affected people in their prime years, at the stage when they normally would be expecting to be firmly launched in their careers and when they were likely to have substantial family commitments. There is much less data over time for men in their forties, because those that were of this age in earlier decades had left the labour market by the time of the survey in the early 1990s. However, it is notable that the overall level was lower among these more established workers and the trend, while still rising, was much less marked than for younger workers. This may reflect the greater job protection for established workers that had emerged from union or shop steward control within the workplace.

Turning to the situation for women, it is notable that women were less severely hit by unemployment than men at each life-stage. This is again consistent with the official statistics on unemployment, but such consistency is less to be taken for granted. The official statistics for female unemployment are known to be problematic, because the rules of eligibility for benefit mean that many women who are without work and wanting a job are not included in the count. However, even when we rely on people's own record of their unemployment experiences, it is clear that women still emerge as less affected by the deterioration of the labour market than men. This presumably reflected the strength of employment growth in the service sector and employers' emphasis on the expansion of part-time work.

Nonetheless, while the overall trend has been towards greater stability in women's employment careers and women have been more protected than men from the vicissitudes of the labour market, it was still the case that the long-term trend has been for a rise in women's vulnerability to unemployment. As with men, this was particularly the case when we compare those in their twenties at different historical periods. In the most recent period, some 30 per cent of young women had experienced a spell of unemployment. The pattern for women differs more markedly from that of men with respect to those in the thirties life-stage. Although there was also a rise in unemployment experience for this group over the longer term, it was much less sharp than for men and there was very little increase at all over the most recent time-period. While in general women in their thirties may be partly shielded from unemployment by the fact that they are spending periods out of the labour market bringing up children, their relatively low vulnerability in the most recent period—when women were more likely to be in the labour market—more plausibly reflects the strong growth in demand for part-time work.

The women whose pattern was most similar to that of men were women in their forties. In good part this was because men of this age were relatively highly protected, with a considerably lower risk of unemployment than that of younger men. It also resulted, however, from quite a sharp rise in unemployment among women in their forties. From a position of exceptionally low rates of unemployment, they came to share a virtually identical position in the most recent period with women in their thirties and were nearly as exposed to unemployment as men in their forties.

Overall, then, it should be noted that, despite the fact that women have become increasingly stable in their employment patterns (in contrast to men), they are also experiencing more unemployment. The decline in non-activity, predominantly time spent in spells of housework or looking after children, is reflecting a tendency for women to be drawn into the labour market where they are finding more stable employment opportunities. However, once women are drawn into the labour market, they at the same time become more exposed to the risk of unemployment.

Towards Dualism?

As was seen in Chapter 1, there has been a growing concern with whether there has been a trend towards the polarization of the employment structure between a core and peripheral sector. With growing product-market volatility, it has been suggested that employers may seek to build up a peripheral workforce that can be easily dispensed with in times of economic difficulty. At the same time, employers may wish to hold on to a core of permanent skilled workers in which they have invested training and resources. These employees are encouraged to stay with the firm by the offer of long-term benefits such as pensions and internal promotion ladders. In some accounts, it has been suggested that this division between core and periphery is closely linked to gender inequalities, with men more likely to benefit from core employment, while women are more likely to find jobs in the periphery (Barron and Norris 1976).

In so far as they involve claims about the distinctive employment conditions associated with different types of non-standard contract, these arguments will be examined in more detail in the next chapter. For the present, our primary concern is with their implications for the trends that should be found in the work-history sequences. The critical characteristic that distinguishes the core from the periphery in these theories is that of job security. If there has been a growing sector of core workers closely tied to one organization, then there should have been a growth over time in the frequency of periods of continuous employment *with the same employer* (for women including maternity leave). If peripheral work is on the increase, then we would expect to see an increase in workers with unstable work patterns, moving in and out of employment. Each ten-year life-stage, then, was examined to see whether there were sequences of continuous employment with a single employer and whether there were sequences in which a person moved from a non-employed status into employment, only to become non-employed again later in the ten-year period.[2]

Figs. 5.7 to 5.10 show the life history analysis of these patterns for men and women in their twenties and thirties.[3] It is evident that there is no simple pattern. Taking first people in their twenties, the incidence of stable work for the same employer (core work) has been on the increase *for women*, but on the decline for men since the 1971–84 period. Conversely, the incidence of 'peripheral' work has been on the increase for men, but on the decline for women. Looking at the thirties life-stage the trends are less clear. There is a general though not steady decline for men in stable core work, but again the converse is the case for women. In particular, in the 1980s there has been a very sharp rise in unstable activity for men. This clearly does not accord with versions of dualist theory which postulate

---

[2] Note that here a stable pattern is defined as work (full- or part-time) for the *same employer for the whole ten-year sequence.*

[3] We include maternity leave in women's employment.

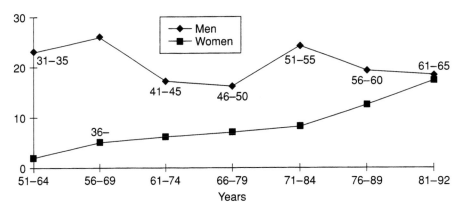

FIG. 5.7 Core work of men and women in their twenties
Note: Figures on points indicate birth cohorts

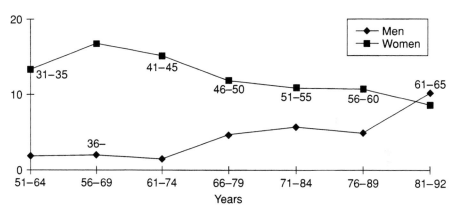

FIG. 5.8 Peripheral work of men and women in their twenties
Note: Figures on points indicate birth cohorts

that it is mainly women who find themselves increasingly drawn into the peripheral sector. If any polarization is taking place, it is not along the traditionally accepted lines. Again the growing disadvantages of being male are apparent here.

These results cast considerable doubt on the view that any general polarization of the employment structure is occurring. It is only men who seem to be increasingly finding themselves in unstable work patterns, but without the corresponding grouping in the core. Women are beginning to benefit from core work. There is a definite trend for young women to stay with the same employer and fewer women have unstable employment patterns. The evidence points to a feminization of the labour market rather than a general core–periphery polarization.

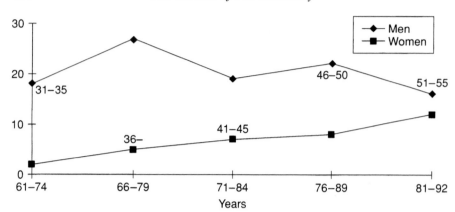

FIG. 5.9 Core work of men and women in their thirties
Note: Figures on points indicate birth cohorts

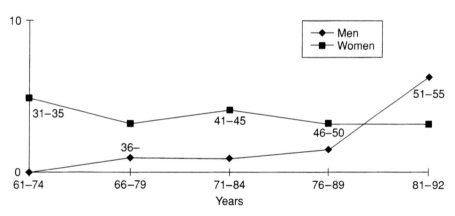

FIG. 5.10 Peripheral work of men and women in their thirties
Note: Figures on points indicate birth cohorts

Women appear to be increasingly able to secure long-term contracts with the same employer and are less likely to find themselves in an unstable position. For men the reverse may be the case: men are increasingly experiencing the fragmentation of their work careers by repeated spells of unemployment and are less likely to have stable employment experiences.

We have shown that the frequency of unemployment events appears to have increased and that men in particular have experienced an increasingly precarious world. For young men especially, security in the labour market has become much less common. We next turn to examine the determinants of stability or instability in people's work histories.

FACTORS ASSOCIATED WITH INSECURITY

We have seen above that the growth of instability in the work histories of men has risen sharply. Also for both men and women the occurrence of a spell of unemployment has become more likely than it was in the past. However, the approach so far has been descriptive and tells us little about the types of influence that can help account for the decline of employment security in Britain. In examining this, we look first at people's work histories, considering the factors that are related to unemployment experience and then turn to consider the sources of current insecurity in work.

Unemployment in the Work Histories

A first point to note is that the risk of unemployment in the work histories was closely related to class. Men in both skilled and non-skilled manual occupations had a much higher vulnerability to unemployment. However, these class differentials could give rise to quite different interpretations. On the one hand, it might be argued that they reflect the structural constraints of the labour market positions in which people happened to find themselves. With rapid technological advance, lower-skilled work has come to be more easily substituted by technical means. Further, the intensification of global competition has led employers to abandon many areas of traditional manufacturing production, where there were particularly heavy concentrations of manual work. An alternative view is that the link between unemployment and occupational class reflects above all early disadvantages experienced by those that occupy lower-class positions. Such people may have been brought up in families that gave their children little encouragement to continue their studies at school or to develop their skills. This early educational deprivation may have marked permanently their later labour market careers.

To what extent were the risks of unemployment related to early background characteristics and to what extent can they be attributed to the structural nature of the labour market positions that people were in? Clearly we need to explore the significance of occupational class for the occurrence of unemployment with controls for factors which might be related to differences in the families in which people were brought up. There are several ways in which we have tried to do this. The first is to take into account people's *class of origin*, that is to say the occupational class of their parents. We have taken as an indicator the father's class when the respondent was aged 14, as this seems likely to be the formative time for people's attitudes to work and the labour market. For women, the mother may have been an even more powerful role model. Given that the mothers of many of the older workers may have not been in the labour market at any selected period in the respondent's childhood, we have taken instead the general extent of the labour market participation of the mother during the person's formative years. People were asked 'While you were at school, to what extent was your mother in a paid job?', with possible responses ranging from 'most of the time' to 'not at all'.

If family socialization has important implications for later labour market experiences, this could be expected to be partly transmitted through parental attitudes to education. A number of questions were asked about how people remembered their upbringing, in particular with respect to their parents' interest in education. To begin with, they were asked 'When you were at school, how much interest would you say your parents took in how you were getting on there?' This was followed by a question asking 'How much encouragement did your parents give you to go on with your studies beyond the earliest school leaving age?' Two further items sought to measure the strength of parental influence over the respondent's views about work and success in life. Finally, given the evidence that lack of resources constitutes an important barrier to educational aspirations, there was a measure of financial deprivation during childhood.

In carrying out the analysis, the individual work histories were divided into five-year segments. The procedure was to use logistic regression to attempt to predict a spell of unemployment in any given five-year segment of a work history, taking as 'explanatory' variables the characteristics of the previous five-year segment (including class and unemployment experience) along with a range of the other variables such as time, age, education, first labour market experience, class origins, and early parental influences. The analysis was based on segments in which people had been in employment at some point in time. The very first segment of each individual's career, however, was excluded on the grounds that very early labour market experiences may be uncharacteristic and obscure the more stable effects. Our analysis, then, focuses on the risk of unemployment once people are clearly launched into their work careers. A one-way causal chain is assumed whereby the origins of the respondent and the respondent's previous experiences in part determine what happens to them subsequently. As in the previous section, the analysis has been carried out using the combined sample, which includes both those who were in employment and those who were unemployed at the time of interview.

Before going on to the results, it must be recognized that there are potential problems with this type of modelling. Recursive models based on data of this type may make strong assumptions about the variables included in the models. This has been well documented in the econometric literature on longitudinal data analysis (see, for example, Allison 1984; Heckman and Singer 1984). First there is the problem of omitted variables. If important individual-specific variables are omitted, then the cases may no longer be independent in a panel study. This can result in artificially low standard errors leading to incorrect conclusions. Second, if explanatory variables are endogenous this can lead to biased parameter estimates (Davies and Pickles 1985). Various ways have been devised to try to get around these problems. Mostly, these involve solving heterogeneity by introducing a second error term of fixed or random effects specific to each individual and estimating using maximum likelihood techniques. Unfortunately, very little software has been developed to date to perform these analyses on large samples and it has been shown that the modelling is extremely sensitive to assumptions about the distributions of the error terms.

In the following we assume that these problems are not a grave cause for concern. We have some justification for this, although the results must be treated with some caution. We are introducing a considerable number of variables into the analysis, including individual-specific origin and social background variables. Hence the possibility of residual heterogeneity through omitted variables is likely to be substantially reduced. Further, we will only be giving interpretative weight to variables that are significant at the 1 per cent level or better in case the models are suffering from the effects of artificially reduced standard errors. This helps protect against the possibility of misinterpreting a variable whose significance level may have been overestimated.

The results of the analysis are presented in Table 5.4. Taking first the data for men, there are a number of indications that people's very early experiences do have a lasting effect on their later labour market careers, in particular experiences in the family in early childhood. It is notable that it was not the class of origin of the parents that was crucial. In our main model, which includes both past and more recent factors, class origins have no significant effect. This could reflect the fact that they translate into later experiences which prove statistically more powerful. But it is notable that the effect was fairly weak even when parents' class was entered on its own. It was only where the parents had been skilled manual workers and foremen that there was a higher risk of later unemployment, perhaps reflecting a stronger tendency of children in these categories to enter skilled manual occupations which later proved particularly vulnerable. A measure of the degree of financial difficulty at home when the person was a child also fails to reach significance.

In contrast, the early factors that proved to have a much more pervasive influence were those relating to the influence of the parents with respect to education. In particular, where people reported that their parents had taken an interest in how they were getting on at school, they were notably less likely to become unemployed later. Even when more recent factors were taken into account, this remained at a very high level of statistical significance. It also should be noted that men who felt that their fathers had had a good deal of influence over their views about work, and whose mothers were in employment for a good deal of the time they were at school, were less likely to become unemployed. The overall pattern points to a significant educational influence of the parents with respect to school studies and, possibly, work values. This interpretation is confirmed by the effect of years of full-time education before entering the labour market. When parental influence variables are included, education fails to reach significance as a predictor of later unemployment, although the direction of the coefficient

TABLE 5.4 *Factors related to unemployment spells in work histories*

	Men		Women	
	Coeff.	Sig.	Coeff.	Sig.
Own class				
Lower non-manual	1.26	n.s.	1.40	n.s.
Self-employed	1.60	n.s.	0.70	n.s.
Technician/foreman	1.36	n.s.	1.60	n.s.
Skilled manual	1.67	**	1.08	n.s.
Semi- and non-skilled	2.20	***	1.17	n.s.
Parent's class				
Lower non-manual	1.70	n.s.	0.73	n.s.
Self-employed	1.33	n.s.	1.26	n.s.
Technician/foreman	1.52	n.s.	0.79	n.s.
Skilled manual	1.25	n.s.	1.60	n.s.
Semi- and non-skilled	0.91	n.s.	1.01	n.s.
Not stated	1.07	n.s.	1.03	n.s.
Size of organization				
Small (1–25)	0.89	n.s.	0.91	n.s.
Large (500+)	0.98	n.s.	0.97	n.s.
Labour market situation				
Historical time	1.09	***	1.04	**
Previous unemployment	3.67	***	2.29	**
Manufacturing	1.23	n.s.	0.92	
Government training scheme	1.00	n.s.		
Period on maternity leave	—		1.39	n.s.
Period as housewife	—		0.17	***
Early influences				
Years of schooling	0.95	n.s.	0.98	n.s.
Parent's interest in schooling	0.95	***	0.73	n.s.
Father's influence on values	0.69	**	0.76	n.s.
Mother's employment	0.72	*	1.28	n.s.
Good living standards	0.70	n.s.	0.76	n.s.

Note: Logistic regression analysis; multiplicative effect on odds. No. of unemployment spells = 4,757 for men, 2,507 for women.

is consistent with a view that longer education reduces the risk of unemployment. However, once the measures of parental influence are removed, the number of years of education moves to a high level of significance. This suggests that much of the influence of the parents is passed through the educational career of the child.

But, even when early educational and parental influences are controlled for, it is clear that the risk of unemployment for men is strongly related to the conditions of the labour market in which people worked. In particular, even allowing for education, the occupational class of the person's job remained a critical factor affecting their vulnerability to unemployment. Men from the skilled manual jobs and even more men in semi-skilled and unskilled manual work had a much higher risk of becoming unemployed.

Two other factors lend weight to a structural interpretation of unemployment risks. First, the probability of unemployment is increasing with time, even after controlling for numerous other factors. It is clear that increased vulnerability to unemployment is to a very substantial extent due to the structural change in the labour market that followed the oil crisis in the early 1970s. Second, it should be noted that the main determinant of future unemployment is a previous spell of unemployment. This was the most powerful factor in all of the analyses. Once a spell of unemployment has taken place, it is much more likely for a man to become unemployed in the future despite other factors. This confirms a finding by Gershuny and Marsh (1993) who analysed work-history data collected in 1986. Given that individual background factors have been controlled for, it seems likely that this in part reflects the fact that people are trapped in particular sectors of the labour market that are characterized by endemic job insecurity. It may also be the case that, once unemployed, people are discriminated against by employers and are obliged to take relatively poor jobs, disproportionately involving temporary or unstable contracts. We will be examining this issue more closely in the third section of the chapter.

Turning to the analyses for women, we find a very different pattern. There is no evidence at all that early background affects the likelihood that people will become unemployed. Neither social class of origin nor the more specific measures of parental influence proved significant. The principal factors that are influential in predicting unemployment relate to later events in the work history. Two of these were factors that were quite specific to women. Interestingly, if a woman had spent time out of the labour market in housework, this led to a *lower* likelihood of becoming unemployed if she had taken a job in the next period. Any explanation of this has to be tentative, given the lack of directly relevant evidence. It may be that women returning to the labour market tended to take jobs primarily in the expanding service sector where the effects of economic restructuring were less harsh. It should also be noted that the most frequent transition of women returners was into part-time jobs. Employers may have been less concerned to dispense with labour which was relatively cheap and provided valuable social skills acquired during periods of childrearing. Certainly the pattern of change in the labour market in the 1980s suggested that employment prospects were increasing particularly sharply for women with children.

The second factor that was distinctive for women was that those that had been on a government training scheme were less likely to become unemployed. This contrasts markedly with the situation for men where such schemes appeared to make no difference at all to unemployment risks. This may point to important differences in the effectiveness of such training schemes for men and for women. Payne's (1991) evaluation of the TOPS/OJTS scheme suggests high rates of successful job acquisition for women coming out of the scheme both in the shorter term and in the longer term (more than two years after the end of the course). Indeed, women coming out of courses found work more rapidly than men and more women than men found work in which they could use the skills in which they had been trained and in which their earnings were higher than in their previous job. Arguably women were receiving training that was better geared into the rapidly changing structure of job opportunities.

The other two factors that had a major impact on women's vulnerability to unemployment were the same as for men. Both of these reflected the structural conditions of the labour market. First, there was an important influence of historical time, with the probability of unemployment increasing over time. Second, a previous unemployment spell was again strongly predictive of a future spell. As in the case of men, previous unemployment is the single most significant factor that leads to future unemployment. However, in contrast to men, there is no evidence of a strong influence of current class.

In short, the notable finding for women is that there is little to support the view that early background characteristics are contributing to spells of unemployment. This difference from the pattern found for men may reflect the much greater generational disjuncture for women in norms about employment and in effective labour market opportunities. The experience of the past generation was a less sure guide at a time of major shifts in patterns of female labour market participation and in the sectoral distribution of women's jobs. The distinctiveness of the rapidly changing nature of career opportunities for women is reflected in the advantageous position of women returning to the labour market after a period out of the labour market and in the greater benefits they received from government training schemes. Nonetheless, it is clear that as with men, women's vulnerability to unemployment was heavily affected by structural factors. The strong association with historical time reflects macro-changes in the labour market, in particular the general growth of job insecurity from the 1970s. The strong influence of a previous spell of unemployment, even after controlling for length of education and early background influences, points to the way in which people become structurally trapped in insecure sectors of the labour market. The results here are consistent with those of Payne and Payne (1993), who have shown that previous unemployment spells have the effect of restricting job choice and reducing the relative likelihood of a recently unemployed individual finding a good job. They are much more likely to return to the labour market in non-standard and/or low-skilled work, thereby finding themselves at risk of yet another spell of unemployment.

All in all, our data indicate that there are several significant differences in the factors that affect male and female unemployment risks. In particular early parental influences, translating into different schooling experiences, had a significant effect for men, but were irrelevant for women. However, the experiences of both men and women were heavily influenced by structural changes in the economy which were largely outside their ability to control. They were both affected by a long-term growth in the inability of the economy to generate sufficient jobs and by the waves of recession in the 1980s and early 1990s. They also experienced in a similar way the problems of structural entrapment whereby once unemployed it becomes increasingly difficult to find stable work.

Current Job Insecurity

Our second approach is to look at the factors that are associated with current job insecurity. We have two measures of people's perceptions of their positions in the labour market. First, to provide a picture of the degree of institutional protection, we asked people how long it would take for someone in their organization to be dismissed if they either persistently arrived late or did not work hard. Second, as a general measure of security, employees was asked how satisfied or dissatisfied they were with their job security.

The dissatisfaction with security and ease of dismissal indicators were scales which were used as the dependent variables in OLS regression models. A wide range of independent variables were introduced into the analysis. Very broadly these sought to examine the influence of individual, occupational, and organizational factors.

At the individual level, we were concerned to examine the effect of age and length of service. At least in an earlier period, there were formal or informal rules in parts of British industry that stipulated that workforce reductions were to be regulated by the 'last in, first out' principle. This is likely to have provided greater security for older workers. However, both the prevalence of such conventions and the extent to which they were eroded during the 1980s remains unknown. In order to examine this, we included not only age, but, as an even more direct measure of such protection, the duration of time that the person had been employed in their current job. The alternative hypothesis is that older workers may feel threatened by younger workers who have new skills and who are possibly more easily trainable for new tasks.

A second type of individual variable that has been included is whether or not the person had been unemployed in the previous five years. It was seen in the previous section that this was a very powerful factor predicting the risks of unemployment in the work histories. In looking at people's current situation, however, there is much richer information on the nature of the organizational contexts in which people were employed. We are in a position, then, to examine whether previous unemployment still affects job security even when a wider range of factors have been taken into account. Since the implications of unemployment may vary depending on the length of unemployment, we created two categories: one of people who had been unemployed for twelve months or less and the other of people with longer durations of unemployment.

With respect to occupation, our main concern was with the effect of occupational class. People with lower skills may have seen a reduction in security, as a result of technological change. We were also interested in the implications of differences in contractual status. Formal employment protection has been markedly weaker for part-time workers, particularly those working less than sixteen hours a week. Even more evidently, temporary workers were in a position which offered very low institutional protection. An important factor affecting security may also be the marketability of skills. To examine this we have included a measure of the ease with which a person could find a similar job elsewhere. People whose skills are in demand may not be too worried about losing their current job or may feel more secure, since they can move to another company relatively easily.

We also examined factors linked to the competitive position of the organizations people worked for. There were grounds for thinking that industry would be important, with the sharp decline of manufacturing in the 1980s leading to widespread job insecurity. To examine this, a variable for manufacturing was included to see whether workers in this sector have been particularly adversely affected. Since it may be the more general phenomenon of exposure to competition that leads to insecurity, we also included a control for whether or not the person was employed in the private sector. Third, there is a measure of whether the respondent feels that his/her organization is in financial difficulty. Finally, given the frequent assumption in the literature that smaller firms face fiercer competition and therefore offer less secure employment than larger firms (e.g. Edwards 1979), the size of the organization has been included.

It was seen in Chapter 2 that there has been a major expansion of the use of new technologies in the last decade and an extensive restructuring of work tasks. Indeed, a majority of the workforce has seen change in the skills that they are required to use. There is a considerable literature in industrial sociology that indicates that work reorganization can generate substantial feelings of insecurity in the workforce. Two variables have been included that reflect the extensiveness of recent changes in work practices within the organization: whether the organization has changed the way work was organized and whether new technology has been brought into the organization in the previous two years.

The organizational factors discussed so far could be expected to intensify insecurity. However, there are also organizational factors which might be expected to alleviate it. It was seen in Chapter 3 that there are signs of the growth of management control systems of a more sophisticated type that sought to influence employee behaviour through target-setting, individual assessment, and merit rewards. Insofar as such systems relied on a reward structure involving opportunities for longer-term career progression, it seems likely that they would be linked to reasonable security of employment. Career incentives would be unlikely to be effective in an organization that could offer little guarantee of a job in at least the medium term. In contrast, the form of control that has been termed 'technical' control, which was associated with the machine-pacing of work and short-term output incentives, has been usually considered to be linked with policies that favour an easy substitutability of labour and therefore limited security. Similarly, systems of strong supervisory discretion have been thought to be associated with 'quick hire and fire' policies. To test the effects of these different types of organizational control system, we have included the indices developed in Chapter 3 for bureaucratic and technical control. We have also introduced a measure of perceived supervisory bias, as an indicator of a system of high supervisory discretion.

With these variables we can build up a picture of the influences on workers' perceptions of job insecurity. It could be reasonably argued that temporary work is definitionally related to job insecurity and therefore problematic as an 'independent' explanatory variable. While we comment upon the results for temporary work, the tables present the model in which the term was omitted to safeguard against any effects from such non-independence.

Ease of dismissal. The first aspect of precarity, the ease of dismissal, is concerned with the degree of institutionalized protection that employees feel that they have in their jobs. The survey included two measures of this: one asking how quickly a person would be dismissed if he/she persistently arrived late at work and the other if he/she persistently did inferior quality work. Overall, taking those who could be dismissed within a month as having relatively precarious employment, 33 per cent reported that this was the case with respect to lateness and 28 per cent with respect to poor quality work. The most insecure group of all, thinking they could be dismissed within a week, represented 7 per cent and 6 per cent of all employees respectively. To facilitate explanatory analysis, the measures have been combined into a single scale of 'ease of dismissal'.

Taking first individual characteristics (Table 5.5), we see that there is no evidence for either men or women that formal protection against job insecurity is linked to age *per se*. However, the length of time spent in the current job does come close to significance for men ($p = 0.07$) and is clearly significant for women. Those who have been employed longer with their organization are less likely to feel that they can be dismissed easily. This is consistent with the view that 'seniority conventions' still have some force in British industry. Our other individual characteristic—previous unemployment experience—did not significantly effect the formal protection of men against dismissal, but women who had been unemployed were clearly more at risk.

It was seen from the work history analysis that there was a strong relationship between class and vulnerability to unemployment. Table 5.5 confirms that those in lower-class positions also were more likely to feel that they could be easily dismissed. For men the technical/supervisory, skilled manual, and the semi- and non-skilled classes are the most vulnerable to dismissal. Routine non-manual and professional workers have significantly greater security. For women, the greater ease of dismissal was only evident among technical/supervisory and non-skilled manual grades. It must be remembered, however, that the sample numbers for female skilled manual workers are very small and hence the estimates for this group of women are very imprecise. In general, it is clear that

TABLE 5.5 *Factors related to ease of dismissal*

	Men		Women	
	Coeff.	Sig.	Coeff.	Sig.
(A) Personal factors				
Age	0.00	n.s.	−0.01	n.s.
Time in job	−0.00	n.s.	−0.00	***
Unemp. 1–12 mths	0.12	n.s.	0.32	*
Unemp. 13 mths+	0.19	n.s.	0.44	*
(B) Occupational factors				
Lower non-manual	0.06	n.s.	0.08	n.s.
Tech./supervisory	0.61	***	0.78	**
Skilled manual	0.61	***	0.12	n.s.
Semi- and non-skilled	0.71	***	0.37	***
Transferable skills	−0.02	n.s.	0.05	n.s.
Part-time work	0.30	n.s.	0.17	*
(C) Organizational factors				
Org. size: Small (1–25)	0.12	n.s.	0.00	n.s.
Org. size: Large 500+	−0.08	n.s.	−0.04	n.s.
Manufacturing	0.10	n.s.	0.21	*
Financial difficulty	−0.02	n.s.	−0.04	n.s.
Perf. mgt. control	0.01	n.s.	−0.00	n.s.
Technical control	0.11	***	0.13	***
Superv. favouritism	0.15	n.s.	0.16	n.s.
Public sector	−0.40	***	−0.45	***
Trade union	−0.32	***	−0.50	***
Advanced tech.	−0.08	n.s.	0.01	n.s.
Change in work org.	0.12	n.s.	0.08	n.s.

Note: OLS regression; no. = 1,511, R^2 = 0.14 for men; no. = 1,490, R^2 = 0.15 for women.

institutionalized employment security was far lower for the working class. Developments in the 1980s are likely to have accentuated this pattern, as a result of the weakening of union strength and of collective bargaining controls.

The growth of non-standard contracts was also associated with lower levels of employment protection. Female part-time workers were more likely to feel that they could be easily dismissed than female full-time workers. It might be thought that this reflects the high concentration of part-timers in small establishments, which have to confront more volatile product markets and have fewer resources to ensure employment stability. However, establishment size has been controlled for, indicating that part-time workers seem to feel more easily disposable irrespective of organizational size. We also carried out an analysis including temporary workers. Male workers on very short contracts (twelve months or less) were significantly more likely to be easily dismissable than permanent employees, but this was not the case for those on longer (one-to-three-year) contracts. Moreover, female temporary workers appeared to be subject to much the same dismissal procedures as their permanent equivalents. While temporary work clearly implies greater job insecurity in the sense of a restricted time-horizon for employment, it does not necessarily mean that people can be more rapidly dismissed on grounds of work performance.

Turning to the characteristics of the organizational environment, there is no evidence that male workers in manufacturing were exposed to easier dismissal procedures, but women in manufacturing were more at risk (although this effect ceases to be significant when temporary work has been taken into account). It is also notable that the degree of financial difficulty of the organization was not associated with easier dismissal procedures; clearly these are of a long-term character and employers cannot easily change them to cope with the vicissitudes of the market.

Were organizations that had been active in introducing new technologies and in restructuring work distinctive in their dismissal policies? Our evidence suggests that they were not. The introduction of new technology was negatively related to speed of dismissal for men and positively related for women, but in neither case was this statistically significant. Recent changes in working organization were positively related for both men and women, but again the relationships were far from significant. There was also no evidence that upskilling was related in any way to dismissal procedures. Overall, it seems that there is little link between work-restructuring and formal protection procedures.

There was more evidence that the system adopted for the control of work performance was related to dismissal procedures. While performance-management systems were not statistically different from others, organizations that had 'technical' control systems did have more rapid dismissal procedures, confirming the view that they tended to be linked to a conception of labour as relatively easily substitutable.

There were two organizational factors that were particularly powerful in reducing insecurity. The first was whether a person worked in the public or private sector. Public sector employees, whether men or women, had much higher levels of protection. The second was whether the employee was a member of a trade union. Trade union membership has a marked effect in lengthening the necessary period for dismissal. It provided substantially greater protection for both male and female employees. Despite the decline in union power over the last decade, it is clear that the unions still exercise a considerable influence at workplace level in ensuring stronger regulation of managerial powers.

Dissatisfaction with job security. The measure of dissatisfaction with job security was a seven-point scale running from completely satisfied to completely dissatisfied. Overall, 15 per cent of all employees were dissatisfied with their job security and 7 per cent were either very or completely dissatisfied. Very high levels of worry were most common among skilled manual workers (13 per cent), technicians and supervisors (9 per cent), and non-skilled workers (8 per cent).

TABLE 5.6 *The class composition of the unemployed, 1975–1992, based on the individual's first spell in each period (column %)*

	1975–9	1980–4	1985–9	1990–2
Prof./managerial	13.3	12.2	19.9	21.6
Intermediate	18.3	23.1	11.9	17.1
Manual workers	68.3	64.7	68.2	61.3

But anxiety about insecurity was also evident among those in non-manual class positions: 6 per cent of professional and managerial employees and 7 per cent of lower non-manual employees reported strong worries about their security.

It is notable that people's satisfaction with their job security was not simply a reflection of dismissal procedures. Many of the factors that were associated with low levels of employment protection did not have significant implications for the way people perceived their personal job security. There is, of course, no necessary reason why the two should go together. Employers may have the power to dismiss employees easily, but have a strong interest in maintaining labour force stability. The time-horizons of the two measures are also very different. The ease-of-dismissal measure is concerned with whether people can be sacked in a fairly short period of time, but people may be very worried about job security even although they would not expect to be thrown out at very short notice. A concern about job security, then, can cover anxieties about the medium term as well as about the short term.

Whereas employment protection differed very substantially by class, worry about job security was fairly evenly spread across different occupational classes. While 21 per cent of skilled manual workers felt that their jobs were insecure, the proportion among professional and managerial employees was also as high as 16 per cent. This might seem surprising given the evidence from the work histories of the greater risks of unemployment for people in manual work jobs.

However, there has certainly been an increase in the proportion of the unemployed coming from professional and managerial jobs over the period 1975 to 1992. In Table 5.6, we have taken the occupational class of the job that the person was in prior to their first spell of unemployment in each time-period. The period covering the recession of the early 1980s (1980–4) saw a slight increase in the proportion of the unemployed who were intermediate workers, but after the mid-1980s, there was a steady increase in the proportion drawn from the ranks of professional and managerial employees. Professional and managerial workers constituted 13 per cent of the unemployed in the period 1975 to 1979 and 12 per cent in the period 1980 to 1984. But their share increased to 20 per cent in the second half of the 1980s and to 22 per cent in the early 1990s.

TABLE 5.7 *Relative risk of unemployment for manual
workers compared to professional and managerial employees
(relative odds)*

Years	1979	1980–4	1985–9	1990–2
Skilled manual	2.6	2.1	2.3	2.6
Semi- and non-skilled	3.1	2.6	2.8	3.0

It is important to bear in mind, however, that this increase in the proportion of the unemployed coming from professional and managerial work did not mean that the *relative* risks had grown greater for this category compared to manual workers. The period had seen a significant shift in the occupational structure, with an increasing proportion of people employed in such higher-level jobs. Even with the relative class risks staying the same, it would be expected that a greater proportion of the unemployed would come from higher-class positions in the later periods compared to the earlier, as these categories expanded in size. Table 5.7 compares over time the risk of a manual worker becoming unemployed, as against having a job, with those of a professional/managerial employee becoming unemployed, as against having a job. This controls for changes in the relative size of classes. It can be seen that, while there was some fluctuation over the 1980s, there is no evidence of a trend towards a higher relative risk for professionals and managers. Class inequalities in unemployment were just as great in the early 1990s as at the end of the 1970s.

Nonetheless, with respect to people's perceptions of insecurity, it is likely that these will be affected more by the absolute increase in the numbers of unemployed coming from higher-level jobs than by the statistical relativities. It is the fact that, in their everyday experience, professional and managerial employees were more likely to meet unemployed people from their own background, or to hear about them, that would tend to affect their view of whether their jobs were secure in times of organizational change.

The evidence again points strongly to the way in which previous unemployment weakens people's later labour market position. Even when class has been controlled for, it is notable that there is still a clear effect of previous unemployment experience (Table 5.8). Both men and women who have been unemployed less than a year are significantly more anxious about their job security and there is an even stronger effect for women who have been unemployed for longer. The direction of the coefficient for longer-term male unemployment also suggests that they feel less secure, although it is not statistically significant. The longer-term male unemployed includes a subgroup that have been unemployed for a very long period of time indeed, and it may be that the relief of having

any type of job partly counterbalances the fact that they are in relatively insecure employment.

Employees on temporary work contracts, as could be expected, also showed
a very high level of concern about job security. Indeed, if temporary work is
added to the model, it is the single most important predictor of job insecurity.
The effect of previous unemployment experience disappears for men and for
women with shorter spells of unemployment when temporary work is introduced,
suggesting that those with previous spells of unemployment feel insecure at least
in part because the jobs that they have acquired are jobs with temporary contracts. We will be looking at more direct evidence about this in the final section
of the chapter. But there was no sign of a more general link between non-standard
employment contracts and a sense of job insecurity. Part-time workers, although
having lower formal protection, did not appear to be more worried about job
security than full-time employees. This was the case not only for men, but also
for women who have been seen as particularly disadvantaged by employment
protection provisions. We will be returning to this issue in the next chapter which
focuses more specifically on the different types of non-standard employee.

An important factor that helped to mitigate the feeling of anxiety about job
security was the person's perception of the labour market for their skills. Those
who felt it would be relatively easy to find another job elsewhere were substantially more likely to feel secure. The benefits of transferable skills, and the
risks of firm-specific skill development, emerge rather clearly.

The nature of the organization was also linked to worry about security. Female
employees in manufacturing, and male employees more generally in private industry, were significantly more likely to be anxious. Worry was particularly sharp
if the organization was in financial difficulty. Insecurity appears to have been
more a result of the collapse of traditional manufacturing processes than of the
emergence of advanced technologies. Despite the speculation that the introduction
of new technologies would reduce job security, through making traditional skills
redundant or reducing the level of staffing needed, there is again no evidence
that it led to greater anxiety about security. Indeed, in the case of men, working
with computerized or automated equipment was linked to a higher sense of security. Separate analyses showed that upskilling, greater responsibility in work, and
even increased flexibility in work practices were neutral in their effects on worries
about job insecurity.

There are again important effects relating to the system the organization had
adopted for controlling work performance. In particular, where supervisors
were seen as able to exercise discretionary power in a way that favoured some
employees over others, there was a stronger sense of insecurity among both men
and women. There was also a very clear effect of performance-management systems for men. This is consistent with the assumption that, given their emphasis
on promotion opportunities as part of the reward structure, such systems normally
guarantee a higher level of employment security at least in the immediate term.

TABLE 5.8 *Factors related to dissatisfaction with job security*

	Men		Women	
	Coeff.	Sig.	Coeff.	Sig.
(A) Personal factors				
Age	0.00	n.s.	−0.00	n.s.
Time in job*	−0.00	n.s.	−0.00	***
Unemp. 1–12 mths	0.30	*	0.37	*
Unemp. 13 mths+	0.29	n.s.	0.56	**
(B) Occupational factors				
Lower non-manual	0.20	n.s.	−0.12	n.s.
Tech./supervisory	−0.16	n.s	0.22	n.s
Skilled manual	0.05	n.s.	−0.41	n.s.
Semi- and non-skilled	−0.11	n.s.	−0.13	n.s.
Transferable skills	−0.41	***	−0.28	**
Part-time work	0.24	n.s.	0.07	n.s.
(C) Organizational factors				
Org. size: Small (1–25)	0.06	n.s.	−0.15	n.s.
Org. size: Large 500+	0.27	**	−0.02	n.s.
Manufacturing	0.11	n.s.	0.42	***
Financial difficulty	0.87	***	0.84	***
Perf. mgt control	−0.09	***	−0.02	n.s.
Technical control	0.01	n.s.	−0.01	n.s.
Superv. favouritism	0.37	***	0.22	*
Public sector	−0.38	***	−0.13	n.s.
Trade union	0.24	**	0.14	n.s.
Advanced tech.	−0.18	*	0.09	n.s.
Change in work org.	0.24	**	0.12	n.s.

Note: OLS regressions; no. = 1,506, R^2 = 0.11 for men; no. = 1,481, R^2 = 0.09 for women.

However, while the coefficient is in the same direction for women, it is not statistically significant. This raises the issue of whether there might be qualitative differences in the forms in which performance-management systems are being introduced for men and women, with possibly a stronger emphasis on promotion opportunities for men.

One particularly puzzling result relates to the effects of trade union membership. It was seen in the previous section that the trade unions appear to have been successful in imposing better formal controls over dismissal. However, at the same time, trade unions members have greater anxiety about job security. There are a number of possible explanations of this. Trade unions may have been effective in the past in preventing manpower reductions, but now, with their power weakened, unionized workforces might be the main targets of management rationalization policies. It may also in part reflect the way unions have access to information about forthcoming redundancies and are often involved in negotiating who will not be laid off. Those in unions may simply have better knowledge of the precarity of their situation in the medium term.

We have seen that a range of factors comes into play to influence perceptions of precariousness. These include the market power of the individual's skills, the nature of contracts, the industry and ownership sector in which people work, and the financial situation of the organization. However, the major changes in work organization and in skill levels that have been noted in previous chapters would not appear to affect job security in themselves. It is the sectoral and competitive position of firms that is important, not the restructuring of work practices.

Two other points emerge with force from the analysis. Despite the substantial inequalities by class in unemployment experience, in the early 1990s job insecurity appeared to have become as substantial a worry for the professional and managerial classes as for the working class. Second, trade unions no longer appear to have the power to provide a strong sense of security in work.

Overall, we have seen that it is important to distinguish between different aspects of job insecurity—in particular, the level of formal protection against quick dismissal and the longer-term sense of job insecurity. The broader set of factors that affect these vary considerably. However, previous unemployment experience had significant effects in increasing both types of current job insecurity.

THE EFFECTS OF UNEMPLOYMENT ON LATER WORK CAREERS

We have seen that one of the major changes in the labour market was the growth of employment insecurity and unemployment. But does unemployment constitute a relatively transitional phase or does it have a lasting impact upon people's careers? One answer to this comes from the evidence that was examined in the last section. An experience of unemployment greatly increases the risk of becoming unemployed again. It would seem that people, re-entering work from unemployment, are much more likely to find a job in an insecure sector of the labour market. But does unemployment also affect other aspects of the quality of the work that people are able to obtain? This issue has been approached in two ways. First, we have examined the effect of unemployment on the types of jobs held by people in the main sample of the employed. This enables us to look at the longer-term effects of unemployment on people's work careers. Second, we have taken an additional sample of people who had recently found work after a spell of unemployment. This gives us more detailed information about the types of jobs that unemployed people find immediately on return to employment.

Previous Unemployment and the Quality of Jobs

In exploring the longer-term effects, a measure was developed from the work histories of people in employment that gave the number of months that each

TABLE 5.9 *Unemployment in last five years among people*
currently in employment and occupational mobility, 1987–1992

	Coeff.	Sig.
All employees with a previous experience of unemployment	−0.18	***
Employees, with a previous experience of unemployment, who were in 1987:		
Prof./managerial	−0.59	***
Lower non-manual	−0.16	n.s.
Self-employed	−0.33	n.s.
Tech./supervisory	−0.35	n.s.
Skilled manual	−0.14	n.s.
Semi- and non-skilled	−0.11	*

Note: OLS regression; no. = 3,556. The results for age and sex
are not shown. The reference group for class comparisons is that
of people who did not become unemployed. Mobility is
measured as change in points on the Hope–Goldthorpe scale.

person had spent unemployed in the five years preceding the interview. It was
then possible to compare the quality of the current jobs of people who had experienced different amounts of unemployment.

Our first, and most general, approach was to look at how unemployment affected
occupational mobility. We took as an indicator of mobility the difference between
the general social standing of the person's job in 1987 and 1992 (as measured
by the Hope–Goldthorpe scale). There is clearly the possibility that mobility
chances vary considerably depending on the stage in the life cycle, sex, and occupational class. So these were controlled for in the analysis. As can be seen from
Table 5.9, there is a very powerful effect of the experience of unemployment
even when these factors have been taken into account. Those who had been
unemployed in the course of the five-year period had significantly lower scores.
Clearly, the extent to which one could be downwardly mobile depended in part
on where a person was initially in the occupational structure. Those in non-skilled
work could fall less far (provided they continued to have employment). We introduced interaction terms to examine the effects of unemployment for specific classes.
Those most heavily hit in terms of the standing of the jobs they were in five
years later were people who had been in professional or managerial work in
1987. In other social classes, the coefficients were uniformly negative indicating
downward mobility compared to people in employment, but it was only among
the non-skilled that the effect was statistically significant. For the non-skilled the
downward descent was less marked than in other classes (presumably because
the floor was reached more quickly), but the effect was more certain.

Second, we looked at how an unemployment experience in the previous five
years affected more detailed measures of job quality in 1992. It must be borne

TABLE 5.10 *Effects of previous unemployment (in months) on current job characteristics*

	Coeff.	Sig.
Work at a place where less than a quarter of the workforce use computerized or automated equipment	1.02	*
Personally use computerized or automated equipment	0.97	**
Had training for the work you do	0.95	***
Likely to get training	0.97	**
Took less than a month to do the job well	1.03	***
Had some choice in current job	0.95	***
On a temporary contract	1.04	***
Have a recognized promotion ladder	0.97	***
Trade union present at work	0.97	***
Works council present at work	0.95	**
Currently member of a trade union	0.96	***
Skill increase in last 5 years	0.97	***
Effort increase last 5 years	0.98	*
Responsibility increase last 5 years	0.96	***
Stress increase last 5 years	0.96	***

Note: Results (multiplicative effects on odds) are drawn from a series of logistic regressions, controlling for class, age, and gender. The sample numbers were 3,159, except for the case of trade union membership (3,001).

in mind that the people who are prone to unemployment tend to be drawn disproportionately from those with lower skills and this needs to be taken into account in comparing the employment conditions of those who previously experienced unemployment with those who had been in stable employment. In order to assess whether unemployment really does penalize subsequent job chances, multivariate analysis is needed to control for the effects of class, age, and gender. If, after these controls, there is still a significant effect from previous unemployment experience, it can be said more confidently that its effects are real. Table 5.10 summarizes the results of a number of logistic regressions. The dependent variables vary in each case, but the independent variables are always the same: class, age, gender, and the number of months unemployed in the last five years.

A number of distinctive patterns emerge. To begin with, unemployment was clearly related to the likelihood that people would be working in a technologically advanced setting. The more unemployment experienced, the less a person was likely to be working with computerized or automated equipment or to be in a workplace where the use of such equipment was common.

Perhaps partly because of this difference in the technical environment of the work, those with longer experiences of unemployment were also less likely to be in jobs where there were possibilities for self-development. They were less likely to have received training relevant to their current job, to have experienced a skill increase, or to have seen the responsibility in the job increase. This was not because of any aversion to training. The desire for training was not significantly associated with the past severity of unemployment. Those who had been unemployed previously wanted training as much as others, but did not appear to get it. The poor quality of the jobs acquired by those who had been unemployed in the past is also highlighted by the measure of how long it took to learn to do the job well. Despite controls for class, unemployment experience was still associated with jobs where there was less need for on-the-job experience. The fact that the unemployed tended to find work in relatively undemanding, technically backward organizations, is also confirmed by the fact that they were less likely to say that work effort or job stress had increased over time.

Finally, it is clear that the unemployed also found themselves in jobs in which employment was more precarious. This confirms the evidence from the work histories discussed earlier, which showed how one spell of unemployment made it much more likely that people would become unemployed in the future. People that had previously been unemployed were less likely to have been able to exercise choice in taking their current job, and they were much more likely to be on a temporary contract, with little prospect of promotion, and without any form of collective representation—either through a trade union or a works council—that could defend workers' rights.

All in all it would appear that becoming unemployed does have serious consequences in terms of later chances of acquiring new skills, using advanced technological equipment, or other forms of self-development that might assist in the future.

Those that had previously been unemployed tended to find themselves trapped in jobs that offered little hope of personal advancement or development, and which were insecure both in terms of the length of contracts and of the presence of effective collective representation of employee interests.

The First Jobs of Unemployed 'Returners'

Our second source of evidence comes from the survey of the 'unemployed'. After the initial sample of the unemployed was selected, some of the people chosen had found work by the time of the interview. There were 122 such 'returners' in the unemployed sample. Clearly, in considering the quality of jobs that these 'returners' were able to obtain it is important to take into account that they were disproportionately people of a manual working background. We have compared, then, the jobs of returners who moved into manual work with those of manual workers from the main sample of employees.

It was clear from all of the indicators that the unemployed had moved into very low-skilled work (Table 5.11). First, it should be noted that only 14 per cent of 'returners' needed any qualifications at all to apply for the work they were currently doing. In contrast, 39 per cent of manual workers in the main sample were in jobs requiring qualifications. Second, a very low percentage of

TABLE 5.11 *Comparison of recently unemployed who had found a job and main employee sample (manual workers only) (cell %)*

	Recently unemployed	Main employee sample
In jobs needing qualifications	14	39
Had training for this type of work	32	41
Using computerized/automated equipment	20	29
gross hourly income (means)	£4.14	£4.88
Anxious re unjust dismissal	30	21
On temporary contract < 12 months	36	5
Very/quite likely to leave job within a year	55	25
Little or no choice in taking current job	77	58
Member of trade union	11	38

Note: The data for the 'recently unemployed' were drawn from a separate sample of 1,000 people from the unemployment register. The analysis is based on the 122 who had found work at the time of the interview.

the returners were in jobs requiring training (32 per cent) compared to 41 per cent of all manual employees. They also tended to find themselves in a less technologically advanced environment. Only 20 per cent of the 'returners' used computerized or automated equipment compared with 29 per cent of all manual workers.

Previously unemployed people also appeared to have found work in organizations that offered relatively low pay. Comparing gross hourly earnings, which controls for the possibility that people moving out of unemployment worked longer hours, we find that 'returners' were on average in jobs that paid £4.14 an hour compared with a manual work average of £4.80 an hour.

The future perspectives of the 'returners' were understandably somewhat bleak. They had relatively high levels of anxiety about the security of their new jobs. As many as 30 per cent were anxious about arbitrary dismissal (compared with 21 per cent of the employee sample). Very few had trade union protection: only 11 per cent were trade union members compared to 38 per cent among all manual workers. Many had only found temporary jobs: 36 per cent had jobs with contracts that lasted less than one year. This was much higher than the figure for the jobs that people in the unemployed sample had been in before becoming unemployed (24 per cent) and it was far above the average for manual workers in the employed sample (just 5 per cent). The very short-term nature of these jobs was evident from the fact that 55 per cent of the 'returners' thought that they were likely to leave their current job within a year, compared with 25 per cent of all manual workers. Indeed, 40 per cent of the 'returners' stated they were already looking for alternative employment.

This evidence strongly confirms the view that the unemployed tended to find low-quality, insecure work that made them vulnerable to further spells of unemployment. It is clear that they took such jobs out of constraint, not choice. Over three-quarters (77 per cent) of the 'returners' said that they had had little or no choice in taking their current job, compared with an average of 58 per cent for all manual workers.

CONCLUSIONS

This chapter has traced the growth of employment insecurity in recent decades. Men's employment careers have become markedly less stable since the 1970s. Although the decline of time spent in full-time housework and an increased use of maternity leave has led to greater overall continuity of employment for women, they too have become more vulnerable to unemployment. While there was evidence that people could suffer from long-term disadvantage due to educational deprivation in their family backgrounds, the decisive factors appeared to be structural. It was the overall weakening of the labour market over time and the nature of the particular sectors of employment in which people found themselves that most clearly determined the risk that they would find themselves unemployed.

The analysis indicated that occupational class still differentiated sharply the extent to which people were vulnerable to job insecurity. Those in manual occupations had a much higher risk of becoming unemployed in the course of their careers. Although professional and managerial employees represented a larger proportion of those entering unemployment in the early 1990s compared to the later 1970s, this simply reflected the fact that these occupations had increased their share of the workforce. There was no change over time in the *relative* vulnerability of manual employees compared to professional and managerial employees. Finally, among those who did have jobs at the time of the survey, it was notable that both men and women in non-skilled manual occupations were much more likely to report that they could be dismissed within a month (and this was also the case for men in skilled manual work). The nature of the employment relationship for different occupational classes, then, still differed fundamentally, in terms of job security.

Finally, our evidence points to a cumulative process. Once people had had an experience of unemployment, this led to a marked deterioration in their future employment prospects. The analysis of those in the employed sample with previous experience of unemployment converges closely with the evidence from the sample of unemployed people who had recently returned to work. Both show that the unemployed have relatively poor prospects of escaping from temporary, low-quality, and undemanding jobs. The unemployed are generally only able to obtain jobs which are insecure and offer little prospect for personal advancement or development. It is clear that unemployment tends to lead to further unemployment.

6

The Flexible Workforce? The Employment Conditions of Part-Time and Temporary Workers

A central theme of the literature on the changing structure of employment has been the rapid expansion of a flexible or peripheral sector of the workforce. This is seen as a consequence of the greater competitiveness and volatility of advanced capitalist economies that resulted from the energy crises of the 1970s, the growth of competition from the cheaper labour economies of the third world, and the pressures of European integration. These changed economic conditions, it is argued, have led employers to abandon the relatively secure employment policies of the early post-war decades in favour of the construction of a two-tier workforce.[1] With advanced technologies there is still a need for a core of skilled employees, with a high level of job security. But to provide the possibility of rapidly adjusting the workforce to changes in the level of demand and of reducing costs in times of economic difficulty, employers are creating a sharp division between this stable core and a peripheral sector of the workforce, which can provide greater flexibility.

In Britain, two groups in particular have been seen as constituting the growing flexible workforce: part-time workers on the one hand, and workers on short-term contracts on the other. Numerically, it has been above all the growth of the part-time workforce that has attracted attention, even if there were signs that this was slowing down in the early 1990s (Dex and McCulloch 1995: 60, 107). In contrast, the size of the temporary workforce was remarkably stable between the mid-1980s and the early 1990s, at around 5 to 6 per cent of all employees, and it was only towards the mid-1990s that any significant expansion began to take place (Watson 1994: 241; Beatson 1995: 7–10).

The employment conditions of those in the 'core' and in the 'peripheral' workforce could be expected to differ in a number of crucial respects. First, the principal rationale for retaining a section of the workforce with high levels of job security lay in the fact that advanced technologies required skills that could not be easily supplied from the external labour market. Employers were therefore obliged to invest in training and upskilling of their own employees. Employees in the core, then, could be expected not only to have generally higher levels of

[1] The most interesting general presentation of these arguments can be found in Berger and Piore (1980) and Edwards (1979). The dangers of such a development were also powerfully made in the report of the US Commission on the Future of Worker-Manager Relations (1994).

skill but to have very good opportunities for skill development. In contrast, those in the periphery were dispensable and there was little motivation for employers to incur training costs which would only be of short-term value to the organization and easily lost to other employers.

The distinction between relatively secure and insecure types of employment is likely to be associated with marked differences in the quality of jobs and in job rewards. Since the main objective of providing 'core' employment is to ensure high levels of motivation and loyalty of workers with key skills, employees in this sector will tend to be given greater responsibility in work, better work conditions, and higher levels of pay. In contrast, the peripheral sector will tend to be composed of those in lower-skilled jobs, where it will be easier to recruit at short notice. Given the assumption that such workers will have a relatively short-term association with the organization, employers are likely to invest little in the work environment and to keep pay rates at the minimum possible level. Since there will be little in the work task to provide strong intrinsic motivation, and employees will have limited familiarity with work practices and procedures in the organization, work performance will be controlled through close supervision.

The creation of two such distinct sectors of the workforce is also thought to have long-term implications for the careers of individuals. There would be relatively little mobility between sectors; rather there are powerful mechanisms locking individuals into particular parts of the labour market. For those that enter the core, employers will offer good career opportunities within the organization in order to increase work motivation and organizational commitment. As has been noted, they would be likely to receive training from their employers, and this places them in a more favourable position for career advancement. To avoid the risk of losing such employees, their employers give them access to an internal labour market which would enable them to satisfy their career aspirations within the organization. Such employees, then, will have a career pattern characterized by employment stability, skill development, and upward mobility. In contrast, employees in the peripheral sector will face frequent job changes and may well be confronted with repeated spells of unemployment. There will be few incentives for employers to invest in the training of such short-term employees and thus those in the peripheral workforce will become increasingly disadvantaged in a context of rapid technical change. At the best then, they will remain confined to this low-skill, poorly rewarded type of employment; at the worst they will be progressively marginalized, eventually falling into the ranks of the long-term unemployed.

A key assumption of those that have developed the thesis of the flexible workforce is that the two main categories of 'peripheral' employee—part-time workers and temporary workers—fulfil rather similar functions for employers and thereby share rather similar employment conditions. However, this has not been justified by any empirical demonstration. This chapter will examine in turn the employment conditions and work attitudes of part-time and temporary workers and then assess the extent to which these two groups can be plausibly seen as sharing a similar work and labour market position as part of a flexible or peripheral workforce.

THE PART-TIME WORKFORCE

Our definition of a part-time worker followed the practice of the Labour Force Survey in relying on people's self-definition. Overall, 21 per cent of the sample were in part-time work and among women the proportion was 40 per cent. These figures are identical to those based on Labour Force Survey estimates (21 per cent and 40 per cent respectively; see Watson 1994: 241). Given the very small numbers of male part-timers, the analysis will focus upon the comparison between male and female full-timers on the one hand and female part-timers on the other.

Part-timers were heavily concentrated in particular sectors of industry. They were primarily in the service industries. In the private services, they were especially in retail (19 per cent of all part-timers) and in finance (11 per cent). In the public services, they were principally employed in education and research, medical services, and welfare. Overall, 41 per cent of part-timers were in the last three of these sectors.[2]

There is also a striking difference in the extent to which full-timers and part-timers are concentrated in small establishments. Whereas 40 per cent of all female part-timers were in establishments with fewer than twenty-five employees, this was the case for only 31 per cent of female full-timers and 25 per cent of male full-timers.

Skill and Skill Development

Theories of the growth of peripheral work have stressed the low skill-level of such work and the limited opportunities for skill development. Moreover, the model implied a progressive divergence in the fate of those in the core and those in the periphery due to the cumulative advantages of the former in terms of experience of new technology and increased employer investment in training. How well do the data confirm this picture of low skill, restricted opportunities for skill development, and increased polarization over time?

A first approach is to examine occupational class position, which is strongly related to a broad range of skill measures. It is immediately apparent that part-time workers are heavily concentrated in the lower-skilled sector of the labour market. As can be seen in Table 6.1, 53 per cent of female part-time workers compared with only 27 per cent of female full-time workers were in semi- and

---

[2] Analyses using the Labour Force Survey 1994 confirm the very high concentration of part-time workers in distribution and other services (Dex and McCulloh 1995: 72; Casey *et al.* 1997: 14–15).

TABLE 6.1 *Part-time work and skill (cell %)*

	Male full-timers	Female full-timers	Female part-timers
Semi- and non-skilled manual	27	27	53
No qualifications required	29	27	56
No training for current type of work	36	38	59
Less than a month experience needed to do job well	16	21	41
Experienced skill increase in last 5 years	66	69	46
Training from current employer	47	50	31

non-skilled work. Women that were part-timers were less likely to be in any other class category than their full-time equivalents. The difference was particularly sharp for professional and managerial work: only 18 per cent of part-timers compared with 40 per cent of full-timers came into this category.

A very similar picture emerges from the more specific indicators of skill. There was a sharp distinction between part-time and full-time workers in the level of qualifications required for the job. Among female part-timers, 56 per cent reported that no qualifications were required of people currently being recruited for their type of work, whereas this was the case for only 27 per cent of female full-timers and 29 per cent of men in full-time work. On the other hand, only 37 per cent of female part-timers said that O levels were needed, compared with 67 per cent of female full-timers and 63 per cent of male full-timers. It is clear that the main division lies not along gender lines, but between women in full-time and women in part-time work. Women in full-time work were virtually indistinguishable from men in terms of the qualifications required of their work. Women in full-time work were also closer to men than to women in part-time work on the training and on-the-job experience criteria of skill (Table 6.1).

On all of our measures, then, part-time workers emerge as occupying jobs of relatively low-skill content. Indeed, even when occupational class was controlled for, part-timers were lower on each dimension of skill.

They were also less likely to have experienced a recent increase in their skills. Taking the overall pattern, only 46 per cent of female part-timers had seen the skill level of their job rise in the last five years, compared with 66 per cent of male full-timers and 69 per cent of female full-timers. This was confirmed by the evidence on the extent to which people had received training from their employer in the last three years. Whereas nearly half of full-time employees (47 per cent for men, 50 per cent for women) had received some training, this was the case for less than a third of female part-timers (31 per cent).

There was, however, a notable industry variation in the skill experiences of female part-timers.[3] In medical services, welfare, national and local government, and education, the dominant tendency was quite clearly towards upskilling. This was the case respectively for 58, 51, 50, and 49 per cent of all part-time employees in these industries, while the proportion reporting deskilling only exceeded 10 per cent in the case of welfare (12.2 per cent). In contrast, in retail and in hotels and catering, there were much higher proportions of people reporting deskilling. In retail, 18 per cent said that their skills had decreased over the last five years and only 37 per cent of part-timers thought they had increased. In hotels and catering, there was a nearly even balance with 28 per cent saying they had been deskilled and 30 per cent that the skill level had increased. The banking and finance sector was a case apart. It had an exceptionally high proportion of part-timers that had experienced a skill increase (62 per cent), but also a substantial minority (17 per cent) that had been deskilled.

Skill differences may reflect a different age distribution between part-timers and full-timers. Whether one takes the proportion of workers over 35 or the proportion over 45, part-time work is more heavily concentrated among older workers. For instance, 68 per cent of female part-timers were aged 35 or more, compared with 51 per cent of full-time male employees and 49 per cent of female full-timers.[4] This could have important implications in that qualification levels in the general population have gone up over time, so that, other things being equal, a younger workforce should also be more highly qualified. The last decade has seen a major expansion in training provision and it is possible that employers targeted this to a greater extent on younger rather than older workers.

However, even when age is taken into account in addition to class, the skill disadvantage of part-timers still stands out very sharply with respect to qualifications required, on-the job experience needed, the experience of recent upskilling, and training from their current employer (Table 6.2). The strength of the coefficient in each case diminishes quite sharply suggesting that over half the effect with respect to qualifications, experience, and training and around one-third of the effect with respect to upskilling, can be accounted for in terms of class position and age. Nonetheless, there remains a highly significant part-time work effect.

Change Over Time in Skill

It is clear that part-time work was associated with major disadvantages in terms of skill. However, has this differentiation in skill position grown greater over the last decade? A number of comparable indicators are available for 1986 from

---

[3] The analysis has been restricted to industries where the sample provided more than twenty-five part-timers in the dataset.
[4] This reflected the fact that women tended to enter part-time work after having had children. Indeed, data from the British Household Panel Study indicates that 70% of women's entries into part-time work follow a period of 'looking after the family' (Dex and McCulloch 1995: 103).

TABLE 6.2 *Effect of part-time work on skills*

	Part-time work effect on skills without controls		Part-time work effect on skills after controls for class and age	
	Coeff.	Sig.	Coeff.	Sig.
Qualifications required	−0.81	***	−0.31	***
Training for type of work	−0.85	***	−0.18	n.s.
On-the-job experience	−0.84	***	−0.39	***
Experience of upskilling	−0.71	***	−0.44	***
Training from current employer	−0.33	***	−0.10	*

Note: Results are drawn from a series of OLS regression with controls for class and age.

the Social Change and Economic Life Initiative surveys. A first approach is to compare the class distribution of part-time work in the two periods.[5] In contrast to what might be expected from stronger versions of core–periphery theory, the signs are of growing equalization rather than polarization. In particular, compared to female full-timers, there has been an increase in the proportion of part-time workers in lower non-manual work (+8 per cent) and a decrease in the proportion in semi- and non-skilled manual work (−8 per cent).

A comparison of the specific skill indicators also suggests that the trend over time has been for the disadvantage of part-time workers to diminish. With respect to qualification requirements, the period saw rather modest increases for men in full-time work, rather more substantial increases for women in full-time work, but above all quite major increases for women in part-time work (Fig. 6.1). For instance, taking those that reported that no qualifications were required, the proportions dropped by 3 percentage points for men, by 7 percentage points for women in full-time work, and by 13 percentage points for women in part-time work. On the other hand, the increases in the proportions saying O-level equivalent or higher was required were respectively 9, 12, and 15 percentage points. A more formal test of change over time using regression analysis with an interaction term contrasting part-time workers in the two time-periods showed that this difference was statistically significant. It must be noted, however, that the change in skill distribution primarily represented a shift of part-timers from non-qualified jobs into jobs requiring middle rather than higher levels of qualification.

A similar pattern emerges for training and the experience required to do the job well (Table 6.3). The proportion of female part-timers that had received no training fell by 19 percentage points between 1986 and 1992, whereas the decrease among women in full-time work was 13 points and among men 8 points. Similarly, the proportion of female part-timers who reported that they needed less than a

[5] To make a more accurate assessment of change, the class categories that have been used are based on the Goldthorpe 1980 classification, since the earlier data were coded into these.

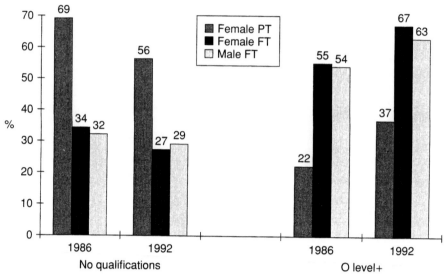

FIG. 6.1 Qualifications required for the job, 1986–1992

TABLE 6.3 *Comparison of women's training and experience needed in full-time and part-time work, 1986–1992 (cell %)*

	No training		1 year training		Less > 1 month experience needed		1 year + experience needed	
	1986	1992	1986	1992	1986	1992	1986	1992
Female PT	78	59	9	14	58	41	10	13
Female FT	51	38	24	26	23	21	27	28
Male FT	44	36	35	33	17	16	45	42

month of on-the-job experience to do their job well declined by 17 percentage points, whereas that for female full-timers fell by only 2 points and for male full-timers by only 1. Regression analysis confirmed that these differences over time reached statistical significance. Again, however, the improvement in the position of female part-timers was primarily in the intermediate categories of training and experience. They improved their position only slightly in the category of those who had had a year or more training or required a year or more on-the-job experience.

Finally, if subjective reports of skill increase are considered, it is again clear that the position of female part-time workers has improved relative to that of full-time workers since the mid-1980s. It can be seen from Fig. 6.2 that while both full-timers and part-timers were more likely to have experienced an increase in

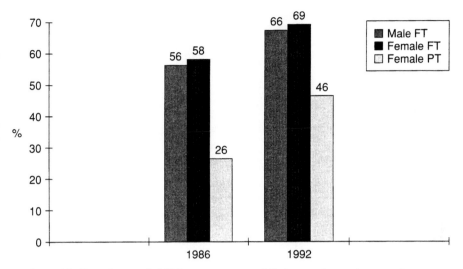

FIG. 6.2 Experience of skill increase among full-time and part-time employees

their skills in 1992 compared with 1986, the change was considerably greater among part-timers. The proportion reporting a skill increase was 10 percentage points higher among male full-timers, 11 points higher among female full-timers, but 20 points higher among female part-timers.

In short, the pattern is very consistent. Whether one takes broad class position or the specific skill indicators, part-timers remain at a severe disadvantage in skill level and skill development compared with full-timers. However, over the last decade, their position has been improving rather than deteriorating.

The Quality of Work

(a) Task discretion

It was seen in Chapter 2 that, among employees as a whole, increased skill requirements were closely paralleled by an increase in the responsibility given to employees for everyday work decisions. It could be expected then that part-timers would have less responsibility in their jobs, but, at the same time, that responsibility would have grown significantly over time reflecting the tendency for the skill gap to diminish.

Our general measure of task discretion certainly brought out the rather lower degree of control over their work that part-timers were able to exercise. Their overall task discretion score was -0.25, compared with 0.06 for male full-timers and 0.09 for female full-timers. This did not simply reflect the fact that they were more likely to be in low-skilled jobs. The disadvantage of female part-timers is evident even when controls are introduced for age and class.

TABLE 6.4 *Part-time work and task discretion (cell %)*

	With normal task control		With new task control	
	1984	1992	1984	1992
Male full-timers	46	52	49	58
Female full-timers	53	63	49	57
Female part-timers	45	50	44	40

But has the disadvantage of part-timers decreased since the mid-1980s? There are two indicators that enable us to compare across time. These were included both in a national survey in 1984 (the Class in Modern Britain Survey) and in the 1992 survey. They distinguish two main aspects of employee control. The first is control over the usual everyday task activities or 'normal task control' and the second is control over new tasks, which could be seen as a rather more demanding form of employee control. The questions were: 'Do you decide the specific tasks that you carry out from day to day or does someone else?' and 'Can you decide on your own to introduce a new task or work assignment that you will do on your job?'

As can be seen in Table 6.4, the lower level of task discretion of part-timers has less to do with the degree of control over established daily tasks than with the ability to introduce changes on the job. With respect to normal task control, part-timers followed the general move towards higher levels of control between 1984 and 1992. A statistical test of the data shows a strong effect of the year (1992), but no effect for the interaction term measuring whether or not the relative position of part-timers had changed between the two dates.

In contrast, with respect to new task control, not only did part-timers start off in 1984 with a disadvantage compared to full-timers, but their position appears to have grown worse over time. Whereas 44 per cent of female part-timers could introduce new tasks on their job in 1984, this had declined to 40 per cent in 1992. In contrast, among both male and female full-timers, there was a substantial rise in the proportion exercising this type of discretion. As a result the period saw an increased difference between part-timers and full-timers in their opportunities for task control. This was confirmed by a statistical analysis that took account of the effects of age and class.[6] It is clear that the general improvement of the position of part-timers on the various skill criteria was not accompanied by a corresponding improvement in the level of discretion they could exercise in their work.

[6] This showed a strong positive effect for year (1992), with a significant negative effect for the interaction representing the position of part-timers relative to others (coeff. -0.52; $p = 0.02$). Even when controls were introduced for age and class, the position of part-timers had deteriorated over the period compared with full-timers (coeff. -0.58; $p = 0.01$).

TABLE 6.5 *Part-time work and job quality (cell %)*

	Male full-timers	Female full-timers	Female part-timers	Sig. (female pt.-time effect)
Str. agree opportunities for self-development	29	30	14	***
Mainly repetitive work	21	25	31	n.s.
Increased variety	70	69	54	***
Able to use previous skills and experience (str. agree)	42	40	31	*
Overall intrinsic work interest score (means)	.05	.16	−.32	***
Increase in work effort	64	66	51	***
Health and safety at risk	35	25	18	***

Note: The significance estimates for the female part-time effect relative to female full-timers in the final column are based on a series of regression analyses with controls for age and class.

(b) Intrinsic work interest

Given their disadvantageous skill position, it would be expected that part-timers would be in jobs that provided less intrinsic interest. A number of measures confirmed that this was the case (Table 6.5). Whereas women in full-time jobs were just as likely as their male equivalents to say that they were in jobs that required them to keep on learning new things (29 per cent and 30 per cent), this was much less frequently the case among part-timers (14 per cent). Part-timers were also more likely to report that their work involved carrying out short repetitive tasks three-quarters or more of the time and they were less likely to feel that the variety of their work had increased over the last five years. Further, they were in jobs that made it more difficult to make full use of their skills. Finally, taking the overall index of intrinsic work interest, it is clear that female part-timers (with a score of −.32) were heavily disadvantaged in comparison with both male and female full-timers (with scores of .05 and .16 respectively).

These disadvantages may reflect primarily the general skill level of the sector of the labour market in which part-time jobs are to be found. To check this, the relationship between part-time status and the various indicators of job quality was tested controlling for class position. This showed that two aspects of job quality—the extent to which the work was repetitive and the ability to use previous skills—were mainly due to the generally low skill-level of the work rather than being inherently linked to part-time work.

However, part-time workers remained distinctive in terms of their limited opportunities for self-development and in the lack of any improvement in the variety of their work in recent years. Thus, while sharing the generally poor quality of work of other semi- and non-skilled manual workers, they had the additional disadvantage in being in a work situation which offered particularly few opportunities for improvement.

But while the intrinsic interest of work was lower for part-timers, there were certain respects in which they had an advantage over full-timers. For instance they were less likely to report a high level of work pressure or an intensification of work effort over recent years. Whereas 64 per cent of male and 66 per cent of female full-time workers considered that the effort involved in the job had increased compared to five years earlier, this was the case for only 51 per cent of female part-time workers. They also had better physical working conditions: they were less exposed to fumes in their work, they worked in a quieter environment, and they were less likely to have to work at very high speed. The combination of these factors appeared to lead not only to a more agreeable, but also to a more healthy work environment. Only 18 per cent of female part-time workers thought that their health or safety were at risk because of their work, whereas, among full-timers, the figure rose to 25 per cent among women and to 35 per cent among men. A logistic regression analysis showed that, when account was taken of age and class position, female part-timers had twice the relative odds of male full-timers of being in a safe working environment.

Extrinsic Rewards

(a) Pay and fringe benefits

Pay is known to be closely related to skill level and, given that part-time work was concentrated in relatively low-skilled jobs, it could be expected to be poorly paid. But was there an additional penalty for those on part-time contracts over and above what could be expected on the basis of skill? Taking gross hourly earnings as our measure for pay, since this provides us with comparability irrespective of the hours worked, there can be no doubt that the pay disadvantage of part-timers was highly significant. On average women part-timers were earning £4.48 per hour compared with £7.13 for men in full-time work and £6.19 for women in full-time work. The low-skill level of the jobs was not sufficient to account for the difference; part-timers were less well paid within each occupational class. For instance, in semi- and non-skilled manual work, part-timers were earning on average £3.70 compared with £5.33 for men full-timers and £4.14 for women full-timers.

Indeed, it was striking how strongly the part-time differential persisted even when a wide range of other factors that were likely to influence pay had been taken into account. The disadvantage of part-timers remained when simultaneous controls were introduced for class, age, and size of establishment (Table 6.6). It was still evident when more detailed account was taken of skill level by introducing the qualifications required for the job, the number of months training that had been required for doing the type of work, and the length of on-the-job experience that had been needed before it was possible to do the job well. Although the introduction of these factors certainly reduced the part-time effect, it by no means removed it. A separate analysis of this data (Lissenburgh 1996) has shown

TABLE 6.6 *Part-time work and pay (log gross hourly earnings)*

	Coeff.	Sig.
Without controls	−0.40	***
+ sex, age, and class	−0.19	***
+ qualifications, training, and experience	−0.18	***
+ industry, firm, and work history characteristics	−0.04	n.s.

Note: Results are drawn from a series of OLS regression analyses.

TABLE 6.7 *Part-time work and access to fringe benefits (cell %)*

Type of benefit	Male full-timers	Female full-timers	Female part-timers
Occup. pension scheme	76	71	47
Special sick pay scheme	67	64	52
Paid time off for domestic problems	64	56	41
Company vehicle	32	16	8
Free transport	25	21	12
Subsidized meals	36	36	32
Provision finance/loans	17	17	12
Accommodation	11	8	5
Life assurance	26	14	6
Private health scheme	28	22	10
Recreation/sports facilities	33	29	17
Extra maternity pay	28	44	30
Child-care assistance	3	5	5
Career break scheme	12	17	8
Overall fringe benefit score	5.3	4.9	3.6

that it is only when detailed information on industry, firm, and prior work history experience has been included that the part-time effect finally ceases to be significant and even then it remains close to the significance threshold ($p = 0.06$).

Moreover, despite the rise in the skills of part-time workers over the decade from the mid-1980s, our data indicate that there has been little change in their relative pay position. In 1986, female part-time workers earned 72 per cent of the average hourly rate for women. Six years later, in 1992, the figure remained virtually unchanged at 71 per cent.

The position of part-timers with respect to fringe benefits also remained markedly worse than that of other employees (Table 6.7). For instance, only 47 per cent of part-timers benefited from an occupational pension scheme, compared with 71 per cent of female full-timers and 76 per cent of male full-timers. Similarly, part-timers were substantially disadvantaged with respect to provision of special sick-pay arrangements, the ability to take time off for domestic problems without losing pay, the opportunity of taking up some type of career break scheme, assistance with transport, access to sporting or recreational facilities, and the availability of life assurance and private health care schemes.

An overall indicator was constructed by summing the number of fringe benefits to which people had access. Male full-timers had the highest number of fringe benefits (5.3), although they were fairly closely followed by female full-timers (4.9). The notable difference is between the advantages available to full-timers of either sex and those available to female part-timers (a score of only 3.6). Again, the disadvantage of part-timers stood out at a high level of statistical significance ($p < 0.000$), even when the potentially confounding factors of occupational class, age, and size of establishment were controlled for. Indeed, as Lissenburgh (1996) has shown, the part-time differential with respect to fringe benefits appears to survive both controls for human capital and controls for industry and firm type.

(b) Career opportunities

Current pay and fringe benefits represent only one aspect of the potential material rewards of employment. A wider assessment also needs to take into account the opportunities available for upward career progression. Where employment offers relatively good opportunities for upward mobility, this should ensure higher levels of financial reward over time. How does the situation of part-time workers compare with that of full-timers with respect to such longer-term advantages?

Two questions were designed to assess career opportunities. The first asked people about the general opportunities that went with the type of work: 'Does your work have a recognized career or promotion ladder, even if it means changing employer to go up it?' The second asked people more specifically about their own chances of being given a significant promotion within their present organization. Given the inevitably approximate nature of such judgements, the scale has been reduced to two categories: those who thought that their chances were 50 : 50 or better and those who thought that they had a lower chance of promotion.

Part-timers emerged quite clearly as having a much more pessimistic view than full-timers about their chances of moving up into higher-level work. Only 43 per cent thought that there was some type of career ladder, compared with 60 per cent of male full-timers and 65 per cent of female full-timers. Further, only 26 per cent of part-timers thought they personally had a 50 : 50 or better chance of getting significant promotion in their organization, whereas the figure was 44 per cent among male full-timers and 41 per cent among female full-timers.

It was noted earlier that part-timers tended to be somewhat older than the average employee and were concentrated particularly heavily in less skilled occupations. Both of these factors could potentially explain their lower promotion chances. The chance of getting promoted is generally higher for younger people

TABLE 6.8 *Part-time work and promotion opportunities (cell %)*

	Those thinking they had a 50 : 50 chance or better of promotion		
	1984	1992	Change
All employees	36	40	+4
Male full-time	43	44	+1
Female full-time	36	41	+5
Female part-time	17	26	+9

and the opportunities are much better in higher-skilled work. Part-timers are also more likely to be found in small establishments and their disadvantage might reflect the simpler control structures of smaller organizations. However, even when these factors were taken into account in a logistic regression, female part-timers stood out as significantly less likely to have promotion opportunities. It is difficult to escape the conclusion that it is the contract type itself that severely curtails women's opportunities.

Has the degree of career disadvantage experienced by part-time workers changed over time? An identical question on promotion opportunities was asked in a national survey carried out in 1984.[7] The overall pattern showed a small overall increase in people's optimism about the opportunities available to them. Moreover, the percentage increase over time was higher among female part-timers (+9 percentage points) than among female full-timers (+5 points) and especially than among male full-timers (+1 point). But this apparent improvement in the relative position of part-timers failed to reach statistical significance. Moreover, the size of the coefficient was sharply reduced when controls were introduced for class, suggesting that it may have reflected primarily the fact that there had been a growth over the period in the proportion of part-time jobs in professional and managerial positions.

Flexibility in Work

The assumption of segmentation and flexibility theories is that part-time workers are valuable to employers in good part because of the flexibility they offer. The notion of flexibility covers rather a wide range of different types of practice. A frequently used typology distinguishes between three main types of flexibility: functional, pay, and numerical (Atkinson 1984). Functional flexibility refers to an employer's capacity to switch employees between different tasks rather than being bound by rigid job demarcations. Pay flexibility concerns the ease

---

[7] The 1984 data are drawn from the Class in Modern Britain Survey.

with which pay can be varied in the light of either individual performance or the difficulties that the organization confronts in the product market. Finally, numerical flexibility refers to the ability of the employer to adjust either the number of people employed or the hours worked.

(a) Functional flexibility

To begin with, were female part-timers distinctive in terms of functional flexibility? Our evidence would suggest that they were significantly less flexible in this respect than either men or women who worked full time. They were less likely to say that they frequently changed the type of work that they were doing. Only 16 per cent were strongly of the view that this was a characteristic of their job, compared with 24 per cent of male and 23 per cent of female full-timers. Nearly half of part-timers (46 per cent) reported that this was not at all the case with their job, compared with fewer than a third of male full-timers (30 per cent) and female full-timers (32 per cent).

Nor was there any evidence that part-timers had been at the cutting edge of organizational changes designed to introduce greater flexibility. Whereas 39 per cent of male and female full-timers were strongly of the view that they were expected to be more flexible in the way they carried out their work than a couple of years previously, this was the case for only 21 per cent of part-timers. Similarly, 38 per cent of part-timers reported that there had been no shift towards greater flexibility in working practices, a proportion at least twice that among male and female full-timers (14 per cent and 19 per cent respectively). This was consistent with the lower proportion of part-timers who reported that there had been changes in work organization over the last two years that had had direct effects on their job (42 per cent compared with 53 per cent among male full-timers and 55 per cent among female full-timers). Similarly, their jobs were less likely to have been affected by recent technical change. In short the evidence points clearly to the fact that part-time work was carried out in a relatively stable organizational environment where there were few pressures on employees for greater flexibility in patterns of working.

(b) Pay flexibility

Pay flexibility refers above all to the ability of the employer to adjust pay levels depending on the level of effort and the quality of work put in by the employee. Were female part-timers more likely to be on some type of effort-related payment system? There were two sets of questions that addressed this issue. The first asked whether or not the payment system contained bonuses that related to performance. People were asked: 'Do you receive any incentive payment, bonus, or commission based on: (*a*) your own pace of work; (*b*) the pace of work of the team or group to which you belong, and (*c*) the results achieved by your organization or your workplace?' As employers can combine different forms of incentive, people were asked to answer each of the options separately.

TABLE 6.9 *Part-time work and flexibility (cell %)*

	Male full-timers	Female full-timers	Female part-timers	Sig. (female pt.-time effect)
Flexible in work				
(a) Frequently change type of work	24	23	16	***
(b) More flexible in work than				
2 years ago	39	39	21	***
Type of pay system				
Individual merit bonus	21	14	7	***
Work-group bonus	8	4	2	**
Organizational bonus	28	18	11	*
Incremental pay scale	39	55	41	***
Rises given to:				
All regardless of performance	69	75	80	n.s.
Those who work hard and				
perform well	44	38	28	n.s.
Those with favoured relation				
with boss	16	16	7	***
Hours				
Often has to work extra hours	36	32	19	***

Note: Significance estimates for the female part-time effect relative to female full-timers in last column are based on regression analyses, with age and class controls.

A first point to note is that such incentives were pretty rare across industry. Overall, only 16 per cent of employees reported that they could receive individual incentive payments and even fewer (6 per cent) that there was a work-group bonus system. The commonest formula (22 per cent) was for a bonus linked to organizational results. However, as can be seen in Table 6.9, the notable characteristic of the payment systems for part-time workers was the rarity of any of these types of performance-related rewards. For instance, they were only half as likely as female full-timers, and only a third as likely as male full-timers, to be on either an individual incentive system or a work-group bonus system.

The existence of one of these incentive schemes does not in itself ensure that pay is related to individual effort or merit, since the bonuses may constitute only a very small proportion of overall pay. To get at people's understanding of the general principles that underlay pay rises in their organization, they were asked: 'Thinking about how pay rises are given where you work, do you think that rises are given to: (*a*) all workers regardless of performance; (*b*) those workers who work hard and perform well; and (*c*) those workers with some favoured relation with the boss.' Again people could mention more than one type of system to allow for cases where there was a genuine mix of systems.

It is again clear that part-time workers were not distinctively subject to a payment system ensuring greater flexibility. The item that comes nearest to the notion of pay flexibility is that indicating reward for hard work and good performance. Yet only 28 per cent of part-time workers said that this was the main principle underlying pay rises in their organization compared with 38 per cent of full-time female employees and 44 per cent of full-time male employees. Quite consistently, part-timers were the most likely to say that pay rises were given to all employees regardless of performance.

Whether, then, one takes the existence of some type of incentive system or people's reports of the general principles underlying the payment system, the conclusion is the same. Far from being characterized by greater pay flexibility, part-time workers were the category most likely to be covered by payment systems that made no attempt to relate pay and individual performance.

(c) Hours flexibility

Even if they were not providing task and pay flexibility, it might be that part-time work gave employers the possibility of varying individual working hours at short notice to reflect unforeseen increases in demand.[8] This might be particularly advantageous in that traditional notions of overtime payments have been traditionally linked to full-time work. To explore this type of hours flexibility, employees were asked how much the case it was that they often had to work extra time over and above the formal hours of the job to get through the work or to help out.

The results show rather clearly that part-timers were distinctive in terms of the regularity rather than the unpredictability of their hours. Only 19 per cent of female part-timers said that it was very true that they often had to work extra hours, compared with 32 per cent of female full-timers and 36 per cent of male full-timers. At the other end of the scale, 42 per cent of part-timers said that it was not at all true of their work, compared with only 27 per cent of female full-timers and 23 per cent of male full-timers. Moreover, part-timers remained very significantly less flexible with respect to hours even when age and class were controlled.

In short, with respect to the 'in work' context, it is clear that there are few grounds for considering part-timers to be particularly flexible. They were less likely than full-timers to make changes in the type of work they were doing, their pay was less closely related to effort, and they had relatively rigid working hours.

Job Security

The most important issue, however, with respect to arguments about the growth of a peripheral workforce is the extent to which employees have job security.

---

[8] On the importance of the growth of this type of flexibility since the mid-1980s see especially Casey *et al.* 1997.

Part-time workers were certainly subject to restrictions which meant that they were less likely to be covered by employment protection legislation than those in full-time work. It is this legal disadvantage which tends to underlie the view that they form part of an insecure secondary or peripheral sector of the workforce. However, this assumes that legal disadvantage translates in a very direct way into practice. Legal rights are likely to be only one factor influencing employer readiness to dismiss workers. It has been seen that the part-time workforce was particularly heavily concentrated in the service sector and this has been the area of the economy that has experienced the strongest growth over the last two decades. How did the conflicting influences of low legal protection and relatively high concentration in an expanding sector balance out in terms of the security that part-timers felt that they had in their current jobs?

Even with respect to the issue of legal protection, it is important not to over-draw the contrast between the position of part-timers and full-timers. A substantial proportion of part-timers did fall within the provisions for employment protection. The rights of part-timers depended on two factors: their hours of work and their length of service. Part-timers who worked sixteen hours or more came under the usual provision of needing at least two years' service to be entitled to employment protection, those who worked for less than sixteen hours a week had to have five years of continuous service with their employer, while those working less than eight hours a week were not covered at all. Using information about the usual hours of work and length of service, it is possible to make an approximate estimate of eligibility. About half of part-time employees in the survey (53 per cent) were likely to have been covered by employment protection.[9] This compares with 74 per cent of male full-timers and 71 per cent of female full-timers.

Part-time workers were aware that they had relatively weak formal protection in their jobs. The survey included two questions designed to see how quickly people felt that a person doing their sort of job would be dismissed if there were problems with their work performance. The first referred to the situation of a person who persistently arrived late and the second to one persistently failing to work hard enough. Taking as an indicator of considerable employment vulnerability those that reported that a person could be dismissed within a month, part-time workers were substantially more likely to report that they could be rapidly dismissed than full-timers in both types of situation (Table 6.10). For instance, with respect to lateness, 42 per cent of female part-timers said that

---

[9] In terms of hours of work, a majority (64%) were working more than sixteen hours a week, putting them onto a similar basis to full-timers, while 29% worked between eight and fifteen hours and only 7% fewer than eight hours. Among those working more than sixteen hours, 69% had been working with their current employer for more than two years and were therefore covered. Among those working more than eight but fewer than sixteen hours, 28% were covered since they had been with their current employer for more than five years. Employment protection distinctions based on hours were subsequently abolished in 1995.

TABLE 6.10 *Part-time work and job security (cell %)*

	Male full-timers	Female full-timers	Female part-timers	Sig. (female pt.-time effect)
Dismissible for poor work within a month	27	26	36	**
Dismissible for lateness within a month	32	28	42	***
Anxious about arbitrary dismissal	21	13	12	n.s.
Job affected by reductions in workforce	50	46	39	**
Job security has decreased last 5 years	40	31	31	n.s.
Mean dissatisfaction with job security score	3.05	2.72	2.71	n.s.

Note: Significance estimates for the female part-time effect relative to female full-timers in the last column are based on regression analyses with controls for age and class.

they would be fired within a month, compared to 32 per cent of male full-timers and only 28 per cent of female full-timers. Similarly, whereas 36 per cent of part-timers said they could be dismissed within this time for poor performance, this was the case for only 27 per cent of male full-timers and 26 per cent of female full-timers.

However, despite this awareness of a relatively low level of regulatory protection, there was no evidence that part-timers were more likely than full-timers to feel that their jobs were in practice insecure. The picture was consistent across a number of indicators (Table 6.10). Together with other female workers, they were relatively unlikely to think that there was a significant risk of arbitrary dismissal or to consider that their jobs had become less secure over the previous five years. They were less likely than either male or female full-time employees to report that there had been reductions in the workforce that had had direct effects on the job. They were the category least likely to think that they might be forced to leave their employer over the next year as a result of redundancies, closures, or termination of contract. This picture of reasonable employment security is confirmed by their relatively low dissatisfaction with job security. Like female full-timers, women in part-time work showed markedly lower levels of dissatisfaction with job security than male full-timers.

This suggests that, while having less formal security, part-timers felt that they had a level of *de facto* security in their current jobs that was as high as that of women in full-time work, and higher than that of men in full-time work. This appeared to be related to the fact that they were in organizations that were undergoing less internal restructuring and that were less threatened by their external environment.

TABLE 6.11 *Risks of unemployment of part-time and full-time workers (all work-history job events)*

	Men	Women
% moves from full-time work into unemployment as a proportion of all moves from full-time work	11	7
% moves from part-time work into unemployment as a proportion of all moves from part-time work	22	6
No. of full-time job events	12,597	7,719
No. of part-time job events	295	2,202

It might be argued, however, that part-timers tend to underestimate their real insecurity, possibly because they attach a lower importance to employment and are thus less sensitive to threats to it. One way of checking this is to examine people's past work histories, looking at the risk of unemployment associated with particular types of jobs. The work histories provide information on all labour market moves since the person first left full-time education. Is there any evidence that a period in a part-time job was more likely to be followed by a spell of unemployment than a period in full-time work? Taking all instances of part-time and full-time work, we calculated the proportion that were followed by an unemployment event as against some other type of event. The pattern this revealed is an interesting one (Table 6.11). First, taking women, there was no difference at all in the risk of unemployment between those in part-time and those in full-time work. Occupancy of a part-time job had been followed by a spell of unemployment in 6 per cent of cases, compared to 7 per cent for a full-time job. This follows very closely the results in the attitudinal data, which showed they were very similar in terms of their relatively low anxiety about arbitrary dismissal and high levels of satisfaction with their job security.

Second, both part-time and full-time female workers were much less likely than male workers to experience unemployment after any given job. This again follows closely the pattern given by the attitudinal data for current perceptions of job security. Third, insofar as there was a link between part-time work and greater job insecurity, it was confined entirely to the rather small proportion of *male* part-time workers. In 22 per cent of cases, men who had held part-time jobs moved into unemployment.

The different sources of data provide a highly consistent picture. Whether one takes attitudinal data about current employment security or work-history data about actual risks of unemployment, it is clear that the great bulk of the part-time workforce, which consists of women, experienced a relatively high level of employment security. This did not appear to be specific to part-time work but rather reflected the employment security of female employment more generally.

It was men that appeared to have borne the main brunt of the job insecurity that flowed from economic restructuring.

There can be little doubt from our data that part-time employees constitute a highly distinctive sector of the workforce. But the picture that emerges is not one that corresponds well with the assumptions underlying theories of a peripheral or flexible workforce. Certainly, part-timers were characterized by relatively low-skill levels, restricted opportunities for skill improvement, low pay, and very poor career opportunities even with respect to others in the same broad occupational class. However, our evidence did not confirm the crucial tenet of these theories that it was a sector of the workforce distinguished by a particularly high level of flexibility. Part-timers were less likely to be flexible in the types of work they did and they were less likely to report increased flexibility in their jobs. With respect to pay, they were less likely to be on payment systems that related rewards to individual effort or performance. Finally, and most crucially, there was no evidence that they suffered from the type of chronic job insecurity that is held to be the defining characteristic of 'peripheral' labour. Certainly, they appeared to be aware that they were less protected in a legal sense, but, in practice, they were more likely to feel that their current jobs were secure than full-timers and the work-history data showed that, for women, part-time work was no more likely to lead to unemployment than full-time work.

THE TEMPORARY WORKFORCE

The second major component of the 'peripheral' workforce, as it is conventionally conceived, is that of workers on temporary contracts. This category has not expanded very rapidly in Britain, but it meets much more directly the criterion of low job security. Relatively little is known, however, about the characteristics of the temporary workforce. Is it the case that such jobs are primarily low-skilled? Are temporary jobs bridges for moving into more permanent work or are people trapped into a cycle of temporary jobs and recurrent unemployment?

Types of Temporary Worker

In examining the characteristics of temporary work, it is important to recognize that there are very different types of temporary work contract.[10] In particular, they vary considerably in their length. The question used to define temporary work status explicitly took this into account. It asked 'Do you think that your job is considered by your employer to be: a temporary job lasting less than twelve

---

[10] Earlier discussions of the potential importance of distinguishing types of temporary work can be found in Dale and Bamford (1988) and Gallie and White (1994).

months, a fixed-term job lasting between one and three years, or a permanent job with no fixed time of ending?' Employees who were on contracts that lasted less time than twelve months will be termed 'short-term temporary workers' and those on contracts lasting between one and three years 'contract workers'. It will be seen that these represent quite fundamentally different groups in the workforce. A good deal of the discussion on temporary work is seriously flawed as a result of the failure to distinguish between them.

Overall, 11 per cent of employees in the survey were temporary workers. Of these, 6 per cent were short-term temporary workers and 5 per cent were contract workers. This figure is considerably higher than that given by the Labour Force Survey, which was 5.5 per cent in 1992 (Beatson 1995: 10). This probably reflects two factors. The survey was carried out over the summer, when the use of temporary work is greater. The question format also differed from that of the LFS in making explicit the classification of medium-length contracts as temporary work. The figure for shorter-term temporary work is very close to the LFS estimate. Both indicators of temporary work provide a broadly similar picture of relatively little change in the size of the temporary workforce since the mid-1980s. The LFS data show virtually no increase from 1984 (when 5.3 per cent of employees were temporary workers). A comparison of our survey with data for 1986 using the identical question, suggests a very slight increase over the period.[11] In the 1986 data, 9 per cent of employees were temporary workers (5 per cent short-term and 4 per cent contract).

For both sexes, temporary workers were likely to be relatively young.[12] Whereas 45 per cent of permanent workers were under 35, the proportion rose to 58 per cent among contract workers and to 52 per cent among short-term temporary workers. Contract workers were distinctive in that a particularly high proportion were in the youngest age category (20–4 years)—24 per cent compared with 15 per cent among permanent workers and 13 per cent among short-term temporary workers. This suggests that contract workers were particularly likely to be at an early stage in their work careers.

The main concentrations of temporary work were in the public and private services: in retail, hotels and catering, and financial services on the one hand, and in education, health, and welfare services on the other. Education was the highest user of all, accounting for 16 per cent of all contract workers and 15 per cent of all short-term temporary workers. This was followed by financial services (13 per cent and 10 per cent respectively). However, there was also a particularly high concentration (14 per cent) of short-term temporary workers (but not of contract workers) in one manufacturing industry—mechanical engineering and vehicles.

[11] The data for 1986 are drawn from the Social Change and Economic Life Initiative Surveys.
[12] This is confirmed by analyses of the British Household Panel Study (Dex and McCulloch 1995: 107).

Such industry concentrations, however, may simply reflect the fact that some industries employ more people of all types than others. If one takes as an indicator of the intensity of use of flexible labour the proportion of all temporary workers employed in an industry compared with its proportion of all permanent workers, there are six industries that stand out as particularly likely to use temporary workers: leisure and personal services, in the private services; health and particularly education, in the public services; and construction and energy/water, in the non-service sector.

The Skill Levels of Temporary Workers

The literature on the flexible workforce invariably depicts temporary workers as having particularly low skill levels. Indeed, it is precisely this that is seen as making them relatively easily substitutable and therefore employable on short-term contracts. This picture turns out to be rather misleading because it fails to take account of the differences between types of temporary worker.

Taking first occupational class position, which is strongly associated with the broad range of skill criteria, it can be seen in Table 6.12 that it was only the short-term temporary workers that had a slightly higher proportion in non-skilled manual jobs than permanent workers (and, even then, the difference was not very substantial—35 per cent compared to 32 per cent). Among the contract workers, the over-representation was at the opposite extreme of the skill hierarchy among professional and managerial employees (45 per cent compared to 35 per cent of permanent workers). Taking the overall distribution by class, what is striking is how similar the contract workers were to the permanent workforce. Whereas there was no significant difference between contract and permanent workers, there was a clear difference between the contract workers and the short-term temporary workers.

The skill distribution varied substantially by sex. For instance, the tendency for short-term temporary workers to be in semi- and non-skilled manual work was entirely accounted for by male workers on these contracts, while the over-representation of contract workers in professional and managerial work was entirely

TABLE 6.12 *Contract type and occupational class (column %)*

	Permanent	Contract	Short-term temporary
Professional/ managerial	35	45	30
Lower non-manual	17	13	18
Technician/ supervisory	5	2	3
Skilled manual workers	11	11	14
Semi- and non-skilled	32	30	35
No.	2,941	206	148

TABLE 6.13 *Effects of contract status on skill*

	Temporary workers		Contract workers	
	Coeff.	Sig.	Coeff.	Sig.
Qualifications required	−0.16	n.s.	0.12	n.s.
Training for type of work	0.19	n.s.	0.02	n.s.
On-the-job experience	−0.08	n.s.	0.06	n.s.
Experience of upskilling	−0.36	***	−0.05	n.s.
Training from current employer	−0.31	***	0.18	**

Note: Results are drawn from a series of OLS regression analyses with controls for sex, age, and class.

due to the positions of women. Indeed, the skill composition of the temporary workforce as a whole was markedly higher for women than for men.

The difference between short-term temporary workers and contract workers was also evident on the more detailed skill indicators. The contract workers were the category most likely to have A-level equivalent qualifications or higher (47 per cent compared with 31 per cent of short-term temporary workers and 37 per cent of the permanent workers) and they were more likely than either category to be in jobs which required more than a year's experience in order to be able to do the job well. These differences in skill requirements were primarily due to the class position of jobs. Once sex, age, and class controls are introduced, neither the short-term temporary nor the contract workers were significantly different from permanent employees (Table 6.13).

The differences between contract and short-term temporary workers also stood out in terms of people's more recent experiences of skill change. The contract workers were indistinguishable from permanent employees in the extent to which they reported that their jobs required more skill than five years earlier, whereas temporary workers were much less likely to report upskilling. Whereas only 41 per cent of short-term temporary employees reported that their skills had increased, this was the case for 62 per cent of contract and 65 per cent of permanent employees. The disadvantage of the short-term temporary workers still stood out clearly even when controls were introduced for sex, age, and class (Table 6.13). An even sharper picture emerges with respect to the training received from employers in the last three years. Whereas short-term temporary workers were significantly less likely to have received any training than equivalent permanent employees, contract workers were more likely to have been trained.

Is there any evidence of either convergence or polarization over time between the skill levels of the temporary and permanent workforce? Comparing with data for 1986, it was clear that the temporary workforce had become more skilled over the decade in terms of each of the indicators of skill.[13] For both short-term temporary workers and contract workers, there was a rise in the proportions in jobs with relatively high qualification, training time, and job experience requirements. Similarly, for both types of temporary worker, there was a rise in the proportion that reported that their skills had increased.

But this has to be seen in the context of the general upskilling of the workforce. For convergence (or polarization) to have occurred, there would need to be evidence of a change in the position of temporary workers *relative* to that of permanent employees. While the statistical tests for this showed a positive coefficient in each case (indicating a tendency for relative improvement), except in one case these failed to reach statistical significance. The exception was with respect to training: the training gap between short-term temporary workers and permanent employees had diminished significantly since the mid-1980s.

The Quality of Employment

Were temporary workers placed in jobs that offered less responsibility and less intrinsic interest? Taking first the level of task discretion, it can be seen in Table 6.14 that both categories of temporary worker have lower task discretion than the permanent workforce. However, the difference in scores between contract workers and permanent workers is relatively small, while the widest gap lies between the permanent workers and the short-term temporary workers. Moreover, it is only the female contract workers that are disadvantaged: the males have the same level of task discretion as their permanent counterparts. A regression analysis showed that, once age and class position had been controlled for, it is only female temporary workers—whether contract or short-term temporary workers—that had lower discretion than equivalent full-time workers.

A comparison of the intrinsic interest of the work can be made using the summary index introduced in Chapter 2. It is based on a range of questionnaire items relating to characteristics of the job: the opportunities that it provided for self-development, the variety of the work, the amount of time spent on repetitive work, the ability to use previous skills and experience, and three items assessing the degree of monotony or interest in the work. This shows in stark form the difference between the experience of the two categories of temporary worker (Table 6.14). The contract workers were even more likely than permanent employees to find their work intrinsically interesting, while short-term temporary workers were much less likely. This was the case both for men and for women. If controls are introduced for class and age, the difference between the contract workers and the permanent employees disappears, but the short-term temporary workers remain very decidedly disadvantaged.

The counterpart to the relatively low level of interest of the work of the short-term temporary workers was that they were in jobs that involved less pressure.

---

[13] The data for 1986 are drawn from the Social Change and Economic Life Initiative Surveys.

TABLE 6.14 *Temporary work, task discretion, and intrinsic job interest*

	Task discretion score	Intrinsic job interest score	Work pressure score
All employees			
Permanent	0.02	0.01	0.01
Contract	−0.07	0.14	0.07
Temporary	−0.28	−0.34	−0.17
Male employees			
Permanent	0.05	0.05	0.04
Contract	0.06	0.14	0.06
Temporary	−0.22	−0.39	−0.15
Female employees			
Permanent	−0.01	−0.03	−0.01
Contract	−0.20	0.15	0.07
Temporary	−0.34	−0.28	−0.19

It can be seen in Table 6.14 that the two temporary-work categories again diverge sharply, with the contract workers having the highest scores for work pressure and the short-term temporary workers the lowest scores. Once controls have been introduced for age and class, the contract workers are indistinguishable from permanent employees, but the short-term temporary workers still experienced significantly lower levels of work pressure.

Consistently with this, the short-term temporary workers were less subject to demands for flexibility in the way they carried out their tasks or in the hours they worked. For instance, only 17 per cent said that it was 'very true' that they often had to change the sort of work they were doing, compared to 29 per cent of contract and 22 per cent of permanent employees. The proportion that were highly flexible with respect to the need to work extra time over and above the formal hours was very similar to that for the other categories of employee (29 per cent compared to 32 per cent for contract and permanent employees). It can be seen in Table 6.15 that, once account has been taken of sex, age, and class, both short-term temporary workers and contract workers are indistinguishable from permanent employees in terms of either task or working hours' flexibility. They were also no more likely to be on a payment system that formally rewarded work effort, although there is some indication that the pay of short-term temporary workers may have been to a greater extent subject to their supervisor's discretionary power.

In short, the two types of temporary worker differed quite sharply in terms of the intrinsic quality of their employment. Contract workers had work that was very comparable with that of permanent employees in similar class positions. Temporary workers, on the other hand, had work that provided much lower levels of job interest, although they were also subject to less pressure in their work.

TABLE 6.15 *Effects of contract status on flexibility*

	Temporary		Contract	
	Coeff.	Sig.	Coeff.	Sig.
Flexibility in work				
Frequently change type of work	0.98	n.s.	1.21	n.s.
More flexible in work than 2 years ago	0.99	n.s.	0.77	n.s.
Type of pay system				
Individual incentive bonus	0.75	n.s.	0.71	n.s.
Work-group bonus	0.95	n.s.	0.99	n.s.
Organizational bonus	0.41	***	0.43	***
Incremental scale	0.50	***	0.96	n.s.
Rises given to				
All regardless of performance	0.65	*	0.82	n.s.
Those who work hard and perform well	0.91	n.s.	0.58	***
Those with favoured relation to boss	1.81	**	0.99	n.s.
Hours Flexibility				
Often has to work extra hours	0.97	n.s.	1.00	n.s.

Note: Results (multiplicative effects on odds) are drawn from a series of logistic regression analyses with controls for sex, age, and class.

Temporary Work and Careers: Bridge or Trap?

The significance of the disadvantages associated with temporary work depends in part on how long people are thought to spend on this type of contract. Is temporary work a transient phase of people's careers, preparing them for a move into permanent employment, or does it lock them on a long-term basis into careers that consist at best of a series of temporary jobs and at worst of sequences of temporary work and recurrent unemployment? The optimistic vision is that it allows people to 'look around' the labour market and to get to know which types of work they like and are good at. If this is the case, it could be expected that people see it as offering reasonable opportunities for future upward career progression. The alternative possibility, underlying a good deal of labour market segmentation theory, is that employers have constructed the temporary workforce as a quite separate category of labour, the primary function of which is to provide the flexibility needed to preserve the privileged employment conditions of the core workforce. In this case, temporary work would offer few opportunities for promotion into better jobs and be associated with chronic job insecurity.

Two approaches were adopted to examine promotion opportunities. In the first, people were asked whether or not there was a career ladder in this kind of job (either in their current organization or by changing employer) and in the second they were asked how good they judged their own opportunities for promotion with their current employer.

TABLE 6.16 *Contract status and promotion opportunities (cell %)*

	In occupation with career ladder	50 : 50 + chance of promotion with current employer
Temporary	49	21
Contract	67	43
Permanent	58	41

TABLE 6.17 *Effect of contract status on promotion opportunities*

	Temporary		Contract	
	Coeff.	Sig.	Coeff.	Sig.
Career ladder	0.93	*	1.04	n.s.
Promotion chances 50 : 50+	0.83	***	0.98	n.s.

Note: Results (multiplicative effects on odds) are from logistic regression analyses controlling for sex, age, and class.

Both of these questions revealed a sharp difference in the circumstances of contract workers and short-term temporary workers (Table 6.16). The contract workers were the group that were most likely to feel that their sort of work had a recognized career ladder (67 per cent compared with 58 per cent of the permanent employees), while the short-term temporary workers were the least likely (49 per cent). Similarly, the contract workers were very similar to permanent employees in their belief that they had a 50 : 50 or better chance of promotion (43 per cent compared to 41 per cent). In contrast, only a small minority of the short-term temporary workers felt that they had a reasonable chance of getting a better job with their current employer (21 per cent). It was seen earlier that the contract workers were more likely than others to be in professional and managerial work and perceived promotion chances were far better in this category than among either lower non-manual employees or manual employees. Once class and age have been controlled for, there is no longer a significant difference between contract and permanent employees (Table 6.17). However, short-term temporary workers still remained much more pessimistic about their chances.

In short, the differences in beliefs about opportunities for promotion reinforce the view that there is a fundamental divide between the two categories of temporary worker. A sizeable proportion of the more highly skilled contract workers see their jobs as offering reasonable opportunities for upward mobility, whereas short-term temporary workers see themselves as trapped in their current positions.

TABLE 6.18 *Effect of contract status on job security*

	Temporary		Contract	
	Coeff.	Sig.	Coeff.	Sig.
Ease of dismissal for poor work	0.46	***	−0.06	n.s.
Ease of dismissal for lateness	0.51	***	0.01	n.s.
Anxious about arbitrary dismissal	0.03	n.s.	0.03	n.s.
Dissatisfaction with job security score	1.63	***	1.07	***

Note: Results are drawn from a series of OLS regression analyses controlling for sex, age, and class.

Job Insecurity

Finally, the critical test of whether temporary work was experienced as a trap rather than as a passage to better things was how far it was linked to job insecurity. While the contractual terms of such work might seem likely to generate a fairly pervasive sense of insecurity, it has been found that the renewal of fixed-term contracts is relatively common in the UK (Atkinson 1993: 3). This, together with a belief that there were opportunities for promotion might mean that perceived security is higher than would be anticipated from the degree of formal protection of the job.

A first indicator of insecurity is how rapidly people thought they would be dismissed if they were poor employees, either persistently arriving at work late or not working hard. The vulnerability of contract workers was very similar to that of permanent employees (Table 6.18). In contrast, the short-term temporary workers were much more exposed to rapid dismissal. Indeed, 56 per cent of the short-term temporary workers said they would be dismissed within a month if they were frequently late and 46 per cent if they were not hard-working. The likelihood of being dismissed at relatively short notice was particularly high among those in lower-class positions. Among workers in non-skilled manual work, 70 per cent of those on short-term temporary contracts thought they would be dismissed within a month if they were persistently late and 66 per cent if they failed to work hard.

While the contract workers were similar to permanent employees in the notice period for dismissal, they nonetheless felt substantially less secure in practice. Both types of temporary worker were more likely than permanent employees to feel that they would be obliged to leave their current employer over the next year for reasons other than their own choice. Whereas only 8 per cent of all permanent employees thought they would be forced out of their job over the year, the figure rose to 21 per cent of contract employees and to 41 per cent of temporary workers. Similarly, both contract and short-term temporary workers were markedly less satisfied with their job security than permanent employees.

TABLE 6.19 *Risks of unemployment of temporary and full-time workers (all work-history job events)*

	All	Men	Women
% moves from non-temporary work into unemployment as a proportion of all moves from non-temporary work	9	10	6
% moves from temporary work into unemployment as a proportion of all moves from temporary work	13	16	10
No. of pairs of non-temporary job events	16,249	9,260	6,965
No. of pairs of temporary job events	6,544	3,599	2,935

Note: Temporary work is defined as any job that lasted less than a year. Employed events include both full-time and part-time work. Self-employment is not included among initial state events, since the focus is on the experience of different types of employee; it is, however, counted as a 'non-unemployment' outcome event.

The dissatisfaction scores rose from 2.74 among permanent employees to 3.77 among contract workers to 4.40 among short-term temporary workers.

The work-history data do not allow a rigorous assessment of the relative risks of unemployment of those on temporary and so-called permanent contracts, since there is no detailed information on the contractual status of past jobs. However, it is possible to look at previous experience of unemployment of those currently in such jobs. Temporary workers were considerably more likely to have had a spell of unemployment in the previous five years. Whereas this was the case for only 11 per cent of permanent employees, the proportion rose to 21 per cent of contract workers and 32 per cent of short-term temporary workers. Contractual status made a particularly sharp difference for men. Among women, 9 per cent of permanent employees had been unemployed compared with 12 per cent of contract workers and 18 per cent of short-term temporary workers. However, among men, whereas 14 per cent of permanent employees had experienced unemployment, the figure rose to 29 per cent of contract workers and to 45 per cent of short-term temporary workers.

An alternative approach is to adopt a proxy for temporary work status in the work-history data by defining any job that lasted for less than a year as a temporary post. It is then possible to assess the likelihood of people entering unemployment on leaving a temporary job as against the likelihood if they left a non-temporary job. Both full-time and part-time jobs have been included in the analysis. It can be seen in Table 6.19 that the results confirm the greater vulnerability to unemployment of temporary workers. Whereas 9 per cent of employees from non-temporary work who moved to another labour market status moved into unemployment, the proportion among those who were temporary workers was 13 per cent. Men generally had a higher risk of unemployment

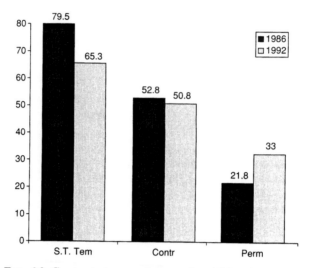

Fɪɢ. 6.3 Contract status and perception of decreased job security

than women, but the implications of temporary work were broadly similar for the two sexes. Among men, 10 per cent of transitions from non-temporary jobs resulted in unemployment, compared with 16 per cent among those from temporary jobs. Among women, the figures were 6 and 10 per cent respectively.

There can be little doubt, then, that the temporary workforce experienced much higher levels of job insecurity, even if there were possibilities for the renewal of contracts and for movement into more permanent jobs, and that this was reflected in a greater experience of unemployment in the past. Has there been, however, a tendency for increased polarization in the experience of insecurity between a 'core' of permanent employees and a 'periphery' of temporary workers?

There is a comparable measure of satisfaction with job security available for both the 1986[14] and the 1992 data that can be used for comparison across time (Fig. 6.3). While, for each year, the dissatisfaction of short-term temporary workers and contract workers is substantially greater than that of the permanent employees, the direction of change is not what would be predicted on the basis of a theory of polarization. There has been a substantial decline in the dissatisfaction of short-term temporary workers and a slight decline in that of contract workers. In contrast, it is among permanent employees that there is evidence of a significant growth of dissatisfaction about job security. The relative improvement in the position of both types of temporary worker over time is confirmed by a regression analysis controlling for both age and class position.

[14] The data for 1986 are drawn from the Social Change and Economic Life Initiative Surveys. The question asks whether there has been a change in job security in the last five years. Responses that job security has decreased are taken as an indicator of dissatisfaction with security.

TABLE 6.20 *Temporary work and the experience of unemployment in the previous five years (cell %)*

	Proportion with a spell of unemployment		
	Temp	Contract workers	Permanent employees
All employees			
1986	37	29	12
1992	32	21	12
Male employees			
1986	51	37	13
1992	45	29	14
Female employees			
1986	24	18	10
1992	19	12	9

Were there parallel changes in people's own experiences of unemployment? In Table 6.20, it can be seen that the broad pattern was very similar for the two categories of temporary worker. A smaller proportion of both temporary and contract workers had experienced unemployment in the previous five years in 1992 than had been the case in 1986. The pattern was very similar for men and women. It is plausible then that the greater optimism of temporary workers in 1992 derived in part from a growth in security. However, the increased insecurity of the permanent workforce would not appear to be related to any marked increase in experiences of unemployment.

In short, there can be little doubt that both types of temporary work contributed to a strong sense of job insecurity. This part of the workforce was genuinely flexible, albeit at a substantial cost to those involved. But there was no sign of increasing polarization over time between the 'permanent' and the 'peripheral' workforce. If anything the short-term temporary workers had come to have a somewhat better view of their prospects (partly reflecting improvements in their experience with respect to job insecurity), while there had been a marked decline in the security of the 'core' workforce.

Is temporary work, then, best understood as a bridge or as a trap? It is clear that the answer is fundamentally different depending on the type of temporary worker. The evidence points to the conclusion that short-term temporary workers experienced a strong sense of entrapment in their labour market position. Nearly two-thirds felt that they had had little choice in taking their current job, only two out of ten thought they had a reasonable chance of upward mobility with their current employer, and four out of ten thought they would be forced to leave their current employer within the year. Although there may have been some improvement in their position over time, they remained quite clearly the most disadvantaged group with respect to job security.

The contract workers, however, were in a rather different position. While they were certainly worried about their long-term security, they were very similar to permanent employees in their beliefs about their chances of upward mobility. Two-thirds thought there was a career ladder for their type of work and 43 per cent thought there was a reasonable chance of being promoted with their current employer. For the contract workers, then, there was a perception of the job as providing a bridge to better work, and for a substantial minority that might be with their current employer.

CONCLUSION

Theories of the flexible or peripheral workforce depict part-time and temporary workers as sharing a broadly similar labour market position. It is characterized by low skill, by poor promotion opportunities, and, above all, by acute job insecurity. It is clear, however, from the detailed analysis of the employment conditions of part-time workers and temporary workers that the tendency to consider them as a relatively undifferentiated flexible or peripheral workforce obscures far more than it reveals. Indeed, even the assumption that temporary workers represent a relatively homogeneous category breaks down on closer inspection. It has been necessary to distinguish two categories of temporary worker—short-term temporary workers and contract workers—that have quite different experiences of employment. While these different types of employee share the fact that they are relatively disadvantaged, the precise nature of these disadvantages and the way they are evolving varies a great deal.

There are grounds for doubt whether part-time employees are in fact flexible employees in the sense usually understood. The defining characteristic of the peripheral sector was that it could be relatively easily disposed of. But part-timers appeared to experience no greater job insecurity than full-timers. It was seen that they also reported low levels of task, pay, and hour flexibility. Much of the discussion appears to involve a confusion of two conceptions of flexibility: aggregate flexibility in the sense of an employer's ability to have more varied patterns of staffing and individual flexibility in the sense of the requirement of individual employees to accept changes in their employment conditions at short notice. Clearly, an employer can acquire greater flexibility over staffing levels by the deployment of part-timers, adjusting the level of staffing more closely to the pattern of work demand. But this is perfectly compatible with a high level of stability in the employment conditions of individual part-time employees. For instance, the fact that an employer decides to increase the workforce during the weekends by employing additional workers as part-timers may lead to a very stable pattern of employment for the individual part-timers involved.

The major disadvantage in the employment conditions of part-timers was not flexibility in the sense of constantly changing tasks and hours or high levels of job insecurity. It was rather the low level of intrinsic interest of the work, the lack of opportunities for training and promotion, and very poor levels of pay and fringe benefits.

In contrast to the part-timers, the temporary workforce clearly did experience relatively high levels of job insecurity. However, with respect to other aspects of employment conditions, there was a sharp divide between contract workers on the one hand and short-term temporary workers on the other. Far from fitting the image of the peripheral worker as low-skilled and entrapped, contract workers tended to have a skill profile that was very similar to that of permanent employees, were involved in work of a similar level of intrinsic interest, and saw themselves as having a career ladder in front of them. These were typically entry jobs that provided bridges into more interesting and stable work.

The short-term temporary workers, on the other hand, did fit more closely the conventional image of the peripheral worker. They were disadvantaged in a wide range of respects. They had limited opportunities for developing their skills, they were involved in work with relatively low levels of responsibility and intrinsic interest, they saw themselves as trapped in this sector of the labour market, with few hopes for getting better work, and they felt that their jobs were highly insecure.

While it is clear that the labour market is highly segmented, the attempt to understand such segmentation in terms of general theories of core and periphery is profoundly misleading. It veils rather than illuminates the distinctive pattern of employment disadvantage associated with specific non-standard contract statuses.

7

Employment Commitment and Job Expectations

Up to this point, we have been looking primarily at changes in the nature of work—in the work task, in forms of control, and in the level of job insecurity. In the final chapters of the book, we focus on employees' responses to these changes. How did the trends towards higher skill levels, more pervasive control, and heightened insecurity affect people's sense of satisfaction and involvement in the work they were doing? What were their implications for levels of work strain and psychological well-being? How did they affect employees' sense of commitment to their organizations?

The impact of changes in work is likely to depend, at least in part, on the centrality of work to people's lives. Studies in the neo-Marxian and 'industrialism' tradition, despite their wide differences about the trends of change, concurred in the view that the characteristics of the work situation had critical implications both for satisfaction with work and for wider personal well-being. Work was seen as the central arena for self-realization and for the establishment of self-identity, and experiences at work were thought to spill over into people's self-confidence in non-work life.[1] There was also support for this view from the major social-psychological study into the relationship between work and personality (Kohn and Schooler 1983).

However, other research challenged these assumptions, arguing that the objective nature of the work situation did not have any necessary implications for personal well-being. Its effects were contingent on the values and orientations that people brought to the work situation, and such values might be primarily affected by out-of-work experiences. Employees should be seen not just as reacting to work situations over which they had little control, but as capable of deliberately selecting types of work which fitted with their personal orientations. Where they are primarily concerned with their family and social lives and view work mainly in instrumental terms, even a work situation as objectively unrewarding as that of assembly-line work could be judged perfectly acceptable (Goldthorpe et al. 1968). Where employment is seen principally as a way of generating the income needed to satisfy out-of-work interests, people may make a conscious bargain to exchange worse work conditions for higher income.

If the importance of intrinsically rewarding work is culturally variable rather than an inherent aspect of human nature, the question is raised of whether the

---

[1] For reviews of the sociological literature, see Rose (1985, 1988).

intrinsic value of work has changed in importance over recent years. One influential view has been that the salience of intrinsic work values has been declining, following a general growth in the cultural importance of personal and family values (Goldthorpe *et al.* 1969). The dissolution of traditional solidaristic communities, higher geographical mobility, the provision of better housing, more leisure time, and higher incomes could all be seen as favouring an increased emphasis on out-of-work activities. In the post-war period, it became more feasible for the working population to develop non-work interests that could be an alternative source of self-identity.

It is also possible that there has been a change in work values as a result of the composition of the workforce. Neo-Marxian and 'industrialism' theories were developed at a time when the bulk of the workforce were men. One of the most important developments in recent decades has been the rise of female employment. But arguably women's commitment to employment is rather different from men's (Hakim 1996*a* or *b*). Given the traditional division of labour, work was understandably the major source of identity for men. Women, on the other hand, were coming into the labour market in a cultural context in which they still had the major responsibility for homemaking. It is possible then that women's primary identity has remained with their roles as mothers or wives, and that employment is viewed very much as of secondary importance. Alternatively, if the increase in female labour market participation was a reflection of deeper changes in women's values, with an increased emphasis on autonomy, equality, and public recognition, then there may have been a marked change over time in women's attitudes to work with the result that there would be no tendency for a decline in employment commitment in the workforce (Dex 1988).

Much of the earlier speculation on changes in attitudes to work derived from a context of relatively full employment. The emphasis on the strategic selection of types of work assumed that the labour market was buoyant and that there was choice available. However, since the recession of the early 1980s, with higher levels of unemployment, this has become much more problematic.

What is likely to have been the impact of higher unemployment on commitment to employment? Again, there could be very different hypotheses. It is possible that, by reducing choice in jobs, unemployment has the effect of reducing people's expectations of finding satisfying work and, as a consequence, their interest in employment *per se*. This may have been reinforced by a reduction of the stigma of unemployment, with high levels of unemployment making it implausible that people are personally responsible for being without work (Kelvin 1980). Alternatively, the experience of unemployment may have made people much more aware of the importance of employment. Jahoda (1982) for instance has argued that many of the advantages of employment for psychological well-being operate largely at a latent level. It helps, for instance, to structure people's use of time and to give them a sense of participation in a collective purpose. Where employment is secure, this is likely to be taken for granted, since there is no way in which people can readily experience the loss of these advantages (except through retirement). However, in the highly unstable labour markets of the 1980s and 1990s, we have seen that a large section of the workforce had at some point personally experienced the deprivations that come with loss of work. And for those that had not, there was a much greater probability of encountering people who had suffered this fate. Unemployment may then have heightened the value that people attach to employment.

An initial problem in assessing the arguments discussed above is that the relevant research to date is often based on rather different indicators of work values. It is important to distinguish at least two distinct dimensions of work attitudes—employment commitment, on the one hand, and job preferences on the other. These differ in their time-horizons and in the extent to which they are likely to be influenced by immediate constraints. We will begin then by examining whether there has been a change in employment commitment and by looking at alternative explanations of the strength of employment commitment. We shall then turn to consider the nature of people's preferences about jobs and the factors that affect these.

TRENDS IN EMPLOYMENT COMMITMENT

In focusing on employment commitment, our central concern is with the importance that people attach to employment on intrinsic grounds, that is to say irrespective of its financial implications.[2] The very notion of commitment implies choice and voluntary consent. It is important, then, to explicitly exclude the possibility that people are expressing an interest in employment uniquely out of financial need. We are fortunate in that we have a measure of non-financial employment commitment, for which there is representative national data at more than one point in time. The pioneer study on non-financial employment commitment was carried out in 1981 by Peter Warr (Warr 1982). Its key measure was a question asking people whether or not they would wish to work even if they had enough money to live as comfortably as they would like for the rest of their lives. This was repeated in the SCELI comparative labour market study in 1986 and in the Employment in Britain Survey in 1992.

In contrast to the view that there has been a long-term decline in employment commitment, a comparison across time shows no overall tendency for nonfinancial employment commitment to decrease (Table 7.1). In 1981, 69 per cent of employed men said that they would wish to continue in paid employment even if there was no financial necessity to do so, whereas in 1992 the figure was 68 per cent. There was, however, some difference in pattern between the

---

[2] See Barbash's discussion of the 'work ethic' (Barbash 1983). On the conceptual distinction and empirical relationships between commitment indicators, see Randall and Cote (1991).

TABLE 7.1 *Non-financial employment commitment,*
1981–1992 (cell %)

Committed	1981	1986	1992
Men	69	67	68
Women	60	62	67
Women full-time	65	63	69
Women part-time	54	59	64

Sources: 1981: SAPU Survey of Non-Financial Employment
Commitment; 1986: Social Change and Economic Life
Initiative; 1992: Employment in Britain Survey. The figures
for 1981 are derived from Warr (1982). The overall figures for
women have been weighted to provide the correct proportions
of full-time and part-time workers for that year.

sexes. The level for men has remained broadly stable. However, among women, there has been a marked change. The proportion highly committed to employment has risen from 60 per cent in 1981 to 62 per cent in 1986 to 67 per cent in 1992. It is clear that this has affected both women in full-time work and women in part-time work. Although women part-time workers are a little less likely to be committed than women full-time workers, it is the part-timers that show the strongest increase in commitment over the period 1981 to 1992. Overall, whereas in the early 1980s there was a clear sex differential of the type that would be expected in a culture with strong traditional gender roles, by the early 1990s the difference in the significance of employment in men's and women's lives would appear to have evaporated.[3]

The conclusion that commitment to employment has not been declining is also confirmed by data from the British Social Attitudes Surveys (Hedges 1994: 41). These used a somewhat differently worded measure, asking people whether they would prefer to have a paid job even if they had 'a reasonable income' (rather than if they could live as comfortably as they would like without working). Far from showing a decline in employment commitment, the evidence shows a tendency for employment commitment to have risen over time. Whereas, in 1984, 69 per cent of employees still preferred to have a paid job, the proportion had risen to 74 per cent by 1993. Again, it is the sharp rise in employment commitment among women that stands out. While in the early 1980s, women showed markedly lower levels of commitment than men, by 1993 they appeared to have even stronger commitment.

In short, employment remains central to people's values and identities and, indeed, for women is probably becoming of increasing relevance. Moreover, that

---

[3] Research indicates a similar convergence in men's and women's commitment to employment in the USA (Bielby 1992).

commitment is not driven merely by financial aspirations, but is rooted in the intrinsic benefits that people feel that they gain from working. Given this persisting attachment to the importance of employment, the quality of work experience is likely to be as fundamental as ever for people's subjective well-being.

DETERMINANTS OF EMPLOYMENT COMMITMENT

It is likely that employment commitment will be a result both of non-work factors and of the cumulative impact of work experience itself. Those that have had a more favourable experience of employment are likely, other things being equal, to be more committed to remaining employed.[4] However, without a longitudinal design, there is no satisfactory way of assessing the significance of such in-work factors. It might be the case that the causal direction works the other way round, with people finding themselves in such an environment (through processes of selection or self-selection) as a result of the fact that they are committed. We focus primarily then on factors that are more likely to be independent in their relation to employment commitment, in particular those relating to early experiences and non-work experiences. The person's current class, however, is retained in the analyses as a broad control for the quality of the employment situation.

Given the research to date, there are three arguments that are of particular interest. The first suggests that the importance attached to employment will be integrally related to a person's self-identity. As such it is likely to have developed at a relatively early age, reflecting the influences of the family and school, and to have been relatively stable across time. The second views the importance of employment as in competition with the importance of activities out-of-work. People can choose either to invest their energies primarily in realizing themselves through their achievements at work or through their social relations and social activities in the household and community. Finally, as has been seen, there is the central issue of the implications for employment commitment of the long-term changes in job security and in the level of unemployment.

Early influences are difficult to capture adequately in cross-sectional work, but there are a number of measures that give a general indication of whether or not there were factors in people's childhood that were conducive to employment commitment. The first is that of educational attainment. An objective of the educational system is to develop a capacity for, and an interest in, creative work. Non-financial employment commitment could be seen as a natural extension of such values, since employment is the gateway to opportunities for creative work over a wide range of activities. In general, then, it would be expected that the

---

[4] A point made well in the debate about the changing work ethic in the USA by Levitan and Johnson (1983) and Zuboff (1983). The importance of work experience has also been emphasized with respect to gender differences (Bielby and Bielby 1989).

TABLE 7.2 *Educational qualifications and employment commitment (cell %)*

| | Committed | | |
Qualifications	All	Men	Women
None	54	54	55
CSE equivalent	66	65	67
O-level equivalent	66	66	66
A-level equivalent	75	74	76
Post-A level (non-degree)	73	74	73
University degree	78	80	75

higher a person's educational attainment, the stronger their non-financial employment commitment.

It can be seen in Table 7.2 that there is indeed a strong relationship between educational achievement and employment commitment. However, since a higher level of education might also have the effect of strengthening employment commitment through the indirect route of providing better opportunities to enter more rewarding types of work, the best indicator of its intrinsic importance is its effect net of current occupational class. As can be seen in Table 7.3, education remains highly significant even when individual characteristics such as sex and age, and aspects of people's current non-work situation have been controlled. The coefficient rises sharply starting with those who took A level or equivalent qualifications. It continues to rise (and becomes statistically significant) among those with post-A level qualifications and in particular, among those with university degrees.

However, while the general link between education and non-financial employment commitment is supported, there is an important difference by sex (results not shown). Although the general trend of the coefficients is very similar for men and women, the effect is less strong for female employees and fails to reach statistical significance except in the case of those with university degrees. Education appears then to be more loosely tied to employment commitment for women. One possibility is that this reflects differences in the types of courses men and women have typically taken. Men are more likely to have specialized in educational courses with a more explicit vocational objective, while women have a greater tendency to select courses in the arts. Such differences in the choice of courses may evidently reflect deeper differences in gender socialization that have an independent effect on employment commitment. However, it is also plausible that different subject specialisms will reinforce and indeed extend such differences.

Two additional measures have been included to capture differences in early socialization that may have led to more general attitudinal differences. The first of these is the importance attached to hard work in general. The particular scale that we adopted was developed by Ho and Lloyd (1984).[5] Its originators conceived of it as a work-ethic scale. The hypothesis is that those who have been socialized into the belief that hard work is inherently good will be more likely to be attached to employment irrespective of its financial advantages. The second measure seeks to capture the extent to which a person feels in control of the environment rather than controlled by it (the locus of control).[6] Those who feel they have a high level of personal control are more likely to be committed to employment on non-financial grounds, since they are more likely to feel that they have the capacity to achieve something at work. The common assumption is that such deeper beliefs will be affected by early experiences. The current evidence for this is rather imperfect, but it does seem plausible that such measures will reflect the cumulative effect of relevant life experience in a wider set of domains than employment. In a broad sense, then, they can be regarded as measures of earlier socialization. Again, it is important to control for current occupational class, to make sure that the effects of the measure of the locus of control do not merely reflect the extent to which people are in work positions that allow for personal control.

The results again support the view that earlier socialization is an important influence on employment commitment. For employees as a whole, both the general work ethic and the locus of control are highly significant. There is again an interesting difference for men and for women. The measure of the work ethic (or significance attached to hard work as an end in itself) is significant for both men and women, but the level of significance is substantially higher for men. The measure of the locus of control is highly significant for women, but fails to reach significance for men. This may suggest that the strength of normative commitment to work in a general sense may be more important for men in determining their attitudes to employment, while for women a more important factor is their self-confidence in their own efficacy. These differences might well be residual effects of traditional patterns of gender socialization.

A further indication of the importance of early socialization is the age pattern. Employment commitment is highest among the youngest category of employee—those that are aged under 25. There is then a remarkably consistent decline in commitment across the successive age groups. The relative odds of

---

[5] People were asked to state, using a 5-point scale, how strongly they agreed or disagreed with the following items: 'Hard work is fulfilling in itself'; 'Nothing is impossible if you work hard enough'; 'If you work hard you will succeed'; 'You should be the best at what you do'; 'By working hard, an individual can overcome most obstacles life presents and makes his/her own way in the world.'

[6] The measure was derived from Spector (1988). People were asked to state, using a 5-point scale, how far they agreed or disagreed with the following items: 'Making money is primarily a matter of good fortune'; 'Promotions are usually a matter of good fortune'; 'To make a lot of money you have to know the right people'; 'It takes a lot of luck to be an outstanding employee in most jobs'; 'The main difference between people who make a lot of money and people who make a little money is luck.'

being committed among those aged 55 or more are only 18 per cent of those of people under 25. A decline in commitment among the oldest group might be explicable in terms of anticipation of retirement; people may begin to shift the centre of their interests so as to avoid too harsh a rupture when they no longer have the opportunity to work. However, the very smoothness of the pattern suggests that there may be a gradual wearing down of motivation as people's early beliefs about the possibilities of self-realization in work encounter the constraints of prevailing employment conditions.

A second general type of explanation of employment commitment focuses on the extent to which people invest their energies in their family and social lives. As was seen earlier, it has been suggested that the growing importance of family and social life has undercut commitment to employment. If this is the case, then we would expect that those who had the strongest family ties and who were most attached to their family and leisure lives would be the least committed to employment for intrinsic reasons. They might well want the money that employment brings to finance their non-work activities, but they would not, to the same extent, look to employment as a source of self-realization.

To examine this, two types of measure have been included. First, we look at family commitments in the form of marriage and the number of dependent children. Second, we look at measures of the degree of satisfaction that people have with their family and leisure lives. The assumption here is that people will primarily wish to invest time and energy in activities that they find satisfying.

The notable feature of the results shown in Table 7.3 is that such non-work commitments appear to have very little effect indeed on employment commitment. Marriage made no difference and further analysis showed this was the case for both men and women. There was also no effect of the number of children in the household. While the direction of the coefficients is consistent with the view that stronger satisfaction with family life was associated with weaker commitment to employment, the effect is very far from statistical significance. The results for satisfaction with leisure activities offer even less support for the view that there is an inherent conflict between work and non-work attachments. Overall, it seems unlikely that either the nature of family responsibilities or the degree of attachment to family and leisure life has strong implications for commitment to employment.

Finally, what was the impact of unemployment? Did it weaken employment commitment or heighten it? It is important to distinguish two rather different ways in which unemployment might affect people's attitudes towards work. It may be that it influences the perception of the local labour market, so that it is the fact of living in an area of high unemployment that is crucial. This, for instance, is the assumption underlying the view that high unemployment may reduce employment commitment by reducing the stigma of being without work. Alternatively, the principal influence on people's values may come from the direct experience of unemployment. This would seem closer to the logic underlying the view

TABLE 7.3 *Factors related to employment commitment*

	Coeff.	Sig.
Age group (ref. 20–4)		
25–34	0.41	***
35–44	0.35	***
45–54	0.23	***
55+	0.16	***
Class (ref. Prof./manag.)		
Lower non-manual	1.01	n.s.
Tech./supervisory	1.01	n.s.
Skilled manual	0.84	n.s.
Semi- and non-skilled manual	0.83	n.s.
Qualifications (ref. none)		
CSE equivalent	0.98	n.s.
O-level equivalent	0.96	n.s.
A-level equivalent	1.23	n.s.
Post-A level non-degree	1.80	***
Degree level	2.23	***
Work ethic	1.07	***
Locus of control	1.03	**
Married	0.98	n.s.
Family satisfaction	0.88	n.s.
Leisure satisfaction	0.99	n.s.
Illness	0.84	n.s.
No. of children	1.16	n.s.
Sex	0.91	n.s.
Part-time worker	1.06	n.s.
Temporary work	0.90	n.s.
Unemployment rate	0.98	n.s.
Unemployment experience	1.38	*

Note: Logistic regression analysis; multiplicative
effects on odds. No. = 2,613.

that unemployment might make people aware of the previously taken-for-granted
benefits of employment. Measures have been included to try to capture both of
these effects. First, labour market data on the level of unemployment in each
local area were attached to the dataset. This is based on the 'official' definition
of the unemployed as registered job-seekers. While this fails to capture a section
of those who are without work and seeking it (in particular married women returning to the labour market after a period bringing up children), the fact that these
are the figures most commonly quoted in the local media makes it likely that
they will be the most directly related to perceptions of unemployment in the
locality. Second, to capture people's personal experiences of unemployment, we
used the information that had been collected in the work histories. The assumption

was that relatively recent experiences of unemployment were likely to have the greatest impact. Hence the measure adopted is whether or not a person had experienced a spell of unemployment (of at least a month) at any time in the previous five years.

As can be seen in Table 7.3, the principal effect comes from the personal experience of unemployment. Once other factors have been controlled for, those that had had a spell of unemployment were more likely than others to be committed to employment. The pattern of the data supports the view that unemployment may lead to greater self-awareness of the benefits of employment for reasons quite other than financial. This is consistent with other evidence. It has been shown that people who are currently unemployed have higher levels of employment commitment than people who actually have jobs (Gallie and Vogler 1993; Gallie *et al.* 1994*a*). The data here suggest that the greater awareness of the benefits of employment that results from unemployment would seem to mark people sufficiently deeply to have a persisting effect even after they have returned to work. This is in contrast to the evidence for psychological well-being which shows that the higher levels of anxiety and depression experienced by the unemployed disappear once people return to work.

In short, the data are consistent with the view that employment commitment is rooted in early experiences, and is affected later primarily by age and by experiences of the labour market. The argument that employment commitment is in some type of competition with family or leisure interests receives little support. Far from showing a decline in employment commitment over time, the evidence suggests that it has remained at a high level for men, and that it has become even stronger among women over the last decade.

EXPECTATIONS OF WORK

The Relative Importance of Job Characteristics

The evidence for employment commitment tells us that people are committed to having a job irrespective of financial reasons. However, it does not reveal the relative importance of intrinsic and extrinsic factors, given the real constraints that people face in their lives. Nor does it tell us the types of non-financial job characteristics that are most important to people. The next section seeks to explore in more detail what people seek from their jobs.

There can be little doubt that, given their real life circumstances, people's most salient concern in seeking work was to secure their livelihood. When asked what was their 'main reason for working', 65 pet cent replied that it was because they needed money for basic essentials such as food, rent, and mortgage. There was a marked difference in the frequency with which this reply was given by men and women. Whereas 76 per cent of men regarded the need to pay for essentials as the principal reason, this was the case for only just over half of women

TABLE 7.4 *Percentage reporting that the need for money to cover basic essentials is main reason for working (cell %)*

Age group	Men			Women		
	All	Married	Unmarried	All	Married	Unmarried
20–4	59	86	50	55	70	45
25–34	78	80	75	62	59	70
35–44	83	84	76	49	41	89
45–54	81	81	80	47	41	82
55+	71	73	62	42	36	61
All	76	81	65	53	48	66

(53 per cent). There is some evidence then that traditional gender role conceptions are still influential. This is reinforced by the differences in age pattern. Among men the need to cover basic expenditure rises sharply from the under-25s to the 25–34 age group. It peaks between the ages of 35 and 55 and then declines among older workers, albeit staying at a high level. Among women, on the other hand, the proportions emphasizing the need to cover essentials is highest among the under-35s and then declines steadily across the succeeding age groups.

A key factor affecting these sex differences in the perceived importance of financial objectives was marital status (Table 7.4). If non-married men and women are compared, they show a remarkably similar pattern across the age groups. The importance of financial need rises sharply in the late twenties and then stays at a high level until the mid-fifties. In contrast, the patterns for married men and women are very different. Married men attach a consistently high importance to the need to meet financial necessities whatever their age group. For all age groups except the 45–54-year-olds, married men are more likely to report the importance of financial pressure than single men. In contrast, married women are less likely to cite financial pressures than single women in all age groups other than the under-25s. The greater importance of household position than sex *per se* is shown by the fact that, while there is virtually no difference between unmarried men and unmarried women, there is a difference of as much as 33 percentage points between married men and married women. It is clear that gendered responsibilities in the household still play a role in determining the significance of financial pressures on people's participation in the labour market.

While the need to bring in an adequate income clearly constitutes the key constraint on most people's labour market choices, this does not mean that people choose jobs simply to *maximize* their income. A wide range of jobs might enable them to meet their essential household bills, allowing considerable scope for other criteria to determine the particular type of work they select. Given that

TABLE 7.5 *Most frequently mentioned job preferences (cell %)*

	Essential			Essential+very important		
	All	Men	Women	All	Men	Women
Secure job	37	41	33	83	85	81
Work likes doing	34	33	35	84	81	87
Good relations with supervisor	29	27	32	79	73	86
Opportunity to use abilities	27	30	25	78	80	76
Good training provision	28	32	23	72	72	71
Good pay	26	31	21	72	75	68

their basic income needs are met, they may give priority to securing additional income, but equally they may look for jobs that give scope to realize their intrinsic work preferences. To examine this, it is necessary to focus more closely on people's expectations of jobs, the relative importance that they attach to specific types of work rewards.

People were given a list of fifteen different job characteristics and asked how important they felt each was in looking for a job. They could judge it as essential, very important, fairly important, or not very important. It is evident that there was no clear dominance of either instrumental or intrinsic expectations (Table 7.5). Taking first those factors that were thought to be 'essential', the most commonly mentioned was job security (37 per cent), followed by work that one liked doing (34 per cent), and good relations with one's supervisor (29 per cent). The opportunity to use one's abilities (27 per cent), training (28 per cent), and good pay (26 per cent) were all of roughly equal importance.

If responses in terms of 'essential' and 'very important' are taken together, it is clear that the majority of employees place a strong emphasis on both extrinsic and intrinsic rewards. The job characteristics that were most frequently ranked highly were: work you like doing (84 per cent), a secure job (83 per cent), a good relationship with the supervisor (79 per cent), and opportunity to use one's abilities (78 per cent). There were differences in emphasis by sex. Men more frequently mentioned job security, training, and good pay as essential in a job; women were more likely to emphasize work that they liked doing, the quality of relations with supervisors, and friendly relations with colleagues.

Income maximization then was not the only, or indeed the most important, criterion in the selection of jobs. Whether one takes the proportion mentioning good pay as essential or the proportion mentioning it as either essential or very important, it comes sixth out of the job characteristics listed. While there was a difference between the sexes, with men placing a stronger emphasis on pay than women, it still ranked only fourth in order of importance among men.

TABLE 7.6 *Most frequent changes in job preferences over last five years (% more important) (cell %)*

	Men	Women	All
A secure job	58.7	48.7	53.9
Good pay	48.4	44.6	46.5
Opportunity to use abilities	39.0	35.2	37.2
Good training	39.9	36.1	38.1
Work likes doing	33.4	36.6	34.9
Opportunity to use initiative	36.8	31.7	34.3

Change Over Time in Job Preferences

How had expectations been changing? Which types of job reward had become more important in recent years? For each of the fifteen characteristics, people were asked whether they had become more important, less important, or were of the same importance as five years previously.

By far the biggest change was in the importance that people attached to having a secure job (Table 7.6). Overall, 54 per cent thought that it had become more important to them. This was particularly the case for men (59 per cent), although it was also the aspect of work which had increased most in importance for women (49 per cent). Security was followed by the increased importance attached to good pay (47 per cent). There is some indication, then, at least over the shorter term, that instrumental factors had become more central. For instance, only 37 per cent attached greater importance than five years earlier to the opportunity to use their abilities. Yet the increased focus on instrumental factors did not imply a *decline* in interest in work that offered intrinsic rewards: only 7 per cent said that the opportunity to use their abilities had become less important to them. Both instrumental and intrinsic preferences have increased in importance, but this has been particularly the case for instrumental preferences. This pattern was the same for men and for women.

However, this was a period in which the labour market had deteriorated sharply. It might be argued that the relative importance of different types of job preference could be expected to fluctuate to some degree with the economic cycle. The increased importance attached to security, for instance, has to be taken in the context of a very sharp rise in levels of unemployment. It might be that there was little change in the absolute level of security that people expected; but that their responses reflected an awareness of the erosion of the type of job security that they had previously possessed. The increased importance attached to instrumental job characteristics may have little to do with deeper-level changes in work values, but mainly with change in the labour markets that people confront.

The only rigorous way of disentangling these processes is to compare the data for the early 1990s with a period in which labour market conditions were similar. The obvious point for comparison would be the early 1980s, which was also a period marked by rising unemployment. Unfortunately there are no data for this period for the overall workforce. However, there are some comparable data on the importance of different job characteristics for female employees (the 1980 Women and Employment Survey).[7] This longer-term perspective confirms several features of the pattern found in people's reports about more recent change. There has been an increase in the importance of job security, with 76 per cent considering it essential or very important in 1980, compared with 81 per cent in 1992. It also confirms that there has been an increase in the importance that women attach to intrinsic rewards. Whereas 70 per cent had regarded the opportunity to use their abilities as important in the early 1980s, the proportion had risen to 76 per cent in the early 1990s. The factor which had most clearly declined in importance was that of convenient hours of work. Whereas 77 per cent of female employees had regarded this as essential or very important in 1980, this was the case for only 41 per cent in 1992. The principal divergence comparing the longer-term trend with the medium term is in the importance attached to good pay. Over the decade, this declined somewhat from 74 to 68 per cent. It must be remembered that the period has seen a substantial change in the composition of the female workforce, in particular in terms of the expansion of part-time work and the increased participation of married women. Overall, the longer-term picture suggests that intrinsic preferences have increased in importance for women at least as much as instrumental.

Job Preferences and Employee Characteristics

While there seems little support for the view that there is any clear trend over the longer term towards more instrumental or intrinsic work values, there may be important differences in job preferences within the working population. Again these may be affected by a range of factors: differences in educational experiences, in the nature of people's current jobs, in their marital status or responsibilities for children as well as in the state of the local labour market.

To simplify the analysis, the various types of job preferences have been grouped. A factor analysis indicated that they fell into three broad categories. The first concerns intrinsic rewards, involving the ability to use initiative in the job, work that the person likes doing, the opportunity to use one's abilities, and variety in the work. The second corresponds to the instrumental or extrinsic dimension, conceived in a broad way. It groups a concern with good pay with an emphasis on promotion prospects, job security, fringe benefits, training provision, and physical work conditions. The third could be labelled a convenience dimension.

[7] The findings of the survey were published in Martin and Roberts (1984).

TABLE 7.7 *Factors affecting types of work preferences*

	Intrinsic		Extrinsic		Convenience	
	Coeff.	Sig.	Coeff.	Sig.	Coeff.	Sig.
Qualifications (ref. none)						
CSE equivalent	0.47	***	0.45	*	0.09	n.s.
O-level equivalent	0.52	***	0.19	n.s.	−0.27	*
A-level equivalent	0.97	***	−0.05	n.s.	−0.34	*
Post-A level non-degree	1.33	***	0.22	n.s.	−0.30	*
Degree level	1.35	***	−0.59	*	−0.13	n.s.
Age group (ref. 20–4)						
25–34	−0.16	n.s.	−0.58	**	0.18	n.s.
35–44	−0.03	n.s.	−1.02	***	0.08	n.s.
45–54	0.01	n.s.	−1.65	***	0.00	n.s.
55+	0.17	n.s.	−1.14	***	0.51	**
Class (ref. Prof./manag.)						
Lower non-manual	−0.22	n.s.	0.49	**	0.54	***
Tech./supervisory	−0.00	n.s.	1.29	***	0.11	n.s.
Skilled manual	−0.24	n.s.	0.20	n.s.	0.62	***
Semi- and non-skilled manual	−0.48	***	0.44	**	0.57	***
Household						
Single men	−0.10	n.s.	0.76	**	0.03	n.s.
Married women	−0.07	n.s.	0.27	n.s.	0.43	***
Married men	−0.12	n.s.	−0.64	*	−1.98	***
No. of children	0.02	n.s.	0.13	*	0.18	***
Sat. family life	0.11	n.s.	0.21	*	0.05	n.s.
Labour market						
Fem. part-time	−0.85	***	−1.34	***	0.75	***
Contract 1–3 years	0.42	*	−0.07	n.s.	−0.26	n.s.
Temporary < 12 months	−0.30	n.s.	−0.40	n.s.	0.10	n.s.
Experience of unemployment	0.27	*	0.14	n.s.	0.03	n.s.
Level of unemployment	−0.02	n.s.	0.12	***	0.07	***
Type of work						
People work	0.61	***	0.18	n.s.	−0.07	n.s.
Assembly-line work	−0.37	n.s.	0.02	n.s.	0.15	n.s.

Note: OLS regression analyses. For intrinsic preferences, $R^2 = 0.14$; for extrinsic preferences, $R^2 = 0.09$; for convenience preferences, $R^2 = 0.16$. No. = 2,620 for all.

It stresses the importance of the hours of work, the ability to exercise choice over hours, and the lightness of the workload. Measures of 'intrinsic', 'extrinsic', and 'convenience' preference types were constructed by adding the scores of the relevant items defining each preference dimension.

The three types of work preference were affected by rather different characteristics. As can be seen in the first panel of Table 7.7, the concern for intrinsic rewards was influenced above all by the level of a person's educational qualifications. Any type of qualification led people to be more concerned about the inherent worth of the job task than among those without qualifications. However, there was a particularly sharp rise in its importance among those with A levels compared with those with lower qualifications, and again among those with post-A level qualifications (both academic and applied). This confirms the view that education develops a preference for more interesting work.

Even when education has been taken into account, however, those in semi- and non-skilled work had weaker preferences for intrinsically interesting work compared to other occupational classes. This may reflect a process in which those with stronger intrinsic motivation are able to achieve higher career positions. Alternatively, the type of work that a person is usually engaged in may mould their expectations. One test of this is to introduce measures designed to tap under- lying belief systems relating to work which are often thought to be formed rel- atively early in people's lives—such as attachment to a work ethic or a belief in personal rather than external control of the environment. When this is done, attachment to a work ethic certainly has a strong positive relationship to intrinsic preferences. But it does not change the earlier relationship between class and intrinsic motivation. This lends some support to the view that the nature of the work that people are involved in on an everyday basis may have independent effect on what they come to value in work.

We also examined whether there were differences in intrinsic motivation between people who did different types of work. Those who were primarily engaged in people-work were more likely to emphasize the importance of the intrinsic qualities of the work task. It has been suggested that those who worked in assembly-line settings might have an instrumental orientation to work, whereby they would be less concerned about job quality and hence about the routine- ness and monotony typically associated with this type of work (Goldthorpe *et al.* 1969). While the coefficient was in a direction that was consistent with this, it failed to reach statistical significance. There was also no evidence that those who were in short-term contracts were less concerned about the nature of their job tasks. Temporary workers on contracts of less than twelve months were not significantly different from others, while contract workers (with contracts for one to three years) were more concerned about intrinsic factors. The one cat- egory of employee on a 'non-standard contract' that was significantly less con- cerned about job quality was that of female part-time workers.

In contrast to education and class position, life-cycle and household factors had no effect on the importance people attached to the intrinsic rewards of work. There was no relationship with age and it made no difference whether people were married or had children. In contradiction to the idea that there may be a trade-off between involvement in family and work life, there was no evidence that those who were highly satisfied with their family lives attached less im- portance to the intrinsic characteristics of work than others. Indeed, although the relationship is not statistically significant, the direction of the coefficient is consistent with the view that they were more likely to consider such factors important.

Finally, it is notable that a previous experience of unemployment had the effect of increasing the importance to people of intrinsic work interest. It was noted in the discussion of employment commitment that unemployment appears to have made people more aware of the previously taken-for-granted benefits of having paid work, irrespective of its financial implications. This is confirmed by the present evidence, which shows that it leads to a greater emphasis when assessing jobs on factors such as the interest of work and the opportunity to use initiative and skills.

Turning from intrinsic to extrinsic job preferences (see the second panel of Table 7.7), there is a notable difference of pattern. In general, education has only a small effect on extrinsic preferences. There is some evidence that those with very low qualifications are more concerned with the material benefits of a job and people with university qualifications place less emphasis on them, but between these two extremes the level of qualification makes no difference. Rather more powerful is social class. Lower non-manual, technical and supervisory, and non-skilled employees are all considerably more likely to emphasize extrinsic factors than those in professional and managerial work. The only group that emerges as similar to professional and managerial employees in this respect is skilled manual workers. However, these class effects are entirely due to the pattern among men; for women, class position appears to make no difference.

In contrast to the results for intrinsic motivation, the importance of extrinsic factors is strongly affected by people's position in the life cycle. The age group most concerned with material rewards is the youngest (those under 25). As people grow older, there is a steady decline in their relative importance up to the 45–54-year-old age group. Marital status is also significant. But, contrary to what might be expected, it is not those who are married or who have children who attach most importance to extrinsic factors. Rather it is the single males. A more detailed analysis showed that this was particularly the case for pay. It is possible that the life-style of the single male is particularly expensive. But it may also be that men who are particularly concerned with financial prosperity delay marriage and family responsibilities.

There is some evidence, however, that is consistent with the view that the demands of, or the importance attached to, family life have implications for the importance given to the material rewards in jobs. Those with more children had stronger extrinsic preferences, and this was also the case for those who were more satisfied with their family lives. Personal investment in the family, then, may lead people to be more concerned about factors such as pay and security. However, this does not provide support for the stronger claim that it produces an instrumental orientation in which intrinsic work preferences are played down in favour of extrinsic. As was seen earlier, household situation makes no difference at all to the importance attached to issues of job quality. People that placed a strong emphasis on material rewards may also remain committed to having interesting work.

There was also little evidence to support the view that people self-select themselves into particular types of work on the basis of the importance they attach to the material conditions of employment. For instance, those involved in assembly-line work were no more likely to stress the importance of extrinsic factors than others. It is also interesting to note that those who worked with people, although having higher expectations with respect to job quality, were no less likely to expect to have jobs that offered good pay and security. It is once more the female part-timers who stand out in terms of their low expectations about material rewards. The striking point about this group is that they are less demanding in their job preferences with respect to both intrinsic and extrinsic factors.

Finally, there is a rather different relationship between unemployment and expectations about the material conditions of employment than had been the case with respect to issues of job quality. This time it was not the personal experience of unemployment that counted, but the level of unemployment in the local labour market. People who were in labour markets where there was a relatively high rate of unemployment were notably more concerned about extrinsic job characteristics. In one respect, this is only to be expected. The measure includes the importance that people attach to job security, and those in high-unemployment labour markets are understandably more concerned about their personal job security. But unemployment also has an influence on the importance of other extrinsic characteristics. A separate analysis of the importance attached to pay shows that this too is affected by the local unemployment rate. It is possible that, where there is pervasive job insecurity, people begin to anticipate possible job loss by maximizing income in an attempt to build up their savings. They may also have to cope with the consequences of loss of income by other household members or with the need to assist local kin who have lost their jobs.

The third preference type was that of 'convenience', the degree of choice over, and the convenience of, hours of work and the heaviness of the workload. The relation of education to convenience preferences was in some ways the mirror image of that with respect to intrinsic factors (see the third panel of Table 7.7). In general the higher the educational level the less weight people put on the convenience of their conditions of employment. When the relationship between education and convenience preferences is considered on its own the pattern is linear. However, once occupational class has been taken into account, it becomes curvilinear. Convenience factors mattered most among the extreme groups: those without qualifications and those with degrees. The pattern for occupational class is more clear-cut. Manual workers (whether skilled or non-skilled) and lower non-manual workers are significantly more likely to emphasize convenience factors than those in professional or managerial work even when other factors have been controlled.

Life-cycle and household factors were also associated with the extent to which people emphasized convenience factors. The oldest category of workers—those over 55—stood out from other age categories in the greater importance they attached to the convenience of the job. There was also a strong effect of marital status. However, this differed significantly between the sexes. Married women were particularly concerned about convenience factors, whereas married men were significantly less likely to think them important. It seems quite likely that in this respect job preferences are still influenced by traditional gender conceptions: with married women particularly concerned about their ability to fit their work hours around their domestic responsibilities.

The only evidence that convenience factors may have been revalent to the selection of specific types of work was with respect to part-time work. Female part-time workers were more likely to emphasize such factors than others. This is the one dimension of job preferences where part-time workers had higher demands than others. It might be that this reflected a concern by part-timers to work full-time hours. But we found no evidence that this was the case. When asked what would be their ideal hours of work if they could choose, only 2 per cent of part-timers wanted to work more than thirty hours a week and part-timers wanted substantially shorter working hours than full-timers. In contrast, neither the short-term temporary workers nor workers with contracts lasting between one and three years differed from other employees.

Finally, it should be noted that the level of unemployment in the labour market was associated with an increased emphasis on this factor. This did not mean that unemployment-prone people were particularly concerned about hours of work. Those with personal experience of unemployment were not more likely to stress this than others. To explore this further we again examined the ideal hours that people would like to work. In contrast to the case for part-time workers, who clearly preferred shorter hours than full-timers, those who wanted to work particularly long hours (more than forty hours a week) were likely to be in areas with a particularly high rate of unemployment. The desire for choice over hours is likely, then, to reflect in this case a desire to work longer hours than usual. This again could reflect an anticipation of future loss of work or additional financial responsibilities due to the unemployment of other family members.

In short, background characteristics, current occupation and life-cycle factors related rather differently to the various types of job preference. The predominant factor affecting the preference for intrinsically rewarding work was education: those with higher levels of education were much more strongly concerned to have jobs that provided opportunities for self-realization and self-development. In contrast, education was only weakly related to extrinsic preferences, which were most notably affected by age. Finally, class position, marital status, and household factors were strongly related to the importance attached to convenience. In contrast to the case for employment commitment, female part-timers stand out quite clearly from other employees, with their job preferences primarily determined by the need to reconcile their family and work roles.

CONCLUSION

Overall, there is little sign from our data that the issue of the quality of employment is likely to become less salient as a result of a decline in the work ethic, the growing importance of leisure life, or a shift towards a more instrumental approach to employment.

In the first place, employment commitment would appear to be at least as strong as it was in the early 1980s. Three factors in particular underlie this. Employment commitment is closely related to education and the long-term tendency has been for education levels to rise. Second, employment appears to have become markedly more important for women's identities over the last decade, reflecting a deep process of restructuring of gender roles. Third, the more frequent experience of unemployment that has accompanied the successive recessions since the 1970s has made people more aware of the previously taken-for-granted advantages of having a job. Structural change has then, if anything, reinforced rather than weakened the view that employment is of central importance for personal development and life satisfaction.

The view prevalent in the 1960s that family and leisure life would become an alternative pole around which people could construct their identities and find meaning in their lives has become much less convincing two decades later. To begin with, there is little reason to believe that an active life out of work is incompatible with a strong interest in achievement in work. Rather the two are likely to be complementary, with work providing not only the income but also the self-confidence and self-esteem needed for a satisfactory life out of work (Kohn and Schooler 1983). But, in addition to this, the increased fragility of family relationships has made the home-centred life strategy a risky enterprise. The cultural shift would appear to have been more towards securing the personal autonomy needed to survive in a world of precarious relationships than to some retreat into the values of domesticity. And it is employment that most obviously provides a basis for personal autonomy.

Nor does our evidence support the view that instrumental values have come to dominate the way in which people relate to employment. Certainly, the majority of people are clear that their primary consideration has to be to assure that their job provides the means for securing a livelihood. However, a wide range of jobs are compatible with meeting satisfactorily this basic need. This provides wide scope for the influence of other factors in determining the way people assess particular jobs. Our evidence suggests that the view held by economists that people are primarily income-maximizers is incorrect. High income was ranked only sixth in importance among job characteristics. People were notably more concerned with job security, with job interest, and with the quality of personal relations with management. Moreover, the ability to make good use of acquired skills and good training prospects were viewed as every bit as important as high income *per se*. Expectations for jobs that are intrinsically worth while have remained then of central importance.

Our analysis of the factors associated with the emphasis on intrinsic job characteristics may help to explain why this has been the case. There is in particular a close link between the educational level a person has reached and his/her attachment to good intrinsic conditions in work. It is not perhaps surprising that a process that is at least in part designed to train people in creative thinking should make them more appreciative of work that enables them to use their skills and take initiative. Since increased technological complexity and the growing importance of organizational skills are likely to boost the demand for greater investments in education over the long term, the very nature of economic development is generating a workforce that will have higher expectations of good intrinsic conditions of work. Rather than declining in importance, it seems likely that such issues will become more central to organizational life.[8]

---

[8] Clearly, as Levitan and Johnson (1983) have emphasized, if job structures fail to adapt to the expectations of a more highly educated workforce, then a growing mismatch between skill requirements and workers' educational attainment could have severe negative repercussions on commitment.

8

Job Involvement, Work Strain, and
Psychological Health

Previous chapters have charted the major changes that have been occurring in the nature of work. It has been seen that there has been a widespread process of upskilling, which has been accompanied by greater responsibility for employees. Intimately linked to these developments have been changes in the type of skills used in work. There has been a striking increase in the use of advanced technology and a shift away from work primarily concerned with the production of objects to work primarily concerned with people. These changes in the nature of work have been accompanied by changes in the organizational and labour market contexts in which people work. Management has experimented with new types of 'performance-management' policies for the motivation and control of employees. There has been, at the same time, a marked rise, at least for male employees, in the insecurity of employment. What have been the implications of these changes for the subjective experience of work: is it likely that they have made work more enjoyable or have they increased discontent?

There are a number of influential theories in the literature about the types of factors that are likely to increase or reduce employee's well-being. In particular, we can distinguish three main approaches which have inspired significant research. The first emphasizes the fundamental importance of a range of 'job characteristics'; it is above all the everyday character of the work task that is thought to influence the way employees feel about their job. The second underlines the significance of the quality of social relations in the workplace. Where people find themselves in a supportive social environment, it is argued, they will be more highly motivated and happier in their work. The third points to the importance of participation in wider decision-making processes. When people can influence decisions, they are less likely to become frustrated in and resentful about their work.

In the light of these general theories, there could be rather differing views about implications of the specific forms of change that have been transforming the world of work. In one respect upskilling could be seen as likely to increase employees' well-being, by enhancing a number of job characteristics that have been shown to have very positive effects. It was associated with greater discretion over the way work was carried out and with more varied and challenging work. There is an impressive body of research that shows the favourable implications of task discretion and job variety for both job satisfaction and psychological health (for an overview, see Warr 1987: ch. 5). However, at the same time, it was found that upskilling and increased responsibility were linked to an intensification of work and higher levels of effort, which may have negative effects for employees' experiences of work.

The development of new performance-management policies also could be seen as potentially two-edged in its effects with respect to the supportiveness of the work environment. Certainly, in part, these policies were designed to enhance social integration by providing for closer, more individualized, relations between management and employees. Management could become better informed about employee aspirations and the organization could make a greater effort to invest in the development of individual skills. This could enhance the image of the organization as a 'caring organization' and give employees a greater sense of recognition. But the other side of such policies was that they were at the same time policies of control. They involved more systematic monitoring and assessment of work performance, in some cases linked to pay sanctions where people were seen as falling short of acceptable targets. The possibility that performance assessments could be seen as unfair meant that such policies might lead to a deterioration of relations with supervisors. Similarly, they might have the effect, as a result of the individualization of rewards, of creating greater competitiveness and mistrust between colleagues. To the extent that they had either of these effects, they would undermine rather than reinforce the sense of being in a 'supportive' work environment.

The third type of change—the growth of job insecurity—has received rather limited treatment in previous empirical research. However, there has been a significant tradition of research on the effects of unemployment which has shown just how severe its effects are for psychological well-being and some work has indicated that, in cases of redundancy, these effects can be experienced before people formally leave their work (Cobb and Kasl 1977; Warr 1987: ch. 11; Warr *et al.* 1988). By extension, it seems likely that job insecurity will be a significant factor undercutting employee well-being. This still leaves, however, the question of the implications of contractually determined insecurity. Are those on fixed-term contracts particularly likely to have negative experiences of their work or does the fact that the term of the job is explicit before it is taken mean that employees have very different expectations and hence responses? It was seen (Chapter 6) that there are important differences between types of temporary worker—in particular, between those on very short-term contracts of less than a year and those on medium-term contracts of between one and three years. This difference was linked to differences in the nature of job tasks and also in perceptions of opportunities for career mobility. How far did these differences in forms of contract work affect employee well-being in work?

An empirical assessment of the factors affecting well-being requires clarification of what precisely is meant by well-being. Peter Warr (1987, 1991) has argued that a serious flaw in many analyses is that they are based upon one-dimensional views of psychological well-being. The approach adopted here is multidimensional, distinguishing three principal aspects of people's experience of work: their degree of involvement in their job, the level of strain or tension in the job, and their level of psychological distress.

The concept of job involvement is related to, but importantly different from, that of job satisfaction (Brooke *et al.* 1988). Job satisfaction has been the major focus of research on employee well-being in the past. There are, however, several grounds for reservation about the concept itself and the way that it is typically operationalized. While the concern with job satisfaction has normally stemmed from an interest in the quality of work experience and, in particular, with whether or not people are positively involved with their work, it is far from clear that, in practice, this is what is being examined. People may be satisfied with an aspect of their work situation not because it gives them a sense of positive interest in their work, but because it makes few demands or because they have very low expectations. In contrast, the notion of job involvement refers to an active interest in and enjoyment of the work.

Job involvement relates to the positive experience of work. For the negative experience, we have distinguished between two states—work strain and psychological distress. The notion of work strain refers to the extent to which a person experiences high and persistent levels of tension or fatigue as a result of their work. High levels of work strain may be experienced both by those with a strong interest in their work as well as by those who dislike it. Clearly, unpleasant aspects of the work environment may be an important source of strain. But it is arguable that some degree of work strain is also inherent in any work of a challenging type. It is likely that the greater the personal investment in a work task, the higher the intensity of work and the greater the tension the person will experience. Precisely the factors that may make work interesting and meaningful to the individual may generate work strain.

Given the potential complexity of its sources, it cannot be simply assumed that work strain is harmful in an enduring way. Rather the extent to which the factors that generate strain lead to more serious psychological conditions must be explored independently. We have termed these more definitely negative consequences 'psychological distress'. This is designed to capture experiences of anxiety or depression which are sufficiently strong to become detrimental to people's ability to act effectively in their everyday lives. Whereas work strain may be an integral part of creative endeavour, psychological distress tends to paralyse active involvement in activities, both with respect to work and non-work.[1]

---

[1] The conceptual distinction between job involvement, work strain, and psychological distress is supported by the statistical differentiation of the measures that we have used. We placed the different measures (which will be described in more detail later) into a common factor analysis. This showed that the items for the different measures load on clearly differentiated factors. Each factor had an eigenvalue of over 2.

We start by examining the factors that encourage job involvement and then turn to the analysis of work strain and psychological distress. Although our primary concern is with the effects on work experience of the factors that have been at the centre of the processes of change over the last decade—skill change, new managerial policies for control, and job insecurity—these have to be assessed within the broader framework of influences that have been thought to be important for people's experiences of their work. The variables in the analysis can be grouped into two main categories. The first relates to the nature of the work task. They are the skill level and responsibility involved in the job, the quality of the work task, and the immediate physical working conditions. The second concerns the nature of the organizational environment: the level of social support from colleagues and supervisors, the use of performance-management or technical control systems, the participativeness of the organization, and the degree of job security. Since there are grounds for thinking that the individual characteristics of age and sex may be independently related to job involvement, these have been systematically controlled for in the analyses (Kalleberg and Losocco 1983; Lorence and Mortimer 1985; de Vaus and McAllister 1991).

JOB INVOLVEMENT

Our starting point was that measures of job satisfaction, taken on their own, are relatively weak indicators of a favourable subjective experience of work and that analysis should focus on the more demanding notion of job involvement. This has the implication that people are not only satisfied with their work, but that the work is found to be interesting and absorbing.

To provide a measure of job involvement, we have brought together four items indicating satisfaction with intrinsic components of work, with three items that focused more directly on the level of interest in the work. The first set of items asked people to assess, on a seven-point scale, how satisfied or dissatisfied they were with the opportunity to use their abilities, with their ability to use their own initiative, with the work itself, and with the variety in the work. The second set asked people how much effort they put into their job beyond what was required, how often time seemed to drag on the job, and how often they thought about their job when they were doing something else. The seven items scaled well, with a Cronbach's alpha of 0.80, and the analysis is based on the combined measure.[2]

Skill and Skill Change

In general, it could be expected that jobs involving higher levels of skill and responsibility would be associated with higher degrees of job involvement. There

---

[2] The measure was based on the factor scores for the first factor.

TABLE 8.1 *Class, upskilling, and job involvement*

	Job involvement score (means)	No.
Class		
Professional/managerial	0.32	1,185
Lower non-manual	−0.09	569
Technician/supervisory	0.10	155
Skilled manual	−0.20	374
Semi- and non-skilled	−0.26	1,091
Skill change		
Increased a great deal	0.38	731
Increased quite a lot	0.19	1,070
Increased a little	−0.22	324
No change	−0.17	942
Decreased	−0.81	300
Type of work		
People	0.22	1,567
Information	−0.06	531
Traditional manufacturing	−0.30	384
Assembly-line	−0.96	93
Driving	−0.32	131

are a number of reasons why this should be the case. Skill and responsibility may be important in themselves, in that people may inherently enjoy more complex work and exercising responsibility. But higher skills may also be linked to other task characteristics that are important for job involvement. As was seen in Chapter 2, the higher the skill level, the higher the quality of the job in terms of its capacity to provide opportunities for self-development, the variety it provides, and the nature of the physical working conditions. Skill level tends to be associated with status in the organization and hence with the quality of communications and the possibility of influencing organizational decisions. Finally, at least in the past, those with higher skill levels have enjoyed substantially greater job security.

Our most general proxy of the level of skill and responsibility in the job is occupational class.[3] As can be seen in Table 8.1, employees in the professional and managerial occupations stand out quite distinctly as having relatively high levels of job involvement. They are followed at quite some distance by those in technical and supervisory jobs. Lower non-manual, skilled manual, and semi- and non-skilled manual workers share a negative scoring on the job involvement index, with semi- and non-skilled manual workers having clearly the lowest levels of involvement.

---

[3] Our measure of class is 'Goldthorpe class'. This is defined by the author in terms of 'employment relations' but correlates particularly well with the range of skill measures used.

The effect of skill level, however, appears to be related almost entirely to its implications for the quality of job tasks. When the main job-quality factors were added, reflecting the intrinsic interest of the work task, nearly all of the differences between occupational classes become non-significant (Table 8.4). Lower non-manual employees are the exception; they still remain less involved in their work than professional and managerial employees.

There was also a strong relationship between the process of upskilling and job involvement (Table 8.1). In general, those that have experienced an increase in their skills showed higher levels of job involvement than those that had seen no change, but this was primarily due to the experience of those whose skills had risen either quite a lot or a great deal. A striking feature of the pattern is the very low level of job involvement of those that had been deskilled. This perhaps helps to account for the fact that deskilling has proved to be a much less common managerial policy than was anticipated by writers who believed there was an inexorable trend towards the fragmentation of skills. Given rising demands for quality, a severely demotivated workforce would have high costs for efficiency. The effect of the experience of skill change proved very robust when successive controls were introduced. It was still highly significant when all of the other variables relating to the work task and the organizational context were taken into account.

The change of work patterns has also meant an increase in the discretion that employees are able to exercise in how to carry out their work. The pattern for task discretion follows closely that of skill development. The direct association is highly significant. Those that had been given greater responsibility over the way their work was carried out were markedly more likely to have a high level of job involvement. Although increased responsibility was closely linked with the process of upskilling (see Chapter 2), it continues to have an independent effect on job involvement even when upskilling has been taken into account. Indeed, it remains highly significant when the full range of controls are introduced (Table 8.4). Overall, the evidence is very consistent that skill development and the growth of task discretion are linked to higher levels of job involvement.[4]

Types of Skill

Our earlier analysis highlighted two important changes in the nature of skills: first the automation and computerization of work tasks and second the growth of people-work. The first of these has led to considerable discussion in the literature, while the second has received relatively little consideration.

---

[4] A similar conclusion emerges from research by social psychologists at the Social and Applied Psychology Unit at Sheffield, using quasi-experimental designs in a field institution (Wall *et al.* 1990). The change from 'specialist' control (where operators were limited to running and monitoring the technology) to 'operator' control (where they were given much broader responsibilities and dealt directly with the majority of operating problems) was associated with greater intrinsic job satisfaction. Similar results were found for autonomous work groups (Wall *et al.* 1986).

Advanced technology. The argument for the beneficial effects of automation for work motivation has been most systematically developed by Robert Blauner (1964). The crucial aspect of automation, he suggests, is that it reverses the historic tendency for an ever-increasing division of labour. It was the process of fragmentation of work that underlay the subjective alienation that characterized industrial settings such as the assembly-line plants of the car industry. Automation reintegrated the work process, thereby providing new opportunities for self-development and meaning in work. In contrast, the 'pessimistic' school contends that automation has broken the traditional relationship between the individual and the machine, thereby undercutting job involvement. Advanced technology, then, takes yet further the degradation of work associated with mechanization and increases alienation rather than job involvement.

Are either of these arguments supported by the evidence? If one takes the bivariate relationship between job involvement and whether or not a person works with automated or computerized equipment, those working with advanced technology have higher levels of job involvement. However, the effect differed considerably between classes. Advanced technology increased job involvement among skilled manual workers and technical and supervisory employees; it was associated with substantially lower levels of job involvement among lower non-manual employees, and it made little difference to job involvement for professional and managerial employees or for non-skilled manual workers. The form that advanced technology takes in these different work settings is clearly very different and it has different degrees of centrality to the work process.

The class pattern is itself differentiated by sex. Although advanced technology is associated with lower job involvement among lower non-manual employees for both men and women, there is a marked difference by sex in its implications for manual workers. Male manual workers working with automated or computerized equipment show higher levels of involvement than other manual workers, whereas the reverse is the case for women. This may be related to our earlier finding (Chapter 2) that there were substantial gender differences in the way work has been organized around advanced technology.

Moreover, the contingency of the effects of advanced technology on wider managerial policies towards work organization is clear from the way they are linked to skill experiences. The positive effects of advanced technology for job involvement are entirely attributable to the way it effects changes in skill.

People-work. The second major change in the nature of work tasks identified in Chapter 2 was a shift away from work involving the manufacture of objects to work that primarily involves dealing with people. How do these different types of work relate to the level of involvement that people are likely to experience in their work?

Work tasks have been classified into five main types: working with people, working with information, traditional manufacturing work (producing or maintaining machines), assembly-line work, and driving. As can be seen in Table 8.1, there was a striking relationship between the nature of people's work and their degree of job involvement. By far the highest level of job involvement was among those whose work predominantly involved people, followed at some remove by those who worked with information. Traditional manufacturing work and driving had roughly similar negative scores. A particularly striking feature is the very low level of job involvement of those in assembly-line work. This confirms the conclusions of the qualitative literature on this type of production process.

However, the implications of both people-work and assembly-line work for job involvement derive in good part from the fact that the two types of work are associated with very distinctive task characteristics and work environments. The positive influence of people-work is linked particularly to higher skill levels, more favourable experiences of skill change, and work tasks that offer greater intrinsic interest.[5] People in this type of work were also better placed with respect to physical work conditions, supervisory support, job security, and organizational participation. When the full range of factors relating to the task and organizational conditions of work are taken into account, the effect of working with people is sharply reduced.

The low level of involvement associated with assembly-line work can be accounted for by a more restricted range of factors. Once skill characteristics, work pressures, and supervisory support are taken into consideration, the initially strong relationship between assembly-line work and job involvement disappears altogether. Again, it is above all task characteristics with respect to skill and the variety of the task that account most strongly for the distinctiveness of assembly-line work.[6]

Thus both advanced technology and 'people-work' are linked to higher job involvement, and assembly-line work to relatively low involvement. These effects are primarily due to the way these different types of work are associated with very different skill characteristics.

Management Control Systems

Chapter 3 highlighted the importance of three contrasting systems of work control. The first was direct supervisory control. The second was 'technical' control, where individual effort was determined by machinery and output-related pay incentives. The third was control through 'performance-management', which relied on the setting of targets, merit pay, appraisal, and opportunities for promotion.

The different types of control system are held to have important implications for motivation. Low trust systems, such as direct supervisory and technical control, are thought to lead to low levels of employee involvement in work (see

---

[5] The initial coefficient (with just sex and age controls) reduces from 0.203 to 0.089 once task characteristics are included.

[6] These reduce an initial coefficient (with age and sex controls) of −0.16 to −0.06.

the influential discussion by Fox 1974). They were likely, then, to be dysfunctional where employees were required to use their discretion. It was this that led Edwards (1979) to suggest that employers would develop a new strategy of control in work settings where they employed highly skilled people. This would focus control after rather than during the work, judging people on their results over the longer term rather than trying to police their everyday activities. It was a system designed to encourage employees to internalize organizational values, in particular through the incentive of promotion opportunities. In many ways, the emerging human resource management policies in the 1980s and 1990s were remarkably close in principle to the types of development that Edwards had predicted.

What has been lacking in the debate in the literature on control systems is any clear evidence about the motivational assumptions that underlie these theories. Is it the case that close direct supervision or technical control systems undercut job involvement? Are performance-management systems any more successful in making employees interested in their work?

Our measure of the closeness of supervision draws on a question asking how much influence the supervisor had over four aspects of the work situation: how hard the person worked, what tasks they were to do, how they were to do the tasks, and the quality standards to which they worked.[7] Taking the bivariate relationship, it is clear that the closer the supervision, the lower the employee's level of job involvement. Further, where supervisory control has grown tighter over the last five years, job involvement is lower. However, earlier research has suggested that the closeness of supervision is likely to be affected by broader organizational and technical characteristics of work and these may have independent effects on involvement (Woodward 1970). It is notable that, once other organizational factors have been controlled, the closeness of supervision no longer makes a statistically significant difference (Table 8.4).

Supervision has also been seen as important in determining the general quality of personal relations at work. This was the factor that was regarded as particularly vital for employee motivation by the theorists of the Human Relations school. It seems likely that one of the qualities that would be most important for maintaining good social relationships would be the supervisor's ability to demonstrate impartiality in the treatment of employees. Failure to do this would generate petty rivalries and jealousies that could create a downward spiral of distrust in work-team relations. It was certainly the case that the view that the supervisor was impartial in treating employees had an important effect on job involvement. In contrast to the 'closeness of supervision', this effect persisted even when the other task and organizational characteristics were taken into account.

---

[7] These items scaled well, with a Cronbach alpha of 0.83. A principal components analysis revealed a single underlying factor, with an eigenvalue of 2.66, accounting for 66% of the variance. The factor scores, then, have been taken as an overall measure of the closeness of supervision.

TABLE 8.2 *Managerial control systems and job involvement*

	Coeff.	Sig.	No.
Technical control			
All	−0.11	***	3,425
Men	−0.11	***	1,783
Women	−0.12	***	1,635
Performance-Management Control systems			
All	0.08	***	3,425
Men	0.11	***	1,783
Women	0.06	***	1,635

Note: Results are drawn from a series of OLS regressions.

Our data also provide support for the view that technical and performance-management controls systems had rather different effects. It can be seen in Table 8.2 that technical control was associated with lower job involvement; moreover, this was the case both for men and for women. In contrast, performance-management systems were associated with higher levels of job involvement for both men and women. These effects cannot easily be explained away in terms of other variables that might be linked to both the forms of control and job involvement. Even with the full range of control variables (Table 8.4), technical control is linked to lower job involvement and performance-management systems to higher involvement. Taken overall, the evidence points strongly to the view that the nature of managerial control systems has important implications for work motivation.

Participation

There was very little evidence that participation in industry increased with the restructuring of employment over the last decade. Indeed, there is some indication that it may have declined. However, this is a factor that has consistently emerged in the literature as a significant influence on job involvement (Blumberg 1968; Brannen 1983). However, the relationship with satisfaction measures is not always easy to interpret. Participation may lead to higher satisfaction because people are more interested in their work or because they can use their influence to reduce work pressure. An assessment of the implications of participation for job involvement is more revealing, since it focuses more specifically on interest in the work.

It was seen in Chapter 4 that there are a number of different ways in which organizations may enhance a sense of participation. The measure of participation used here was designed to cover the range of these by asking people 'How much influence do employees have over decisions in this organization?' Participation proved to be one of the strongest variables for predicting levels of psychological involvement in the job. It was highly significant for both men and women, and within each occupational class. Further, its effect proved very robust when other potential factors affecting job involvement are taken into account (Table 8.4). Even when the full range of task and organizational characteristics were included, it remained highly significant. Our evidence, then, confirms earlier research that employee participation is a major factor contributing to more positive involvement in work.

Job Security

Theories of internal labour markets have placed considerable emphasis on the importance of job security in building up the type of highly motivated workforce needed to cope with more complex and technologically sophisticated production systems. Yet a major feature of the period since the 1970s has been the lower level of job security, especially for men. Is it the case empirically that low job security tends to undercut people's involvement in their work? We have focused in particular on two issues: the implications of non-standard employment contracts and the effects of more general anxiety about job security.

Taking first the nature of contracts, there was a strong negative association between working on a short-term contract and involvement in the job. However, if one distinguishes within the category of temporary worker between those on very short contracts (twelve months or under) and those on contracts lasting between one and three years, it is notable that the really important contrast is between those on very short-term contracts and all others. Whereas those in permanent work had an average job involvement score of 0.03, and those on contracts of between one and three years a score of –0.01, those on contracts lasting less than a year had a score of –0.43, indicating a very low level of involvement.

What was it about temporary work status that lowered involvement? It is possible to assess this through looking at how the strength of the effect of temporary work changes when various characteristics of the work situation are controlled. For this, employees have been classified into those who had contracts for less than a year and all others, and then controls have been added cumulatively to the regression. Table 8.3 shows the effects of each set of controls on the coefficients and on the significance levels. There is little reduction in the

TABLE 8.3 *Temporary work and job involvement*

	Coeff.	Sig.
No controls	−0.45	***
+age and sex	−0.44	***
+job security	−0.05	n.s.
+skill characteristics	0.10	n.s.

Note: Results are drawn from a series of OLS regressions.

TABLE 8.4 *Factors related to job involvement*

	Coeff.	Sig.
Age	0.01	***
Sex (men)	−0.03	n.s.
Task characteristics		
Lower non-manual	−0.16	***
Technicians/supervisors	−0.07	n.s.
Skilled manual	0.04	n.s.
Semi-/non-skilled manual	−0.08	n.s.
Upskilled	0.12	***
Task discretion	0.27	***
People-work	0.08	**
Works with advanced technology	−0.08	*
Work pressure	0.14	***
Change in effort comp. to 5 years ago	0.07	**
Health and safety at risk?	−0.06	n.s.
Uncomfortable work position	−0.04	**
Organizational context		
Supervisory influence high	−0.02	n.s.
Change in tightness of supervision	−0.03	n.s.
Sup. treats some better than others	−0.06	***
Friendliness of people worked with	0.23	***
Performance-management policies	0.03	***
Technical control	−0.03	**
Dissatisfaction with job security	−0.15	***
Contract wkr 1–3 years	0.20	***
Temporary wkrs under 12 months	0.13	n.s.
Female part-time	0.00	n.s.
Employee influence in organization	0.12	***
Unionized	0.03	n.s.

Note: OLS regression analysis. No. = 2,655; R^2 = 0.5.

effect of temporary work when age and sex controls are added. However, the coefficient falls sharply and becomes non-significant when account is taken of job insecurity. Short-term temporary work, then, is associated with lower levels of work motivation and this is primarily due to the insecurity it involves.

In the conditions of the 1990s job insecurity had spread beyond the ranks of those whose formal contracts were short-term to affect a significant proportion of employees with supposedly 'permanent' contracts. Taking as our measure people's dissatisfaction with their job security, it was clear that insecurity reduced involvement even when temporary contract status was discounted.[8] Indeed, dissatisfaction with job security undercuts job involvement even when all the other task and organizational variables were controlled. It was clearly one of the

[8] The measure asked people to assess on a seven-point scale how satisfied or dissatisfied they were with their job security.

most powerful determinants of the extent to which people felt involved in their work and indicates that the increased precariousness of employment over the last two decades may have had very serious consequences for people's attitudes to their jobs.

WORK STRAIN

The next aspect of people's experience of work that we are concerned with is that of work strain—the extent to which people experience high and persistent levels of tension or fatigue as a result of their work. The view that the changing nature of job tasks is leading to increased tension at work has been developed primarily by those who have argued for a general tendency for the quality of work to decline. For instance, it has been argued that with advanced techno-logy the worker is subject to sharply increased work pressure arising from the requirement to be more flexible in terms of work practices and hours of work (see for instance Naville 1963). Some aspects of this scenario have been confirmed in our earlier analyses. In particular, there was evidence of an increase in levels of work pressure, although these were linked to a process of upskilling rather than deskilling of work. What implications did this intensification of the work process have for the level of strain that people experienced as a result of their work?

Our measure of work strain was derived from a four-item question asking people how they felt at the end of the workday, using for each item a six-point scale ranging from 'never' to 'all of the time'.[9] The emphasis on the duration of symptoms of strain is of central importance. Persistent work strain is far more likely to be damaging than a merely transient spell, and it was such persisting forms of strain that we wished to capture through the measure. The items scaled well and for most of the analysis we have used the overall scale of work strain based on the four items.[10] The wording of the items was as follows:

'Thinking of the past few weeks, how much of the time has your job made you feel each of the following?'

- After I leave my work I keep worrying about job problems.
- I find it difficult to unwind at the end of a workday.
- I feel used up at the end of a workday.
- My job makes me feel quite exhausted by the end of a workday.

A first point to note is that in terms of these measures a substantial propor-tion of the British workforce would appear to have experienced substantial work

[9] The measure was taken from Warr (1990).
[10] The items scaled with a Cronbach's alpha of 0.82. A principal components analysis produced a single factor, with an eigenvalue of 2.63, accounting for 66% of the variance. The factor score has been taken as the measure of work stress.

TABLE 8.5 *Work strain (row %)*

	Much of time	Some of time	Occasionally	Never	No.
Worries about job after work	11	22	39	28	3,321
Finds it difficult to unwind	13	23	34	30	3,320
Feels used up at end of workday	21	27	37	15	3,321
Feels exhausted at end of workday	22	29	36	13	3,325

strain. The basic distributions for the four items are given in Table 8.5. Taking the items separately, each symptom of work strain was experienced by at least of third of all employees at least some of the time and, depending on the particular item, between 11 per cent and 22 per cent experienced it regularly. Each of the items expresses a fairly strong manifestation of work strain. If one takes as an indicator of the prevalence of strain the proportion of those reporting that they experienced at least one of these symptoms much of the time, 31 per cent of all employees experienced a high level of work strain.

Finally, there was a widely prevalent view among employees in British industry that the level of strain involved in the job had increased over recent years. Our indicator here is a simple one-item question, asking people whether the stress involved in the job had increased, stayed the same, or decreased over the last five years. Overall, 53 per cent said that stress had increased, 34 per cent that it had stayed the same, and 12 per cent that it had decreased. Increased stress was particularly common among those in professional/managerial work (67 per cent) and among those in technical/supervisory jobs (66 per cent). Even among those in lower non-manual work a majority (55 per cent) reported that the stress involved in their work had increased. In contrast, this was the case for only a minority of skilled manual workers (44 per cent) and semi- and non-skilled manual workers.

Skill and Responsibility

It was noted above that higher-skilled work and skill development were strongly related to job involvement. What, however, were their implications for work strain? Our best proxy for the overall *level* of skill and responsibility is people's occupational class. There was a highly significant bivariate relationship between class position and work strain. The basic pattern can be seen in Table 8.6. In particular, those in professional and managerial work had much higher levels of work strain than any other category. Lower non-manual, technical/supervisory, and skilled manual work seemed to involve rather similar levels of strain, while semi- and non-skilled manual work was the least stressful. The pattern remains remarkably stable even when the full range of variables affecting work strain is taken into account (Table 8.8).

TABLE 8.6 *Class, upskilling, and work strain*

	Work strain score (means)	No.
Class		
Professional/managerial	0.33	1,155
Lower non-manual	−0.08	554
Technician/supervisory	−0.05	149
Skilled manual	−0.11	358
Semi- and non-skilled	−0.28	1,059
Skill change		
Increased a great deal	0.33	702
Increased quite a lot	0.09	1,033
Increased a little	−0.17	320
No change	−0.23	919
Decreased	−0.20	293

A more direct indication that the rise in skill levels may have increased work strain comes from the measure of change in skill experience. Those that had increased their skill over the last five years had significantly higher levels of work strain. As can be seen in Table 8.6, those whose skill level had remained the same had the lowest levels of work strain, followed by those who had been deskilled and those whose skills had increased in a relatively modest way. High levels of work strain were mainly typical of those who had experienced a marked increase in their skill level. It is likely that this is due to the fact that upskilling was associated with an intensification of work effort. When controls were introduced for the intensity of work and for change in work effort over the previous five years, upskilling itself no longer had a significant effect.

A number of studies have suggested that more highly automated methods of production are directly linked to increased work strain. They point in particular to the much greater (and sometimes catastrophic) consequences of error in highly automated settings such as chemical plants. Further, some accounts have suggested that the introduction of new technology into non-manual work is associated with an intensification of work effort, in part by making it easier for management to record the level of work effort. However, there are also arguments that suggest that automation could have the opposite effect of reducing work strain. For instance, it takes away much of the physical drudgery of manual work and has been generally found to be linked to better work conditions.

The view that advanced technology is associated with higher work strain would appear to receive some support. Taking the bivariate relationship, we find that those working with computerized or automated equipment had a higher work strain score.[11] However, there were substantial differences between occupational

[11] Those working with automated equipment had a score of 0.10 compared to −0.13 for those who were not ($p < 0.000$).

classes. While the literature has focused principally on the possible links between advanced technology and work strain for manual workers, our evidence suggests that it is most marked for lower non-manual employees. In contrast, there is no indication that it increases strain for professional and managerial employees, or for skilled manual workers.

What is it about such work that increases strain? The most common view is that it results from an increased intensification of work. If this is correct, then the association could be expected to disappear once measures of general work pressure are introduced. However, the link between advanced technology and work strain remained highly significant when a variety of controls were introduced for work effort. Rather what did appear to account for the effect was skill experience. The use of advanced technology was associated with higher demands for skill and responsibility, and these, in turn, were related to greater work strain. As can be seen in Table 8.8, once skill variables have been controlled for, the effect of advanced technology disappeared.

There also has been a strong emphasis in the literature on the detrimental effects of highly repetitive work in manufacturing industry, in particular in assembly-line production. But although it was seen earlier that assembly-line work was related to low levels of job involvement, there was no evidence that it involved particularly high levels of work strain (Table 8.7). In contrast, there was a very clear effect of being involved in work that was primarily concerned with people. Of the various types of work (people-work, traditional manufacturing, assembly-line, information work, and driving), the highest level of work strain was experienced by those engaged in people-work, followed at a considerable distance by those whose work primarily involved driving and by those working with information.

Moreover, the effect of people-work proved very persistent. Not only was it still highly significant for the level of strain when other skill related factors were introduced, but it remained significant even when a much wider range of determinants of strain have been included (Table 8.8). This suggests that it may be intrinsic to the type of work. Perhaps the vulnerability of those engaged in such jobs may be linked to the fact that work with people is likely to involve an even less predictable work environment than work with objects. Moreover, given the interactive nature of such work, it is likely to require high and sustained levels of concentration, while errors may lead to bruising encounters with clients or customers who feel that they have been unfairly treated.

Yet, it should be noted that there was considerable variation in work strain between industries with relatively high levels of employees involved in people-work.[12] The highest level of work strain was to be found in education, followed by the financial industries (banking, finance, and insurance) and then medical

---

[12] We have excluded two industry categories where there are less than twenty respondents, since the data for these are likely to be of a low level of reliability: agriculture and repairs.

TABLE 8.7 *Work strain, type of work, and industry*

	Work strain scores (means)	No.
Types of work		
People-work	0.13	1,510
Information	−0.03	513
Manufacturing	−0.17	373
Assembly-line	−0.25	92
Driving	0.04	126
Industry		
High work strain		
Education	0.37	277
Financial industries	0.18	263
Medical services	0.15	168
Welfare	0.07	127
Energy	0.07	46
Low work strain		
Leisure	−0.25	45
Personal services	−0.24	37
Retail	−0.17	222
Wholesale	−0.16	86
Catering/hotels	−0.13	63

services (Table 8.7). These were followed at some remove by welfare and the energy industries. However, at the other end of the spectrum, there were relatively low levels of strain in leisure and personal services, followed by the retail, wholesale, and catering industries. These are industries with high levels of people-work, but of a far more routine type. This suggests that the critical factor that generates strain may be more specifically the degree of responsibility for people that is involved in the work. It is when people-work involves substantial responsibility for others that it is likely to generate exceptionally high work strain.

Work Pressure and Work Strain

It was seen earlier that employees reported a marked increase in the intensity of work effort over the last decade and there is considerable research evidence that the pressure of work is a major factor determining work strain. This is strongly confirmed by our own data. As can be seen in Table 8.8, even when other factors had been included in the analysis, work intensity proved a highly significant influence. The link with the intensification of work is even clearer if one takes a measure of change in work effort. A rise in the level of work effort over the previous five years was directly associated with increased strain, even when the current level of work intensity had been taken into account.

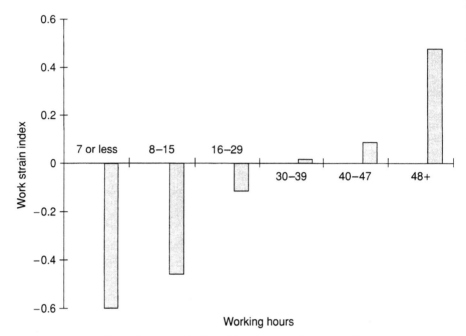

FIG. 8.1 Length of working hours and work strain

An influential argument has been that the impact of job demands varies depending on the level of task control available to employees (Karasek and Theorell 1990; Marmot *et al.* 1991). Where a person can control the work task, it is suggested, a high level of job demand is not as likely to lead to strain. The situation that is most damaging to health is one where a high level of job demand is associated with a limited capacity to take decisions about the work. To examine this, we introduced into the analysis an interaction term reflecting the effect of job demands at different levels of task discretion. However, in contrast to the expectations of the argument, the interaction term was not significant. Our evidence then confirms the view that work intensity is a major factor leading to work strain, but there is no sign that the impact of job demands is any less where people have a higher level of task discretion.

The length of working hours is also likely to be an important component of work pressure. There is certainly a strong relationship between overall weekly working hours and work strain, with strain rising the longer the hours worked. The direct association between working hours and the level of strain can be seen in Fig. 8.1. The categories of working hours have been selected in part to reflect different groups of part-time workers, defined in terms of their eligibility with respect to various types of legal protection. Taking thirty hours as the conventional cut-off between full-time and part-time work, it is clear that part-time workers are less likely to experience work strain than any of the categories of full-time worker. But perhaps the most notable feature of the data is the strong linearity of the relationship between work hours and strain within both the part-time and full-time groups. The underlying pattern is quite simply that the longer the hours worked, the greater the strain, although the effect was particularly pronounced among those working 48 hours or more.[13]

Finally, we also examined whether strain was affected by the physical conditions under which people carry out their jobs. A number of aspects of physical working conditions were associated with work strain when taken on their own. But the two features of physical working conditions that showed persistently strong effects were those relating to the discomfort of work positions and, above all, to the safety of the work environment.

The Organizational Context

A wide range of research has demonstrated the importance of the supportiveness of the social environment in mediating the effects of deprivations in the work setting. As was seen earlier, the role of the supervisor may be of central importance in determining the extent to which people perceive themselves to be in a supportive work environment. While the closeness or looseness of supervision was not in itself linked to the level of work strain, a factor that was clearly significant was again the impartiality of the supervisor in handling personal relations (Table 8.8). There was also a clear effect of the perceived overall supportiveness of the work environment. Those who felt they were in a more friendly environment were less likely to experience work strain, even when the level of work demand was held constant.

The managerial control policies that have been at the centre of the analysis are the use of technical and performance-management systems of control. It has been seen that these had strongly contrasting effects with respect to job involvement. Technical control was associated with lower levels of job involvement, while performance-management systems were linked to higher involvement. What were their implications for work strain? There are fairly clear expectations with respect to technical control. A distinctive feature of this system is its reliance on mechanical forms of control, whereby work pace is determined by machinery. It is also a form of control that allows very limited discretion to employees and typically involves a fragmented and repetitive work process. The evidence from qualitative studies would certainly lead to an expectation that this type of control would accentuate work strain.

The implications of performance-management systems are less easy to predict. From one perspective, they involve a closer, more individualized, relationship

---

[13] A similar conclusion was reached by Clark (1996) in an analysis of job satisfaction based on the British Household Panel Study.

TABLE 8.8 *Factors related to work strain and psychological distress*

	Work strain		Psychological distress	
	Coeff.	Sig.	Coeff.	Sig.
Age	0.00	*	0.00	***
Sex (men)	−0.11	**	−0.13	***
Task characteristics				
Lower non-manual	−0.26	***	−0.01	n.s.
Technicians/supervisors	−0.32	***	−0.05	n.s.
Skilled manual	−0.39	***	−0.06	*
Semi-/non-skilled manual	−0.38	***	−0.04	n.s.
Upskilled	0.03	n.s.	−0.02	**
Task discretion	0.01	n.s.	−0.02	**
People-work	0.12	***	0.04	*
Works with advanced technology	0.01	n.s.	0.02	n.s.
Work pressure	0.33	***	0.04	***
Change in effort comp. to 5 years ago	0.08	**	0.00	n.s.
Health and safety at risk?	0.26	***	0.08	***
Uncomfortable work position	0.07	***	0.02	*
Organizational context				
Supervisory influence high	−0.01	n.s.	0.01	n.s.
Change in tightness of supervision	0.03	n.s.	0.00	n.s.
Sup. treats some better than others	0.04	*	0.03	***
Friendliness of people worked with	−0.04	*	−0.05	***
Performance-management policies	0.00	n.s.	0.00	n.s.
Technical control	0.06	***	0.01	n.s.
Dissatisfaction with job security	0.04	***	0.04	***
Contract wkr 1–3 years	−0.02	n.s.	0.07	*
Temporary wkrs under 12 months	−0.04	n.s.	−0.10	**
Female part-time	−0.11	*	0.00	n.s.
Employee influence in organization	−0.07	**	0.00	n.s.
Unionized	−0.01	n.s.	−0.03	*

Note: OLS regression analyses. For work strain, no. = 2,578, R^2 = 0.27; for psychological distress, no. = 2,567, R^2 = 0.14.

between employers and managers, which potentially provides better channels of communication and might help to resolve frustrations at an early stage. However, it is also a system that involves much closer monitoring and assessment of individual work performance and this may be linked to much greater work pressure.

The empirical evidence with respect to technical control confirmed the expectations from the qualitative research literature. Technical control was associated with higher levels of strain, and this effect persisted strongly when other factors were controlled (Table 8.8). It was also the case that performance-management systems were linked to higher work strain. But while the direct relationship was at a high level of statistical significance, this was no longer the case once other task and organizational factors had been taken into account. While technical control systems aggravate strain over and above their implications for work pressure, the negative consequences of performance-management systems appear to be primarily due to the fact that they are linked to an intensification of work effort.

Our evidence again confirms the findings of earlier research that participation in organizational decision-making has major benefits for employee well-being. Where the organization provided opportunities to participate in wider decisions, employees were less likely to suffer from work strain (Table 8.8). The effectiveness of participation in reducing work strain may partly reflect the fact that it gives employees a stronger sense of control over their work environment. Problems and sources of frustration can therefore be resolved more easily through active intervention. Another factor that may be important is that participation is likely to lead to a greater transparency of the organization's activities. The social-psychological literature has stressed the importance of environmental clarity for psychological well-being. This is often couched in terms of the significance of having clearly defined performance goals. But an important structural factor underpinning employees' ability to understand the objectives of their organization, and the relationship between these and their own work tasks, may well be the extent to which they are informed about, and feel that they can influence, organizational decisions.

Finally, our evidence once more underlines the critical effect of job insecurity for employees' experience of work. If the clarity of goals and the predictability of the environment are increased through participation, they are heavily undercut by job insecurity. While direct threat of job loss is clearly the most severe and disorientating form of insecurity, it is also likely that the extensive changes in work organization that have marked the period have accentuated a pervasive sense of insecurity. It has long been recognized in the literature that organizational change is frequently perceived as highly threatening by employees. Certainly, the measure of perceived insecurity was strongly related to the level of work strain. The effect remained with a high level of statistical significance, even when the full range of task characteristics and organizational factors were taken into account.

However, despite this general effect of insecurity, there was interestingly no overall tendency for those on short-term temporary contracts to show higher strain. This lack of relationship between temporary work status and work strain may seem surprising given the disadvantages associated with this type of work (Chapter 6). However, it must also be recalled that temporary work jobs were characterized by a markedly lower level of work pressure. The lack of relationship between contract status and work strain is likely to result from the countervailing effects of higher insecurity but lower work pressure.

PSYCHOLOGICAL DISTRESS

Work Strain and Psychological Health

The final stage of our analysis turns to the relationship between changes in work and psychological distress. Work strain and psychological distress are clearly not synonymous. The notion of psychological distress relates to a general psychological state of people, not specifically to their work-related experiences. It is likely, then, to reflect a wider range of influences. It may be that there are countervailing factors in people's non-work lives that allow them to support high levels of work strain, without it leading to psychological distress.

The measure of psychological distress that has been adopted is based on a validated measure of minor psychiatric morbidity—the General Health Questionnaire (GHQ). The version used consists of twelve items, from which an average score was computed for each individual.[14] While there are different ways of deriving an overall measure, it has been shown that the 'Likert' version has superior statistical properties and this has been adopted here.

There can be no doubt that there was a strong overall association between work strain and psychological distress. There was a correlation of 0.38 for both men and women. But considering the broad types of factors that contributed to work strain, it is clear that some had no discernible impact on psychological health, while others were strongly associated with it.

It was particularly the factors associated with skill level and skill change that had very *different* implications for work strain and for psychological distress. While there was a very clear gradient between occupational class and work strain, with professional and managerial employees standing out as having far higher levels of work strain, there was no evidence that this was the case for psychological distress. The results with respect to skill change are also revealing. Those who had experienced an increase in the skill requirements of their job over recent years were more likely to suffer from work strain, but they were less likely than others to experience psychological distress. Upskilling then was linked to higher work strain but it appeared to provide protection against psychological distress.

However, there were factors which did increase both work strain and psychological distress. This was the case for work pressure and for working conditions that were poor with respect to safety and comfort. Supervisory favouritism also had a consistently negative influence. Finally, dissatisfaction with job security strongly accentuated both work strain and psychological distress.

Overall, then, the determinants of work strain appear to divide into two rather different categories: those relating to skill level and skill change that do

---

[14] The original thirty-item measure was developed and validated by Goldberg (1972, 1978). For an assessment of the twelve-item version, see Banks *et al.* (1980).

not undermine psychological health, and those such as work pressure, poor physical working conditions, and job insecurity which do appear to have a spill-over effect for more general psychological health.

Work Strain, Out-of-Work Life, and Psychological Health

The concept of psychological distress is more general than that of work strain, in the sense that it captures the influence of experiences in a range of domains of people's lives. The level of psychological distress may depend to a considerable degree on out-of-work factors. It may also be the case that the quality of people's non-work lives may mediate the impact that work strain has on psychological stability. Our enquiry was not specifically focused on this, and there are only a very limited number of indicators of non-work experiences that can be considered. The results here are necessarily very tentative; they show, however, a pattern of some interest.

To begin with, there is an indication that work strain may have had detrimental consequences for people's out-of-work lives. The survey included three indicators of people's satisfaction with different aspects of their lives: their family life, their social life outside the home, and their leisure life. As can be seen in Table 8.9, those with higher levels of work strain reported lower satisfaction with each of these life spheres. The link was particularly strong for social and leisure life. The pattern was similar for both men and women. The main difference was that work strain seemed to have a sharper impact on satisfaction with family life for women than it did for men. This would suggest that the pressures of work generated a level of anxiety and tiredness that made it more difficult for people to enjoy their families and their social lives.[15]

Does work strain continue to have a clear impact on general psychological distress even after account has been taken of satisfaction with certain other life domains? The effect of work strain was examined taking account of non-work

TABLE 8.9 *Work strain and satisfaction with non-work life*

	Men		Women		All	
	Coeff.	Sig.	Coeff.	Sig.	Coeff.	Sig.
Family life	−0.06	**	−0.15	***	−0.11	***
Social life	−0.19	***	−0.22	***	−0.21	***
Leisure life	−0.23	***	−0.23	***	−0.23	***

Note: Correlation coefficients.

[15] For a discussion of the literature on the negative spill-over of work on non-work life, see Brannen *et al.* (1994).

TABLE 8.10 *Effect of work strain on general psychological health*

Variables	Men		Women		All	
	Coeff.	Sig.	Coeff.	Sig.	Coeff.	Sig.
Age	0.00	***	0.00	***	0.00	***
Illness	0.03	n.s.	0.06	*	0.02	n.s.
Married	−0.05	*	0.00	n.s.	−0.02	n.s.
No. of children	0.02	**	0.00	n.s.	0.01	*
Satis. with family life	−0.04	***	−0.08	***	−0.07	***
Satis. with social life	−0.04	***	−0.02	*	−0.03	***
Satis. with leisure life	−0.01	n.s.	−0.01	n.s.	−0.01	n.s.
Financial anxiety	0.07	***	0.09	***	0.08	***
Work strain	0.12	***	0.11	***	0.11	***
Job involvement	−0.08	***	−0.07	***	−0.08	***
Job involvement × work strain	−0.03	***	−0.01	***	−0.02	***
Sex (men)					−0.12	***

Note: OLS regression analyses. For men, no. = 1,650, R^2 = 0.31; for women, no. = 1,502, R^2 = 0.36; for all, no. = 3.073, R^2 = 0.31.

factors (Table 8.10). The factors that have been considered include: age; people's family situation (whether they were married and the number of children they had); their satisfaction with their family, social, and leisure lives; and finally the frequency with which they worried about money. The model also takes into account people's level of job involvement. As sex has a powerful independent impact on psychological distress, with women showing substantially higher vulnerability than men, the results have been given both for the sample as a whole and for men and women separately.

Several of the non-work variables proved to be important for general psychological health. While people's marital status and the number of children they had were not significant for the sample as a whole, they did make a difference for men. Married men showed better psychological health than single men, but the more children they had the greater their psychological distress. Marriage, on the other hand, did not appear to bring women the same advantages, and they were unaffected by the number of children. But, while family status variables had a gender-specific effect, the measures of satisfaction with different spheres of non-work life showed a rather more similar pattern for men and women. People that were more satisfied with their family and social lives showed lower levels of psychological distress, even when work-related strain had been taken into account. The link was particularly strong for family life, although this mainly reflected its importance for women. In contrast, social life outside the home appeared to be more important for men than for women. Finally, the degree of financial difficulty people experienced had a very strong effect on psychological distress for both sexes.

Yet even when family status, satisfaction with family and social life, and financial deprivation have been controlled, the impact of work strain on general psychological health still comes across very clearly. Overall, it had the strongest effect of the factors that we were able to consider. This was the case for both men and women, although its effect was somewhat stronger for men.

When interaction terms were introduced to see whether the relationship between work strain and psychological health changed depending upon family status or satisfaction with non-work life, the results were consistently non-significant. It would seem that neither family nor social life can easily buffer the impact of work experience.

The only significant factor we were able to discover that did mediate the impact of work strain on more general psychological health was that of job involvement. As can be seen in Table 8.10, people who were involved in their jobs were less likely to experience psychological distress in the first place. But, over and above this, when an interaction term was introduced, it was found that, when people who experienced work strain were involved in their jobs, the strain of work was less likely to spill over into psychological distress.

Overall, then, the effects of work strain on psychological well-being persist even when account is taken of people's satisfaction with their family and social lives. While the latter have an important direct effect on psychological health, there is no evidence that they blunt the impact of stresses from work. Negative experiences at work appear to have harsh psychological effects, irrespective of people's contentment with other aspects of their lives. The only factor that does appear to mediate the implications of work strain for more general psychological distress is that of involvement in the work. Hence, the more intrinsically interesting the work, the more is should shield people from the damaging effects of work strain.

CONCLUSIONS

It is clear that the different processes of change in the nature of work and the employment relationship have sharply contrasting implications for the psychological experience of work. In this analysis we have distinguished between three aspects of this experience: job involvement (which captures satisfaction with a work task that is experienced as intrinsically interesting), work strain (which refers to persistent work-related tension or fatigue), and finally psychological distress (which refers to a vulnerability to anxiety or depression).

The process of upskilling was strongly related to higher job involvement. This reflected in part the fact that it led to work that provided greater variety and opportunities for self-development. But the persistence of the effect through the full range of controls suggests that people in general enjoy the challenge of more complex work and become motivated by it. The negative side is that upskilling also helped to produce higher levels of work strain, through its effects in increasing work pressure. As we have seen, upskilling tended to be accompanied by a decentralization of responsibility to employees. The combination of meeting the requirements of new skills and increased responsibility seem to have taken their toll in greater tension at work. But our evidence did not suggest that, at least in the short term, upskilling led to psychological ill-health.

Our analyses also indicate that, irrespective of these broad patterns of change in the nature of work, management organizational policies can have significant effects on employees' experience of work. In particular, they have highlighted three organizational factors that have proved consistently important. First, they provide support for earlier views about the importance of patterns of supervision. The ability of supervisors to provide supportive relationships was a major factor making for higher involvement. Conversely, arbitrary forms of supervision led to higher work strain. Technical control systems, with their emphasis upon mechanical control of performance and output-related incentives, were associated with lower levels of involvement in the job and greater work strain. Performance-management control systems, on the other hand, appear to have encouraged greater employee interest in the job. However, they also were associated with higher levels of work strain, primarily as a result of the way they led to increased work pressure.

It should be noted that although there has been no evidence of a growth of participation in recent years, our results provide strong support for the conclusions of earlier research on participation. It had consistently favourable effects for the experience of work. Where participation was higher, employees were more likely to feel involved in their work tasks, and they were less likely to feel strain in their work, even controlling for the level of work pressure.

Finally, the results concerning the effect of job insecurity on psychological well-being were remarkably consistent. Lower job security reduced job involvement, since, presumably, people were reluctant to invest psychologically in work projects that they might never see to fruition. At the same time, it was a major source of work strain and this seems to have had a clear follow-on effect in terms of poorer psychological health. The decline in job security seems, then, to have had deeply negative effects both for work motivation and for people's personal psychological stability.

The three aspects of change that were highlighted in earlier chapters—upskilling, new management policies for the control of work performance, and heightened job insecurity—all had marked implications for employee's experience of work. Upskilling and the use of performance-management systems were ambivalent in their effects. On the positive side, the rise in skill levels and of more indirect forms of control contributed to higher levels of personal involvement in the job. However, at the same time, they led to substantially greater work strain, in part because they were linked to an intensification of work effort. It was, however, the impact of job insecurity that gives the greatest cause for concern. The effect of job insecurity was consistently negative. It seemed to lower people's involvement in their jobs and to lead to higher levels of work strain. But, over and above this, it appeared to contribute directly to more pervasive psychological distress.

9

Organizational Commitment

Previous chapters have demonstrated the varied ways in which employers in the 1990s were demanding more of their employees. They required more responsibility and discretion to be taken, more control and pressure to be accepted, more flexibility to be displayed, more uncertainty and stress to be absorbed. In short, employers sought employee commitment to the organization. Indeed, human resource management (HRM) has been described or advocated as a movement from a strategy of control to a strategy of commitment (Walton 1985a; Lundy and Cowling 1996).

An obvious obstacle to organizational commitment is the growth of insecurity which was analysed in Chapter 5. Can a high level of commitment be expected from employees if employers cannot or will not offer them long-term security—given competitive pressure and the policies of flexible workforces? Not a few theorists have answered this question in the negative, and reasserted the critical importance of security and long-term employment guarantees (Brown *et al.* 1993; Kochan and Osterman 1994). Others, however, have suggested that the problems issuing from insecurity and uncertainty can be contained through policies on recruitment, training, and skills. Useem (1993), for example, argues that there are effective responses to 'downsizing' in terms of basic entry-level training. It has also been proposed that security of employment can be substituted by the offer of 'employability', that is, training, development, and qualification leading to better chances of *external* career moves (Waterman *et al.* 1994).

The issues of commitment, however, transcend the managerial agenda. The commitments made by individuals represent the values they hold and the priorities among which they distribute their choices and actions over long periods of their lives. Elizabeth Anderson (1993) has written of the importance to people of having values which shape their long-term priorities, since it is these values which bind together their lives and make it possible for them to have a self-esteem which is based on being a certain kind of person. Thus, people need to make commitments, and commitment to organizations (in which one is likely to spend a large part of one's active life) is one of the salient opportunities for satisfying this need. But organizations can only be satisfactory objects for individual commitment if they are worth while in terms of the values they themselves represent. Thus Walton (1972), in one of the earliest statements of the high-commitment strategy, argued that organizational commitment depended partly on the intrinsic interest of work, but also on the 'human dignity offered by management' and the 'social responsibility' demonstrated in the organization's business. But how far have organizations, in the Britain of the 1990s, developed and communicated values, whether of these types or others equally strong, to their employees?

This question relates to one of the main changes taking place in Britain in the 1980s and 1990s, namely the progressive replacement of public service organizations by privatization and marketization. Supposedly this has led to the introduction of market and competitive values among public service employees (Whittington *et al.* 1994). But how far have market values stimulated organizational commitment in practice? And how far has the erosion or loss of public service values had adverse effects on commitment?

VALUES AND COMMITMENTS

The relationships between organizational policies and commitment, and the contrasts between market (or competitive) and public services values, will be central to the analysis in this chapter. Perhaps the most profound observations on just these relationships were provided by Philip Selznick in his short book on leadership (1957). The focus there was on the role of the organization's top management in creating the conditions for commitment. Selznick saw, as the central function of leaders, the building and preservation of the values which give an organization its coherence. Selznick argued that this could only be achieved through *making commitments*, whether these were commitments to the provision of certain services for the community, or, more importantly, to qualities or standards in the provision of services. Of course, effective leadership does not stop short at the framing of committed policies, but must go on to put those policies into practice. This necessarily involves getting employees to identify with the organization's commitments. So Selznick's model is one of individual commitment through association with the organization's own commitments.

Most of Selznick's examples were drawn from public services and their commitment to the community. Commercial organizations attempting to develop commitment along similar lines have often depended upon product quality or customer service to provide their ethos. There are, however, several potential difficulties or limitations with this approach in the commercial sector. The most obvious is that public-benefit services tend to have quite a different character from commercial services, and often involve altruistic values. This is most apparent in medicine, where practitioners to this day take an ancient religious oath of conduct. Education, too, has religious and civic associations which have continued through to the present day; indeed, the historian Corelli Barnett (1986) argued that the prevalence of religious and moral objectives had been a significant obstacle to the modernization of education in Britain to recent times. Again, in the armed forces, police, or fire service, employment involves an understood commitment to the sacrifice of one's own life or limb, if necessary, for the protection of the community. This willingness to sacrifice oneself in the public benefit is quite different from commercial jobs where there may be an *accidental* risk which is hardly less serious, as in the case of the drivers of goods vehicles.

Further, an employee's commitment to a particular type of service, or a particular service quality, does not necessarily give an organization all that it wants. Such a commitment may be too narrow, or too rigid, for complex organizations which continually have to adapt their objectives to changing external pressures, a current reality for the public as well as the market sector. It is common to hear managers stating that they need committed employees to cope with competition and to adapt to a regime of continuous change. The suggestion is that the committed employee is not merely attached to certain fixed priorities, but is ready and willing to serve the as yet unknown future priorities of the organization.

When they underline this point, organizations are indeed drawing near to the meaning of commitment as it is understood in spheres of life beyond work. If two people enter a committed relationship, they accept the need to absorb the unexpected problems of the future, and more generally to care for each other's well-being without knowing in advance precisely what that may involve. So it should be with committed employees, whose willingness to contribute according to the organization's needs is the basis of the latter's competitive flexibility.

But the analogy of personal relationships suggests the need for commitment to be reciprocal. Do not organizations have to show, by their personnel policies and practices, a commitment to the needs of employees which balances the commitment which it seeks from them? The less the organization can draw upon its external service commitments as a motivational source, and the more it needs employees' commitment to give it flexibility, the more it may have to exhibit commitment to the well-being of its own employees.

A notable discussion of the internal policies by which organizations can produce commitment is that of Amitai Etzioni (1975). Etzioni argued that 'utilitarian' (that is, commercial) organizations were generally best served by using 'remunerative power' in order to obtain a 'calculative' commitment from employees. His argument was that such organizations had rather precise and often quite complex production objectives to satisfy, and these usually involved the coordination of many people each of whom had to perform a specific task which fitted in with others. Payment systems were the best method of getting each person to do exactly what was required of her or him. However, Etzioni recognized that many employees had a 'moral' rather than a calculative involvement with their work, and he also recognized that when an organization needed an intense, open-ended kind of commitment, then remunerative power was less effective, and the organization had to rely instead on this moral commitment.[1] The kinds of organization in this latter position, he thought, were chiefly public-benefit

---

[1] In Etzioni's scheme, moral commitment is linked with the 'normative power' of the organization, and the latter is identified with status symbols, honours, and other extrinsic rewards of a non-financial type, rather than the intrinsic worth of the organization or its goals. Yet Etzioni offers no explanation as to how such apparently feeble rewards as status symbols can generate the intense, sustained effort of the morally committed individual. The approach in this chapter makes no use of the notion of 'normative power'.

or civic organizations: examples he gives are education, research, medicine, the arts, and religion.

Etzioni points out a fundamental problem for commercial organizations. Remunerative power only connects with a calculative response *of low intensity*: it generates just as much effort as is paid for, and no more.[2] As the organization seeks a response of higher intensity from its employees—commitment rather than compliance with a contract—then it needs to tap into moral commitment. Yet that may require an altogether different approach.

THE MEASUREMENT OF ORGANIZATIONAL COMMITMENT

The growing interest in organizational commitment has resulted since 1980 in the emergence of a specialist literature, mostly in the USA. Much of this work is concerned with the development of reliable and valid measures of commitment, and useful steps have been taken in showing that commitment is indeed separate from measures of job satisfaction or job involvement (Brooke *et al.* 1988), and that organizational commitment is different from work commitment or career commitment (Mueller *et al.* 1992).

The most frequently used measure has been provided by Mowday's Organizational Commitment Questionnaire or OCQ (Mowday *et al.* 1982). The OCQ has also formed the basis of our measure, but we have used items from it selectively and the measures we have derived are perhaps untypical. Several studies have shown that the OCQ contains distinct concepts, including a somewhat passive concept of attachment, or low capacity to leave the organization, rather than a more active concept related to high performance or flexibility (Meyer and Allen 1984; Mueller *et al.* 1992). The more active concept is clearly what was meant by Selznick's commitment to service values, by Etzioni when speaking of moral commitment, and by Walton with his strategy of high commitment. It is expressed by Kalleberg and Berg (1987) when they define organizational commitment as 'the degree to which an employee identifies with the goals and values of an organization and is willing to exert effort to help it to succeed'. Our selection of items from the OCQ attempted to follow this definition as far as possible, with the aim of identifying commitment in a strong sense.

The set of six questions on organizational commitment from the OCQ to be discussed in this chapter is as follows (their labels identify them in Table 9.1).

(*a*) 'I am willing to work harder than I have to in order to help this organization succeed.'
(*b*) 'I would take almost any job to keep working for this organization.'
(*c*) 'I would turn down another job with more pay in order to stay with this organization.'

---

[2] This point has been extensively demonstrated by psychologists: for a review of the evidence, see Lane (1991).

TABLE 9.1 *Indicators of organizational commitment (row %)*

	Strongly agree	Agree	Disagree	Strongly disagree	Don't know
(a) Work harder	20	60	15	2	4
(b) Take any job	5	20	50	22	4
(c) Turn down pay	6	22	39	26	7
(d) Similar values	8	55	25	5	7
(e)* Little loyalty	3	16	49	30	2
(f) Proud	14	61	18	4	4

* Item (e) is scored in reverse order, with strongly agree = lowest and strongly disagree = highest.

Notes: 'Don't know' includes 'not stated'; the questions were not presented to respondents in the order shown above. No. for all items = 3,458.

(d) 'I find that my values and the organization's values are very similar.'

(e) 'I feel very little loyalty to this organization.'

(f) 'I am proud to be working for this organization.'

Statistical analysis showed that these six questions were intercorrelated[3] and can be regarded as a coherent set. There are, however, some apparent differences of meaning among the questions. Item (a) in the table is the only one which refers specifically to *effort* as a means of showing commitment. Two items (b and c) are particularly interesting in the light of the current debates concerning *flexibility*. They express a willingness to be flexible to the point of some personal sacrifice, and can be regarded as particularly strong expressions of commitment. The remaining three items (d–f) are rather weaker expressions of identification with the organization and its *values*.

There are both advantages and disadvantages in treating all six items as a single scale of organizational commitment. The advantages are those of presentational simplicity and statistical reliability. The price to be paid for this, however, is a loss of information about possibly different facets of organizational commitment. At the present early stage of investigating this concept, there seems more to be gained by respecting apparent differences of meaning among the questions than by using a single composite measure. Accordingly the analysis which follows keeps separate effort commitment, which is represented by the single item (a) in Table 9.1; flexibility commitment, consisting of the sum of

[3] The questionnaire sequence on organizational commitment contained two additional items, 'The success of my organization depends a lot on how well I do my job' and 'People usually notice when I do my job well.' These items do not possess much face validity as aspects of organizational commitment, but seem rather to represent possible influences on organizational commitment. They were included chiefly for purposes of comparison with the American General Social Survey of 1991, in which all eight items were used. In principal components factor analysis, the six items of Table 9.1 loaded highly on a single factor accounting for 43% of the variation, with the other two questions remaining distinct.

items (*b*) and (*c*); and value commitment consisting of the sum of items (*d*), (*e*), and (*f*).[4]

Table 9.1 also reports the distributions of replies to these questions about organizational commitment. There are no previous British results with which these can be compared, but they at least provide some impression of the current state of commitment. Perhaps the most surprising, and for management worrying, results were those for 'flexibility commitment'. Only 25 per cent express willingness to be flexible over the job they do, and only 28 per cent would stick with the organization in the face of a better pay offer. Indeed, only about one in twenty *strongly* assert their willingness to do these things. So, on these measures at least, an intense organizational commitment seems to apply only to a minority.

The picture was, however, considerably more positive on the effort commitment question. Here 80 per cent express a positive attitude, and 20 per cent express it strongly. Willingness to work above the normal level therefore appears to be a weaker form of commitment than willingness to be highly flexible over job and pay.

On the questions about value commitment the majority expressed positive views, especially in asserting some loyalty to their organization. This confirms the view that the flexibility questions are a more severe test of commitment than the value questions. But only 8 per cent *strongly* agreed that their values and the organization's were very similar, while 30 per cent disagreed to some extent. On this simple evidence, there seems to be some unattractiveness in the organizational values communicated to many employees and there appears to be considerable scope for increases in organizational commitment.

CONTRASTS BETWEEN THE MARKET AND SOCIAL SECTOR

An important issue, in the light of the opening discussion, is how far underlying values actually differ between employees in different sectors of employment. Do some prefer to work for the public benefit, while others feel more comfortable in a commercial climate?

The first step in examining this question is to define the public-benefit sector, or henceforth (for the sake of brevity) the social sector. Education, health, and welfare services form its largest and most obvious constituents. It was decided to include, alongside these, all the other functions of central and local government. The argument against including them is that many of these occupations are purely administrative and are not directly concerned with the well-being of individuals or the community. On the other hand, people working in the public sector

---

[4] The two items for flexibility commitment had a reliability (Cronbach alpha) of 0.55. The three items for value commitment had a reliability of 0.74.

TABLE 9.2 *Perceptions of the social utility of organizations in the social and commercial sectors (column %)*

	Social sector	Commercial sector
1 (Not at all)	1	9
2	3	12
3	7	27
4	17	26
5 (Very much)	73	26
No.	1,173	2,154

Note: The table excludes self-employed respondents, and 131 people who did not reply to the question.

(whatever their personal jobs involve) may identify with the idea of their organization giving service to the community.

Included in the definition of the social sector were the small numbers who worked in the police and fire brigade, childcare workers, trade union officials, charity administrators, and clergy. Also included were creative people in the artistic and literary professions, because these too can be expected to have distinctive creeds which cannot be reduced to commercial terms.

On this composite definition, just over one third (35 per cent) of the sample —excluding self-employed people—worked in the social sector. A check on the validity of the definition is provided by answers to the question, 'How much does your organization do something useful for society?' The answers are shown in Table 9.2 and confirm a massive difference between the social sector and the commercial sector. In addition, people working in the social sector were substantially more satisfied with their 'chance to do something worthwhile in life', with 67 per cent describing themselves as very satisfied or satisfied, by comparison with the commercial sector where the corresponding proportion was 48 per cent.

There is, then, evidence enough to support the view that the social sector is different in terms of the values of many of those who enter it. In addition, employees in the social sector contrast with those in the commercial sector in terms of some important background characteristics. Higher qualifications are required for entry to medicine, nursing, teaching, and a number of other occupations in the social sector. Accordingly, much higher proportions in the social sector than in the commercial sector have attained higher educational levels. Another important difference is gender, with women playing a much larger part in the social sector. This again is as one would expect, given that teaching and nursing have traditionally constituted such important careers for women, and given also that women have for so long been the chief carers in family and community life.

INFLUENCES ON ORGANIZATIONAL COMMITMENT

The central focus of the chapter is on how organizations influence commitment. The analysis assesses two main ways in which commitment may be shaped by the organization: (*a*) through its personnel practices, and (*b*) through the culture of values which it communicates to its employees.

The distinction between these two forms of influence merits a brief discussion. The first group consists of the more formal or tangible steps taken by employers, such as systems of control, supervision and reward, channels for communication, and programmes of change. The second group consists of the subtler attributes of the organization, sometimes called climate or style as well as culture. It is hard to pin down how these attributes are communicated by the organization or how they come to be perceived by the employees; they may in part be inferred from the behaviour of managers and supervisors, or the priorities which they display over time. Of course, the formal practices installed by an organization may also help to convey its culture—for example, whether it has rigid controls, or extensive opportunities for participation, must affect employees' perceptions. But the assumption of the analysis is that the cultural values of an organization are more diffuse and cannot be wholly reduced to formal practices.

The plan of analysis is at two levels, corresponding to the preceding distinction. The first level concentrates on the policies and practices and ignores the perceived values or culture of the organization.[5] The aim here is to assess the total influence of each practice—for example, open communications—on organizational commitment. Practices were selected to represent themes that were particularly relevant to organizational commitment. The five main thematic groups were: (i) practices relating to job security or insecurity; (ii) practices demonstrating a paternalistic relation to employees; (iii) control systems; (iv) practices relating to individual development and upskilling; and (v) practices relating to communications and participation.

The second level of analysis retained the organizational practices, but added the perceived cultural values of the organization, which became the chief focus of attention. This expanded analysis measures the influences of both practices and perceived values net of each other. What is of particular interest here is whether the perceived values or culture of the organization help to explain employees' organizational commitment, even when the influences of the main personnel

---

[5] An accepted principle of path analysis or structural analysis is that intermediate variables can be omitted from the model without biasing the estimates of more basic or lower-level variables. In the present case, the organizational practices are at a more basic or lower level than the perceived organizational values or culture, because the former influence the latter whereas the latter do not influence the former. Under these assumptions, the model provides unbiased estimates of the total effects of the organizational practices on commitment.

practices have been controlled. Two thematic groups of organizational values are considered: (i) market values of efficiency, innovation, and reward; and (ii) social values of caring, service, and participation. These will be further explained at a later point.

The second level of analysis also brings various aspects of individuals' job satisfaction into the picture. If job satisfaction was not taken into account, the importance of organizational values or culture might be exaggerated. This is because individuals' views about the organization's values are likely to be affected by the rewards or inducements they have gained from the organization, rather than being a true or pure judgement of the organization as such. Measures of individual satisfactions or rewards are needed as controls against this bias.

One might also be interested in how far satisfactions or rewards themselves shape organizational commitment, but such an assessment cannot be achieved in a cross-sectional survey. Rewards or inducements could be (indeed, sometimes undoubtedly are) given to employees precisely because they have already demonstrated that they are committed, and are therefore valued more by their employer. So one cannot be sure how much the analysis says about the influence of satisfaction on commitment rather than the influence of commitment on satisfaction. The important point is that, by taking account of individual satisfaction in the analysis, one makes sure that the importance of organizational values for commitment is not overstated.

All the analyses reported in the chapter applied the same set of standard controls: gender, social class, length of service with the present employer, and size of establishment.

Change, Insecurity, and Organizational Commitment

The climate of change and insecurity, so prevalent in the 1990s, led to profound doubts about the objective of building a committed workforce. Accordingly, one of the main themes of the analysis was to examine how sensitive commitment was to the experience of change and insecurity.

The first step was to identify which organizational practices had the most adverse impact on satisfaction with job security. This was already considered in Chapter 5, and those earlier results were used to guide the selection of factors included here. The most important of these were: whether there had been cuts in staffing, affecting the respondent's own work area; whether the overall size of the organization's workforce had been reduced; whether the individual was employed on a temporary or fixed-term, as opposed to a permanent, contract; and whether he or she felt anxious about the possibility of unreasonable dismissal—which was presumed to reflect the prevalence of arbitrary hire-and-fire practices. All of these were powerful influences on satisfaction (or dissatisfaction) with job security, even when the general characteristics of the individual (such as social class) and of the organization (notably size) had been controlled.

The most general influence on insecurity, from within the organization, could be the contraction of its total workforce, reflecting policies of retrenchment or rationalization. In fact, 42 per cent of the sample reported that numbers employed by the organization had fallen during the past five years, so this was a common experience. There were several indications that commitment was undermined where this had taken place, and these are summarized in Table 9.3. In the social sector, decreasing employment was linked to both lower levels of value commitment and lower levels of flexibility commitment, although effort commitment was unaffected. Both these influences were highly significant. In the commercial sector, a reduction in the organization's numbers only had a negative impact on value commitment, and this was at a lower level of significance than the corresponding result in the social sector. Commitment in the social sector, therefore, seemed more sensitive to declining employment. It should be borne in mind that a reduction in numbers employed may have a dual meaning to social sector employees: not only a reduction in security, but a limitation on the service which can be provided to the community.

If a fall in overall numbers reduces commitment, one might expect a still stronger effect on the individual when a reduction of the workforce directly affects the area of work in which she or he is located. Some 41 per cent of respondents in the social sector, and 51 per cent in the commercial sector, stated that reductions had taken place within the past two years affecting their own area of work. These proportions underline the degree of retrenchment being experienced. Remarkably, however, there was *no evidence* of staff reductions at the individual's job level reducing organizational commitment, on any of the three measures of commitment considered, or in either the social sector or the commercial sector. So it seems to be the general state of the organization, rather than what is happening in the individual's immediate section of work, which makes an impact on organizational commitment.

Other local changes, apart from staff cuts, were considered in the analysis: the introduction of computers and automation, the introduction of other new equipment, and changes in work organization. As with the question on staff cuts, these questions focused on changes over the past two years which affected the individual's own job. Although none of these changes was directly related to feelings of job insecurity, they might in other ways create an unsettled atmosphere which weakened commitment to the organization. However, individuals experiencing work reorganization within the previous two years had neither lower nor higher organizational commitment than others. The introduction of computers and automation did have some negative impact, but only in the social sector, and only on value commitment (not on flexibility commitment or effort commitment). Conversely, the introduction of other types of equipment (by implication, involving more traditional technology) tended to *increase* value commitment in the social sector, while leaving other aspects of commitment unaffected.

TABLE 9.3 *Effects of job security on organizational commitment*

(a) Social sector employees

	Flexibility commitment		Value commitment		Effort commitment	
	Coeff.	Sig.	Coeff.	Sig.	Coeff.	Sig.
Change in size:						
increase	−0.15	n.s.	−0.19	n.s.	0.03	n.s.
decrease	−0.42	***	−0.44	**	−0.09	n.s.
Changes affecting work:						
staff cuts	0.00	n.s.	−0.04	n.s.	−0.12	n.s.
work organization	−0.11	n.s.	−0.12	n.s.	0.01	n.s.
computers/automation	−0.08	n.s.	−0.33	**	−0.03	n.s.
other new equipment	0.15	n.s.	0.27	*	−0.01	n.s.
Anxious of dismissal	0.28	*	−0.30	n.s.	0.25	*
Contract						
temporary	−0.24	n.s.	−0.21	n.s.	−0.19	n.s.
fixed 1–3 years	−0.01	n.s.	0.22	n.s.	0.03	n.s.

(b) Market sector employees

	Flexibility commitment		Value commitment		Effort commitment	
	Coeff.	Sig.	Coeff.	Sig.	Coeff.	Sig.
Change in size:						
increase	0.01	n.s.	−0.01	n.s.	0.10	n.s.
decrease	−0.10	n.s.	−0.24	*	−0.01	n.s.
Changes affecting work:						
staff cuts	0.03	n.s.	0.02	n.s.	0.07	n.s.
work organization	0.02	n.s.	0.02	n.s.	0.05	n.s.
computers/automation	−0.06	n.s.	−0.02	n.s.	0.02	n.s.
other new equipment	0.02	n.s.	−0.03	n.s.	−0.03	n.s.
Anxious of dismissal	−0.11	n.s.	−0.14	n.s.	−0.02	n.s.
Contract						
temporary	−0.16	n.s.	−0.10	n.s.	−0.09	n.s.
fixed 1–3 years	−0.14	n.s.	−0.44	**	−0.29	*

Note: OLS regression analysis for flexibility commitment and value commitment; ordered probit analysis for effort commitment. No. = 985 for social sector, no. = 1,885 for market sector. Only part of analysis is shown. For social sector employees, $R^2 = 0.14$ for flexibility commitment, 0.19 for value commitment; for market sector, $R^2 = 0.12$ for flexibility commitment, 0.24 for value commitment.

On the whole, then, there were only weak indications of individuals' commitment being adversely affected by changes affecting their work group or workplace. A more personal threat may be posed by arbitrary hire-and-fire policies, and where these prevail one might expect commitment to be weak. A question which should pick up the full range of threats or risks is worded 'In your job, do you feel any anxiety about the possibility of being dismissed without good reason?' About one in six of respondents expressed anxiety on this score. It was particularly unexpected to find that those saying they were anxious about dismissal did *not* have below-average organizational commitment. In the commercial sector, there was no link between this aspect of insecurity and any measure of organizational commitment. In the social sector, those feeling insecure in this way actually had *above-average* flexibility commitment and effort commitment. Perhaps this paradoxical result (shown in Table 9.3) might be interpreted through selective attention: those who are insecure are more aware of the value of their jobs and so feel more committed.

The final aspect of insecurity considered was the individual's *employment contract*—that is, whether he or she was a temporary worker, a contract worker, or a permanent worker. After excluding self-employed people, 6 per cent of the sample were on short-term temporary contracts and 5 per cent on contracts of one to three years; the earlier analysis (Chapter 5) confirmed that they had particularly low levels of satisfaction with job security. One would reasonably assume that for this reason, among others, a temporary or contract worker would be less committed than a permanent worker.

The results here were particularly surprising. Having a short-term temporary job, the most insecure position of all, had no independent effect on any aspect of organizational commitment in either the social sector or the commercial sector. Being on a fixed-term contract of one to three years also had no effect on commitment in the social sector, but did adversely influence both value commitment and effort commitment in the commercial sector. The lack of any adverse effect of fixed-term contracts in the social sector is understandable if many of these contracts are training or probationary periods leading towards more permanent professional posts; this is likely to be much less the case in the commercial sector. But while the results for fixed-term contracts were not unreasonable, those for the very precarious short-term temporary posts were so unexpected as to need further investigation.

In a variant analysis, therefore, all the other variables about personnel practices were omitted (but controls for individual and organizational background characteristics were retained). Looked at in this simpler way, people on short-term temporary contracts did emerge as having lower organizational commitment. So the commonsense observation that short-term staff are on average less committed is correct, even though the assumption that this is *because of the employment contract* is incorrect. The explanation which reconciles the results of the two analyses is that those on short-term contracts have less access to many of the practices and policies which are applied to longer-term employees. When these practices and policies are equalized, those on short-term contracts can be as committed as permanent staff. It may, however, be difficult in practice to apply some kinds of policies, such as training for enhanced skills, to employees who will be retained only for short periods. This, however, is a quite different disadvantage of temporary employment from insecurity as such.

As a whole, the specific practices which influence insecurity have surprisingly limited impacts on organizational commitment. Yet these specific practices are not the whole story concerning insecurity and commitment. In the second stage of analysis, information about the individual's *satisfaction with security* was introduced, and its influence on commitment assessed while controlling for specific organizational practices. Such an analysis evaluates the importance, for commitment, of those aspects of security which are *not* measured through the specific workplace practices. The most important of these are likely to be first, the particular situation of the individual in his or her job, which might be secure or insecure relative to that of other workers, and second, the effects of wider economic or industry conditions in generating a pervasive sense of security or insecurity.

The results of this further analysis were sharply contrasting between the social sector and the commercial sector. In the social sector, satisfaction or dissatisfaction with job security had *no influence on organizational commitment*, on any of the three commitment measures used. Indeed, the estimated effects were very close to zero in each case. In the commercial sector, however, dissatisfaction with security was linked to a significant reduction both in flexibility commitment and in value commitment.[6] In relation to 'effort commitment', however, its influence remained near zero even in the commercial sector.

The difference between the sectors is plausible. One might expect that workers in the commercial sector on average would be more exposed to insecurity of a personal or job-specific nature, while workers in the social sector might be more affected by collective changes. Again, the commercial sector would perhaps be more exposed to the uncertainties of the general economic climate and the market place.

In sum there was some influence of insecurity on organizational commitment, but far less than expected. The strongest effects were of overall dissatisfaction with security, but these applied to the commercial sector only, not the social sector. Also, as explained earlier, it is probable that this type of analysis, explaining commitment in terms of satisfaction, will overestimate the influence by failing to take account of reverse causation. In other words, committed employees may be highly valued by employers and so be less exposed to insecurity.

The more reliable estimates are those relating corporate and workplace practices and policies to organizational commitment. Internal practices and policies are also, naturally, much more within the control of employers than are the external economic or business conditions which may affect overall feelings of

---

[6] Satisfaction with security had a coefficient of 0.07, significant at the 99% confidence level in the extended model of flexibility commitment. It had a coefficient of 0.09, significant at the 99.9% confidence level in the extended model of value commitment.

insecurity. What made a marked difference to commitment was total workforce reductions: other policies and practices had few impacts. On the whole, then, employees seemed remarkably capable of absorbing change and uncertainty without detriment to their organizational commitment.

Yet even if insecurity and change have become accepted as inescapable realities, employees may remain sensitive to the organization's treatment of employees who are adversely affected. A later section of this chapter considers the importance attached by employees towards the caringness of the employer. The results there give a different perspective on the issue.

Paternalism, Welfare, and Working Conditions

Paternalistic management, in some respects a forerunner of the 'human relations' movement and of current human resource management ideas, emphasized the provision of a range of benefits which protected the employee from various risks, while exacting an unquestioning loyalty in return. With a history stretching back to the crusading work of Robert Owen (Polanyi 1957), paternalism was developed to a remarkable degree by Quaker capitalists such as the Cadbury family, and reached its zenith as a widespread movement among major British employers in the interwar years (Fitzgerald 1988). Although the idea of paternalism has become discredited, many of the protective policies remain widely installed in industry. These aspects provide a counterpoint to the analysis of insecurity of the previous section. Could it be that paternalist or welfarist policies have a renewed effectiveness in building commitment in the riskier environment of recent years?

The survey included detailed questions about *fringe benefits*, first establishing whether each of fifteen types of benefit was present in the employee's organization (to her or his knowledge), and then checking whether or not the available benefits influenced the individual's joining or leaving decision. For the analysis of commitment, this information was condensed to a count of the benefits available. Also included in the analysis were measures of whether any benefit had influenced the individual's joining decision, and whether any benefit was influencing retention. Thus, for each sector, there was a total of nine estimates of the effects of benefits on commitment: three benefits measures, by each of the three measures of commitment.

In the social sector, only one of the nine possible relationships was significant: the number of benefits available was associated with increased levels of value commitment (see Table 9.4). In the commercial sector, the effects of benefits were more extensive, but they were negatively as well as positively related to commitment. Where an individual stated that benefits were an important reason for staying with the firm, this did have a positive influence on flexibility commitment, which from earlier results seems a particularly strict measure of commitment. In the great majority of cases, the benefit which constituted the reason for remaining with the firm was an *occupational pension*. It therefore seems that attractive pensions can be an effective way of increasing flexibility commitment in the commercial sector. The lack of such an effect in the social sector is probably because pension provision there is much more uniform and is to some extent taken for granted. The other significant results in the commercial sector were that benefits influencing joining decisions were linked to higher value commitment, while the number of benefits provided were linked to *lower* effort commitment.

Another provision regarded as characteristic of the paternalistic firm, and of the public sector generally, is the *incremental pay scale*. By rewarding length of service overtly, one might assume that incremental pay scales would help to increase commitment. However, this turned out not to be case, either in the social sector, where 72 per cent reported being on incremental scales, or in the commercial sector, where the proportion was 30 per cent. There was, however, a significantly *negative* effect on effort commitment in the commercial sector, with those on incremental scales less willing to accept abnormal workloads. This last result can be placed alongside the similarly negative effect, in the commercial sector, concerning the number of fringe benefits (see above). It seems that commercial employers whose pay and benefits systems cocoon their employees from financial risk, obtain relatively low effort commitment. This could be either because the pay and benefit systems themselves remove incentives, or because of incidental characteristics of the firms which can afford such provisions.

Recent years have witnessed a rapid increase in corporate *profit-sharing and share-option schemes* for employees, and their effects on motivation have been fiercely debated (Estrin *et al.* 1987; Florkowski 1994). Few of these schemes are closely dependent on business results, and so they can be regarded as closer to welfare benefits than to pay incentives. Some 23 per cent of employees in the commercial sector could participate in a profit-sharing or share-option scheme, while the proportion was only 3 per cent in the social sector. In the commercial sector, the existence of such a scheme made no difference to any of the three kinds of commitment. Where it did make a difference was in the social sector: the very small minority with access to such a scheme had considerably above-average flexibility commitment, and also some tendency towards above-average effort commitment. But the numbers involved were so small—and the circumstances presumably so untypical—that there was little practical significance in this finding.

The final aspect considered under this heading was *working conditions*. Although there was quite a range of questions on working conditions which might have been considered in the analysis, it was decided to focus on one particularly strongly worded item: 'Do you think your health and safety is at risk because of your work?' The same proportion (29 per cent) felt that they were at risk in both the social and commercial sectors. This aspect produced a rather clear result, with dangerous working conditions having no effect on organizational commitment

TABLE 9.4 *Effects of paternalistic policies on organizational commitment*

(a) Social sector employees

	Flexibility commitment		Value commitment		Effort commitment	
	Coeff.	Sig.	Coeff.	Sig.	Coeff.	Sig.
No. of benefits	0.02	n.s.	0.05	*	0.01	n.s.
Benefits led to joining	−0.13	n.s.	−0.00	n.s.	−0.01	n.s.
Benefits prevent leaving	0.13	n.s.	0.17	n.s.	0.11	n.s.
Profit-sharing	0.76	**	0.11	n.s.	0.46	n.s.
Incremental pay	0.07	n.s.	0.23	n.s.	0.11	n.s.
Unsafe/unhealthy working conditions	−0.03	n.s.	−0.14	n.s.	0.05	n.s.

(b) Market sector employees

	Flexibility commitment		Value commitment		Effort commitment	
	Coeff.	Sig.	Coeff.	Sig.	Coeff.	Sig.
No. of benefits	0.01	n.s.	−0.01	n.s.	−0.03	**
Benefits led to joining	0.07	n.s.	0.24	*	0.09	n.s.
Benefits prevent leaving	0.25	**	0.06	n.s.	−0.02	n.s.
Profit-sharing	−0.01	n.s.	−0.07	n.s.	−0.01	n.s.
Incremental pay	−0.07	n.s.	−0.06	n.s.	−0.12	*
Unsafe/unhealthy working conditions	−0.23	***	−0.56	***	−0.15	*

Note: OLS regression analysis for flexibility commitment and value commitment; ordered probit analysis for effort commitment. No. = 985 for social sector, no. = 1,885 for market sector. Only part of analysis is shown. For social sector employees, $R^2 = 0.14$ for flexibility commitment, 0.19 for value commitment; for market sector employees, $R^2 = 0.12$ for flexibility commitment, 0.24 for value commitment.

in the social sector, but having a substantial negative effect on all three aspects of organizational commitment in the commercial sector.

Overall, the results (summarized in Table 9.4) indicate that paternalistic policies are of doubtful value from the viewpoint of developing organizational commitment. The positive impacts are somewhat limited, and may be offset, in the commercial sector, by some negative impacts. The clear exception to this is safe working conditions (or the elimination of dangerous working conditions), which appear to have important implications for commitment in the commercial sector. The results on corporate welfare are consistent with those concerning the effects (or lack of effects) of insecurity. Insecurity does less to undermine organizational commitment than might have been expected; and policies to reduce or protect against insecurity, through welfare benefits, do less to increase organizational commitment than might have been hoped.

Control Systems and Human Resource Practices

In exploring the effects of insecurity or of paternalistic security on commitment, the positive findings have been rather meagre. However, there are other ideas to be considered. Two main themes which recur in the literature are, first, the importance of practices which develop the individual and increase self-realization in the work (Allen and Meyer 1990; Randall and Cote 1991; Brown *et al*. 1993); and second, those which foster mutuality through consultation, participation, or shared responsibility (Walton 1985*b*; Kochan and Osterman 1994). Our analysis of personnel practices was guided by these themes.

Additionally, the analysis examines the influence on commitment of formal systems of control, which apportion rewards to individuals for the contributions they are deemed to make. Some HRM theorists regard these control-and-reward systems as problematic for a high commitment strategy (Beer *et al*. 1984; Walton 1985*a*; Kohn 1993; see review in Wood 1996). However, in Chapter 3 we showed that the assumed opposition between control and task discretion did not operate in practice; whether there is an opposition between control and commitment is an issue which we now test.

The two control systems represented in the analysis have played a part in several of the previous chapters (see especially Chapter 3 for details). Technical control consists of a combination of work-pacing, short repetitive tasks, and individual incentive pay. Performance management (also referred to as bureaucratic control) is based on individual appraisal, target-setting, merit pay, and internal career structures.

Chapter 3 showed that the two systems have very different implications for organizational conflict. Do these differences carry over into their effects on organizational commitment? The answer is that they do, but much more so in the commercial sector. For the social sector, it makes rather little difference to organizational commitment whether a technical control system, a performance-management system, or neither, is applied. The sole exception is a fairly weak effect of performance-management systems in increasing value commitment.

In the commercial sector, however, the effects of performance-management systems on all aspects of commitment are positive and highly significant (Table 9.5). Technical control systems also—in the commercial sector only—have a significant effect, but it is a *negative* one: value commitment declines with increasingly intensive technical control.

The lack of effects in the social sector may reflect a relatively recent, and as yet less far-reaching, adoption of performance management by comparison with the commercial sector. One simple indicator of this is the scale score for the two sectors, which is 2.65 in the case of the commercial sector but 2.43 in the case of the social sector (a statistically significant difference). Thus, the lack of impact in the social sector may change with time. However, it is also possible that, because of the different values which lead many people to enter the

TABLE 9.5 *Effects of control systems and human resource practices on organizational commitment*

(a) Social sector employees

	Flexibility commitment		Value commitment		Effort commitment	
	Coeff.	Sig.	Coeff.	Sig.	Coeff.	Sig.
A. Control						
Performance management	0.02	n.s.	0.06	*	0.04	n.s.
Technical control	−0.07	n.s.	−0.04	n.s.	−0.01	n.s.
B. Supervision						
Supervisor coaches	0.33	***	0.55	***	0.30	***
Supervisory norms	0.06	**	0.11	***	0.06	***
C. Participation						
Open communications	0.28	**	0.41	***	0.25	**
Work-improvement group	0.19	*	0.21	n.s.	0.02	n.s.
Decisions affecting work	0.07	n.s.	0.35	***	0.15	n.s.

(b) Commercial sector employees

	Flexibility commitment		Value commitment		Effort commitment	
	Coeff.	Sig.	Coeff.	Sig.	Coeff.	Sig.
A. Control						
Performance management	0.05	**	0.11	***	0.05	**
Technical control	−0.02	n.s.	−0.07	*	0.02	n.s.
B. Supervision						
Supervisor coaches	0.32	***	0.44	***	0.24	***
Supervisory norms	0.02	n.s.	0.05	**	0.04	***
C. Participation						
Open communications	0.20	**	0.42	***	0.19	**
Work-improvement group	0.04	n.s.	0.20	*	0.16	*
Decisions affecting work	0.25	***	0.71	***	0.43	***

Note: OLS regression analysis for flexibility commitment and value commitment; ordered probit analysis for effort commitment. No. = 985 for social sector, no. = 1,885 for market sector. Only part of analysis is shown. For social sector employees, $R^2 = 0.14$ for flexibility commitment, 0.19 for value commitment; for commercial sector employees, $R^2 = 0.12$ for flexibility commitment, 0.24 for value commitment.

social sector, performance-management systems (in common with financial rewards in general) tend to be less effective with them.

Performance management is based largely on appraisal and target-setting processes which take place between managers and their subordinates. More generally, current management thinking tends to place emphasis upon the role of the line manager in creating commitment and high performance, especially by encouraging individual development. Accordingly, in considering developmental practices, our focus is upon the supervisor.

An assessment of the personal role of the supervisor in fostering commitment poses some difficulties: one needs to watch out for influences running the other way. For example, supervisors may keep a close check on employees with weak commitment while lightly supervising those who are highly committed. In that case, the degree of commitment is influencing the nature of supervision rather than vice versa. For this reason, questions which concern the direct relationship between the individual employee and his or her supervisor are best avoided in an analysis of commitment.

To assess the supervisor's role in coaching and developing employees, a question was posed in a way which avoids the trap just described. The question asked: 'How true is it that your supervisor helps employees to learn to do their jobs better?' Another suitable group of questions concerned the importance which the supervisor attached to employees performing in various ways (through time-keeping, attendance, effort, and quality). Summing the replies to this group of questions, one obtains a scale which was labelled supervisory norms (see Chapter 3 for details).

These two measures both proved to be strongly and positively related to organizational commitment (Table 9.5). Strong supervisory norms increased effort commitment and value commitment in both sectors, and increased flexibility commitment in the social sector although not in the commercial sector. Furthermore, supervisory help with learning significantly increased all three types of commitment in both sectors. These findings support the view that supervision, or line management, plays a crucial role in developing organizational commitment. They also suggest that processes of employee development delivered through line managers are important for commitment.

The importance of the supervisor's development role leads on to the issue of upskilling, which has been so prominent in previous chapters. Upskilling is not so much a corporate policy as the result of the whole set of policies to do with development. It therefore seemed too broad a measure to be included with the 'personnel policy' analysis, but it was brought into the more complete analysis which will be described in the next section. In that analysis, it was subjected to a much more extensive set of statistical controls, and its effects on commitment were therefore tested in a particularly stringent way. Upskilling was represented by a five-point scale.

The results of this aspect of the analysis are summarized in Table 9.6. The main effect of upskilling was to increase effort commitment, which it did significantly, and to a similar degree, in both the social sector and the commercial sector. It also increased value commitment, but only in the social sector, where its impact was large. Upskilling had no effect on flexibility commitment in either the social sector or the commercial sector.

Overall, the analysis of upskilling reinforces the results concerning the personal role of the supervisor. Together the results indicate a positive impact of development opportunities on employees' organizational commitment.

TABLE 9.6 *Effects of upskilling on organizational commitment*

	Flexibility commitment		Value commitment		Effort commitment	
	Coeff.	Sig.	Coeff.	Sig.	Coeff.	Sig.
(a) Social sector	0.005	n.s.	0.10	**	0.07	*
(b) Commercial sector	−0.007	n.s.	0.03	n.s.	0.06	**

Note: Results are from analyses shown also in Table 9.8; see these for details.

The final aspect of employer practices and policies to be assessed concerns emphasis on open communications and on individual participation. Three items on this theme were included in the analyses:

(a) 'Does management hold meetings in which you can express your views about what is happening in the organization?'

(b) 'Are you involved in a quality circle or similar group?'

(c) 'Suppose there was going to be some decision made at your place of work that changed the way you do your job. Do you think that you personally would have any say in the decision about the change or not?'

Item (c) is open to some criticism, as a policy influence on commitment, because of its rather personal nature. Perhaps the more committed employees would also be more trusted by management, and would therefore expect to be consulted to a greater degree. This potential weakness must be borne in mind in the following discussion. Items (a) and (b) seem better, as they refer to general aspects of organization which go wider than the individual.

It turned out that the importance of communications and participation did not depend excessively on item (c). In fact, the results showed that all three aspects of communication and participation were strongly connected with organizational commitment. The most consistent effects were those of open communication (item (a) above). In both the social sector and the commercial sector, those who could express their views at meetings had substantially higher levels of all three types of commitment (see Table 9.5). This set of findings is particularly interesting, from a practical viewpoint, because the majority of employees felt that they did enjoy open communications: 74 per cent in the social sector, and 60 per cent in the commercial sector. Here, it seems, there is a practice which is in widespread use and also enhances commitment.

Work-improvement groups or quality circles (item (b) above), which applied to only 25 per cent in the social sector and to 19 per cent in the commercial sector, had less consistent but still appreciable effects on commitment. In the commercial sector, participation in such groups was connected with significantly increased effort commitment and value commitment, although not to flexibility commitment. Conversely, in the social sector, there was only one significant effect of work-improvement groups, and this was on increasing flexibility commitment.

Turning finally to participation in changes (item (c) above), one finds a marked difference in the results for the social sector and the commercial sector. In the social sector, this kind of individual participation affected only value commitment. In the commercial sector, however, those who participated by giving their views about work changes had much above-average commitment on all three measures. It might be thought that the difference between sectors results from consultation and participation being more a matter of custom in the social sector (especially in view of high levels of public-sector unionization), and more a matter of individual trust in the commercial sector. But this is not borne out by the proportions reporting that they participated to some degree: at 56 per cent in the social sector and 48 per cent in the commercial. If managers consulted over changes only with selected trusted individuals, one would expect the latter proportion in the commercial sector to be lower. On balance, despite some reservations about potential bias in the question, one can reasonably conclude that participation in decisions has a positive impact on commitment, especially in the commercial sector.

Overall, these results provide substantial evidence that communications and participation are important for commitment. This applies to both sectors, but the value of the practices is particularly pervasive for the commercial sector.

Overview of the Effects of Workplace Policies

The first main conclusion from this section of the analysis was that insecurity, *in terms of the changes or policies* which produce it, did relatively little to reduce organizational commitment. Most remarkably, short-term temporary employment contracts did not, *in themselves*, reduce the level of organizational commitment (although longer fixed-term contracts had a negative effect in the commercial sector). The findings do not mean, however, that security is unimportant. Indeed, other evidence from the survey shows that employees regard security as very important (see Chapter 8). And in the present analysis, there were adverse effects on commitment—albeit in the commercial sector only—from *overall dissatisfaction with security*. The key point here is that the dissatisfaction with security, as measured here, reflected either purely individual job problems, or reactions to uncertainty in the external economic climate rather than systematic personnel policies which the organization could choose and control.

In view of the foregoing results concerning insecurity, it was not surprising to find that policies of the protective, paternalistic kind had little positive impact on commitment. An important exception, however, and one which tends to have been rather neglected, was the safe working conditions of employees. This was particularly important to employees in the commercial sector.

The policies and practices which had a positive impact on organizational commitment were, on the whole, those which concerned personal development and participation. The continuous development of employees, both through performance-management systems, and through supervisory support for learning, was an important influence. Line management evidently played a major role, both through the development processes already mentioned, and through the transmission of norms about performance. Finally, employee participation, and especially open two-way communication, seemed particularly effective in building organizational commitment.

An important limitation on these findings, however, was that the impact of workplace policies and practices on commitment was in part different between the commercial sector and the social sector. Safe working conditions, performance-management systems, and quality circles produced positive effects to a greater extent in the commercial sector. Participation in change decisions had some effect in the social sector, but a much stronger one in the commercial sector. The converse applied to supervisory norms. The only policies with consistently positive influences across both sectors were supervisory support for learning, and open two-way communications.

Organizational Values and Commitment

The next and final section of the analysis considers links between organizational commitment on one hand, and broad organizational values or 'culture' on the other. Organizations—as it was argued by Selznick—generate commitment to the extent that they embrace and transmit values with which the individual employee identifies. An analysis which directly assesses the relations between organizational culture and commitment seeks to test this notion.

The questions posed by the survey did not explicitly refer to values or culture, but rather to how the organization was seen to behave towards its tasks and its employees. Values or culture, it is assumed here, shape choices and actions and so are behavioural in effect. Employees infer values from the consistency of choices and actions rather than from what is said in organizations' mission statements or policy documents. If therefore anyone prefers to substitute a phrase such as behavioural patterns for values or culture, in the following account of analyses and findings, they will find it possible to do so.

The series of questions was introduced as follows in the interview:

'The next series of questions is about how you see the organization where you work. We would like you to think about the *whole* organization, not just the part where you work. You may not have precise information about some of these questions, but we would still like to have your personal impressions.'

Each of the questions following this introduction asked the respondent to rate the organization on a five-point scale with the end-points defined, for example ranging from 'very much' to 'not at all', or from 'very true' to 'not at all true'.

The analysis of organizational commitment used six items from the series that was introduced in this way. These consisted of two groups of three items each,

TABLE 9.7 *Questions on organizational values*

	Mean scores (with standard deviation in brackets)	
	Social sector	Commercial sector
(a) Market values		
How efficiently is work carried out	3.59	3.70
in this organization?	(0.93)	(0.91)
How keen is management on new ideas	3.91	3.93
and improvements in this organization?	(0.99)	(1.03)
How well does this organization	3.10	3.18
pay, in general?	(1.02)	(1.08)
(b) Social values		
How much influence do employees have	2.64	2.45
over decisions in this organization?	(1.09)	(1.08)
How much does this organization care	3.47	3.39
about employees' well-being?	(1.10)	(1.20)
How much is this an organization that	4.61	3.56
does something useful for society?	(0.92)	(1.40)

Note: Higher scores represent more positive replies.

labelled the market and the social values of the organization. The selection of these items was a priori, based on the meaning of the items, and will be tested by the following analysis. The actual wording of the questions is shown in Table 9.7 along with the average scores recorded by employees in the two sectors.

The first group of three items is intended to reflect the kind of values which are frequently stressed in debates on competitiveness and orientation to the market place. Superiority in terms of efficiency and innovation is very commonly claimed as one of the main advantages of competitive market societies. Accordingly firms are exhorted to be competitive through efficiency and innovation, and these attributes are increasingly also urged upon public sector services. A higher standard of living for employees and their families is similarly offered as recompense for the increased effort and adaptability which a more competitive society demands.

After labelling the first group of items as market values it is natural to label the contrasting second group as social values. As with all labelling, this is to some degree arbitrary. It is certainly the case that many commercial organizations have wanted to stress the importance of some or all of these values to achieving competitive success. Nonetheless, there is a contrast between these values and the more materialistic views of market society. These social values are also similar to the values present in the 'support culture' described by Harrison (1972, 1987) in his typology of organizations, whereas the market values are more akin to the 'achievement culture' which he describes (see also Pheysey 1993).

Organizational Culture and Commitment: The Findings

A set of six analyses was carried out, just as in the previous part of the chapter concerning organizational policies, seeking to explain each of the three measures of organizational commitment first within the social sector and then within the commercial sector. All the control variables used in the previous block of analyses were retained. As explained earlier in the chapter, controls were also introduced for individual satisfaction or individual rewards, which might bias individuals' judgements of the organization. All six measures of organizational values or culture were included together in each analysis, so that the relative strength of their effects on organizational commitment could be gauged.

Market Values

The results concerning the three measures of market values are summarized in the upper panel of Table 9.8. They are rather simple. Only one value from this set is an important influence on organizational commitment, and that is *efficiency*. The perception that work was carried out efficiently throughout the organization was linked to a higher level of all three aspects of commitment, in the commercial sector. Perceived efficiency was also linked to both value commitment and to effort commitment in the social sector, although not to flexibility commitment.

There was no evidence that *innovation* (defined in terms of 'new ideas and improvements') was important either in the social or the commercial sector. However, the analysis controlled for a range of policies or practices relating to innovation or productivity (notably, the use of quality circles and the introduction of new technology). It was shown in the preceding set of analyses that the existence of quality circles did, indeed, influence commitment in the commercial sector although not in the social sector. So specific innovative practices do have some impact on commitment; what seems to be lacking, however, is a wider culture of innovation.

The most surprising of the results here concerned pay. It might have been expected that employees in the commercial sector, in particular, would be influenced by whether or not the organization was perceived to pay well. However, the analysis controlled for the individual's satisfaction with pay. When pay satisfaction is removed from the analysis, views about the organization's pay levels do appear to be positively related to commitment, in the social sector as well as the commercial sector. It is when individual pay satisfaction is introduced that these apparent influences disappear. So in this respect, perceptions of the organization are dominated by what the individual receives.

TABLE 9.8 *Effects of organizational values on organizational commitment*

(a) Social sector employees

	Flexibility commitment		Value commitment		Effort commitment	
	Coeff.	Sig.	Coeff.	Sig.	Coeff.	Sig.
A. Market values						
Organization efficient	−0.00	n.s.	0.15	**	0.10	*
Management keen on new ideas	0.01	n.s.	0.04	n.s.	0.04	n.s.
Organization pays well	0.02	n.s.	0.02	n.s.	0.06	n.s.
B. Social values						
Employees influence decisions	0.12	**	0.10	*	0.09	*
Organization cares about employees	0.16	***	0.33	***	0.10	*
Organization useful for society	0.05	n.s.	0.15	**	0.17	***

(b) Commercial sector employees

	Flexibility commitment		Value commitment		Effort commitment	
	Coeff.	Sig.	Coeff.	Sig.	Coeff.	Sig.
A. Market values						
Organization efficient	0.08	*	0.16	***	0.09	**
Management keen on new ideas	−0.01	n.s.	0.01	n.s.	−0.00	n.s.
Organization pays well	−0.05	n.s.	−0.02	n.s.	−0.02	n.s.
B. Social values						
Employees influence decisions	0.11	***	0.16	***	0.14	***
Organization cares about employees	0.20	***	0.36	***	0.15	***
Organization useful for society	0.05	*	0.06	**	0.06	**

Note: OLS regression analysis for flexibility commitment and value commitment; ordered probit analysis for effort commitment. No. = 935 for social sector, no. = 1,885 for commercial sector. Only part of analysis is shown. For social sector, R^2 for flexibility commitment = 0.21, for value commitment = 0.38; for commercial sector, R^2 for flexibility commitment = 0.27, for value commitment = 0.48.

The results concerning social values in the organization are summarized in the second panel of Table 9.8. In contrast to market values, the social values had an extensive and highly significant influence on organizational commitment. Looking first at the commercial sector, one finds that all aspects of commitment were responsive to all aspects of social values. Since each of the value measures was scored in the same way, it is reasonable to compare the size of the effects across them, although such comparisons are not precise. Commercial organizations seen as to a high degree *caring for the well-being of employees* tended to have particularly committed employees, on all three measures of commitment, but especially value commitment. Commercial organizations with a *participative culture* on average scored well above average on all three measures of commitment. If they were seen as highly *useful for society*, commercial organizations again had higher commitment scores than average, but the effects were somewhat smaller in magnitude than for the other two organizational value measures.

In the social sector, the strongest influence on commitment again appeared to be a high level of caringness towards employees. Having a participative culture was rather consistently linked with all aspects of commitment. Being seen as useful for society was, in the social sector, strongly linked both with effort commitment and with value commitment—the magnitude of the effects, here, was noticeably higher than in the commercial sector, as one would expect.

Strangely, however, usefulness to society was *not* significantly related to flexibility commitment in the social sector. The explanation is probably that this aspect of commitment was particularly associated, in the social sector, with satisfaction with work. In a variant analysis with the satisfaction measures omitted, the organization's perceived usefulness to society *was* significantly related to flexibility commitment.

Overall, there can be no question that social values in organizations are very important to all aspects of organizational commitment. Further, these values are at least as important for employees' commitment in the commercial sector as in the social sector.

Three further points of interpretation or emphasis are worth making. The first concerns the value of participation. Employees on average rated their organizations lower on participation than on any other value (see Table 9.7), both in the social sector and in the commercial sector. It would seem, then, that relatively few organizations have successfully incorporated a participative ethos into their culture. To the extent that they have, however, there are highly consistent positive effects across all aspects of commitment.

The second point refers back to the issue of security or insecurity. The earlier analyses concerning workplace policies and practices indicated that those affecting security had surprisingly little impact on organizational commitment. However, the present results indicate that perceptions of the organization as *caring for employees' well-being* were of the greatest importance for commitment. Individuals seem to persist in their organizational commitment despite policies or practices which increase insecurity, but at the same time they are highly sensitive to an uncaring culture, and this may well involve how insecurity is handled by the organization.

Finally, the perceived usefulness of the organization to society has a bearing on organizational commitment, even after controlling for the background attributes of the individual, for the policies and practices of the organization, for individual satisfactions of various kinds, and for other types of organizational value. The organization's own commitment to service therefore remains an independent influence on employee commitment, over and above the development of internal policies or the allocation of individual rewards.

CONCLUSIONS

The present study is the first to assess organizational commitment in Britain on a nationally representative sample.[7] The commitment of employees to their organizations seemed rather weak, but this was an impression which will need testing in future research. We are at this stage on firmer ground with conclusions about the nature of the influences on organizational commitment. The prescriptive literature on employee commitment has tended to stress the importance of two types of personnel policy: those to develop employees in their work and those to foster mutual relationships of consultation and participation. The findings gave considerable support to these views. In a number of respects, though, these were policies which employers had adopted only weakly and selectively, thereby potentially limiting the development of high levels of organizational commitment. This was most obvious in the case of participation, a powerful influence on commitment but seen by employees as particularly low in the order of their organizations' values.

The findings in connection with control systems were less in accord with the prescriptions on commitment. A developed system of performance management contributed to higher commitment, at least in commercial organizations, so there seems to be no necessary opposition between control and commitment. It is possible, however, that this would not remain true if organizations were seeking to stimulate much higher levels of employee commitment than those observed currently.

There was also rather limited support for the view that job insecurity is incompatible with organizational commitment. There were some weakly adverse effects of some aspects of insecurity on some aspects of commitment, but on the whole employees' commitment appeared remarkably unaffected by employers' policies which undermined security or by paternalism which buttressed it.

Here, however, one should consider perhaps the most important finding of this chapter: that organizational commitment is less a matter of personnel policies and more a matter of broad values. Even after controlling for a wide range of policies, and for various facets of job satisfaction, it was these values which had by far the strongest relationships with commitment.

So, while individuals' organizational commitment was little affected by employers' policies which undermined security, commitment was much reduced where the organization was seen as not caring for the well-being of its employees. Perhaps employees see insecurity as externally imposed, on their employers as much as on themselves. But they do not see external conditions as giving employers a licence to be uncaring.

---

[7] Subsequently, Guest *et al.* (1996) studied organizational commitment, among other matters, in a national quota sample of 1,000 British employees. Their measure of commitment was a three-item scale. Two of these items were similar to our 'value commitment'; the third concerned leaving intentions and so did not correspond to our concept of commitment.

The sub-plot of this chapter has been to explore differences between two sectors of employment labelled the social and the commercial. This has not been a merely academic exercise, for the social sector of employment—including education, health, and welfare services—has been under particular pressure in recent years to adopt practices and values from the commercial sector and so become marketized. The results of this chapter suggest that this has been a one-sided and ill-considered process.

Social sector organizations, as one might expect, are perceived as being far more useful to society than those in the commercial sector, and their employees have on average considerably higher levels of satisfaction with their opportunities for doing something worth while. This suggests that the policies or practices which are effective in fostering commitment in the commercial sector may not be equally effective in the social sector, and so it appears to be. The analyses indicated many differences between the sectors in the impacts on commitment of a wide range of policies and practices. Except for some aspects of supervision, developmental practices appeared to have a markedly greater positive impact on commitment in the commercial sector than in the social sector.

On the other hand, there was little difference between the sectors in the importance of organizational values for commitment. Moreover, as has already been stressed, perceived organizational values had far more impact on commitment than did any collection of policies and practices. But among the market values, it was only efficiency which stimulated commitment, while the social values of participation, care for employee well-being, and benefit to society, were all powerful influences on commitment. So overall it was the social values which dominated. They dominated as much in the market sector as in the social sector of employment. The substance of the findings, therefore, does not support those who argue that the market must be the teacher and the public sector the pupil. Rather, the findings suggest that the market itself has much to learn from the social sector. The social sector may well benefit from greater commitment to efficiency, but at the same time it is social values which are needed to generate employee commitment in the commercial sector. Organizations which are not themselves committed to the communitarian values of participation, care for employee well-being, and useful service to society, have limited scope for stimulating high levels of employee commitment.

10

Labour Cost and Performance

The main focus of previous chapters has been on how employees experienced the workplace changes taking place in the 1980s and early 1990s. But what are the implications for the performance of organizations? Do the significant changes which have been identified in previous chapters also result in reductions in labour costs or increases in employees' productive performance? Answering these questions will bring two additional dimensions to the findings. First, it will extend our view of the consequences of change in a more behavioural sense. Second, it will suggest how far the changes taking place are propelled by their contribution to the profitability or efficiency of organizations.

The role of organizational commitment will be of particular interest. Is it true that more committed employees are also more effective, and if so in what respects? The value of employee commitment is certainly a central assumption of human resource management (HRM), which has played such a prominent role in the debate on corporate personnel policy in recent years. The analysis will provide a test of the practical importance of organizational commitment, and hence also of the effectiveness of a high commitment strategy. Indeed, some pointed criticism of the HRM approach has been based on its lack of demonstrated impact on performance (Guest 1992).

The evaluation will also consider the impacts of the more detailed policies and practices adopted by employers. The analysis will take a fresh view of those policies and practices which, in previous chapters, have been found important as influences on organizational commitment, job involvement, stress, and employee satisfaction. Those outcomes are of great concern to employees, but perhaps less so to employers. Employers, are likely to judge their own policies largely in terms of their effectiveness in reducing costs or increasing performance. There is no reason to expect that this will be less so in the case of employers adopting the policies of HRM, for the leading thinkers and advocates of HRM have continually stressed that these policies are justified and made necessary by market competition and the drive for high levels of economic performance (e.g. Walton 1985*a*, 1987; see review in Lundy and Cowling 1996).

Finally, if the policies most effective in these terms are most likely to survive and be imitated, then the findings of this chapter may provide some sense of which changes are likely to continue in the future.

INDICATORS OF EMPLOYEE PERFORMANCE

Organizations have difficulty in measuring or assessing the performance of employees in an objective or reliable way, even with all the resources that they have. It may seem foolhardy to try with the limited information collected by a survey. Surveys can only ask simple questions, are reliant on the respondents to provide a true picture of their behaviour, and lack external corroborating evidence. Yet surveys also have advantages. Respondents do not stand to gain or lose promotion or a pay rise when replying to surveys, and so have no incentive to bias their replies. They may of course still want to say something socially acceptable to the interviewer, but there are survey techniques to minimize this problem. The most important advantage which surveys can tap into is that respondents know many things about themselves better than is available from any other source (for general review of subjective and objective performance measures, see Bommer *et al.* 1995).

The quality of a survey measure of performance can often be gauged through the results it produces: 'the proof of the pudding is in the eating'. One of the performance measures we examined, lateness for work, failed this test. The frequency of an employee's lateness was poorly predicted by a whole range of potential explanations including commitment, satisfaction, stress, and nature of supervision. There could be various reasons for this failure. Lateness is hard to define (how late do you have to be?), and the key factors may not be relevant to performance: for example, the uncertainty of the transport system. Whatever the reasons, lateness was evidently not a reliable performance measure. However, the three other cost and performance outcomes to be discussed—absence, leaving intentions, and work quality—were all strongly influenced by a variety of the organization's personnel practices, though sometimes in surprising ways.

The following sections of the chapter will present each of the cost and performance outcomes in turn. For each outcome there will be some further discussion of how well we can measure it, then a description of the kind of analysis carried out, and finally the results. An overview and discussion of all the findings will follow at the end of the chapter.

ABSENCE

A prominent theme emerging in employers' policies in the late 1980s was the need to tighten control over employee absence (Industrial Relations Review and Report 1987). This was not because absence rates were higher than before, but because the labour force was being cut and therefore (with reduced slack) absence had a more immediate impact on production than before (Edwards and Whitson

1989; 1993). Increased emphasis on the control of absence can be seen as part of an increased drive for productivity, or as part of the growth of control systems, as discussed in earlier chapters.

Absence is undoubtedly an important source of labour costs or of lost revenues for most organizations. The organizational consequences of unscheduled absence include instability in the supply of labour to the firm, disruption of scheduled work, and loss or underutilization of productive capacity (Steers and Rhodes 1978; Mowday *et al.* 1982). To avoid loss of sales revenue or shortfalls in service levels from absences, employers either have to carry a permanent labour slack, or temporarily hire additional labour (while continuing to bear some costs for the absent workers), or put on overtime which often involves premium pay rates.

Organizational commitment is particularly central to the issue of employee absence or attendance. It is often assumed that high organizational commitment leads to a high level of employee attendance. But exactly why this should be so is seldom made explicit; commitment is likely to be just one of many factors involved (Guest 1992). Early studies of employee attendance looked at one or two factors at a time, and lacked an integrated approach (Mowday *et al.* 1982). The first attempt to view absence behaviour systematically, and to portray how various influences combine, was provided by Steers and Rhodes (1978). They examined more than a hundred previous studies, searching for variables which had some apparent bearing on absenteeism. The model of absence which they proposed as a result distinguished voluntary from involuntary absence, and suggested that to explain attendance or absence one had to consider both *ability to attend* and *motivation to attend*. An employee's motivation to come to work represents the primary influence, provided that there are no external barriers to attendance. Among those factors which affect the motivation to attend are the employee's own feelings about the job (for example, job satisfaction) and various pressures to attend, for example from management or co-workers. Job satisfaction is affected by the details of the job situation, but also by the employee's expectations of the job, which are in turn affected by the personal characteristics of the employee such as education. Ability to attend, however, depends on such things as illness and family emergencies.

A framework of this type can incorporate measures of organizational commitment alongside job satisfaction, as the motivational factors influencing attendance or absence. Job absence should be related to commitment, because commitment represents agreement on the part of the employee with the goals and objectives of an organization. If an employee firmly believes in what an organization is trying to achieve, she or he should be more motivated to attend in order to work towards these goals.

The concept of commitment also suggests a growth process where the individual comes to perceive that her or his goals and values are consistent with those of the organization, and so becomes increasingly involved in the service of those goals and values (Steers 1977*b*). In this way, organizational commitment is considered even more stable than such factors as job satisfaction in predicting job absence (March and Simon 1958; Steers and Rhodes 1978; Mowday *et al.* 1982).

Despite the existence of this framework for analysing absence, little systematic analysis began until a decade later. One study which made a substantial effort in this direction was that by Brooke and Price (1989). It tested the Steers and Rhodes model with a sample of 425 full-time employees from a single public sector organization. Among other findings, the authors found little support for the supposed link between organizational commitment and subsequent job absences. This study and a longitudinal study of hospital employees conducted by Price and Mueller (1986) have created an impression that there is little link between commitment and absence (as concluded by Guest 1992). If this is so, then either absence is not an outcome of employee motivation, or organizational commitment is not a good indicator of employee motivation. However, the studies cited were confined to special rather narrow groups of employees, which may not be typical of employment as a whole. The nationally representative sample obtained in the Employment in Britain Survey provides a sounder test.

Measuring Absence

There is no universally accepted measure of absence. The potential measures include the number of spells of absence within a period (ignoring length of absence), the total amount of time absent from work (ignoring the number of spells), and whether ever absent during a stipulated period (usually recent and short, so as to minimize recall problems). There is some evidence that the number of spells of absence (absence frequency) exhibits higher reliability than other measures (Mowday *et al.* 1982), but the decision on which measure to use depends on the aims of the study and availability of data.

In this study, the primary interest lies in employee voluntary absence, which is related to issues of personal motivation. It is also assumed that employers want to curtail voluntary absence in order to reduce the labour costs referred to earlier. Frequency of absence seems a more suitable measure than total days of absence, since it will be relatively unaffected by long periods of serious illness, which are unrelated to motivation, and also cause less disruption.

In the survey, respondents were asked the following question: 'How many times have you had to be away from work at any time due to sickness during the last year?' It might seem that, by referring to sickness as the reason for absence, the question excludes absences taking place without good reason, while it includes spells of absence which result from ill-health rather than low motivation. However, sickness is often stated in a purely customary way as the supposed reason for voluntary absence. There are also many minor illnesses or ailments —colds or sore throats, for example—which will be ignored by a highly motivated employee even though they are an adequate reason for being absent.

TABLE 10.1 *Self-reported frequency of absence, by gender (column %)*

Frequency of absence in past year	Men	Women	All
Never	46	35	41
Once	30	30	30
Twice	14	20	17
3–5 times	8	12	10
6+ times	2	3	2
No.	1,721	1,577	3,298

The effect of substantial or persistent ill-health can be 'netted out' from the results by including information about this factor in the analysis.

To reduce the risk of biased answers intended to impress the interviewer, the question about absence was placed in a section at the end of the interview, which the respondents completed privately. In addition, several predefined categories were provided for replies, to help convey the idea that absence was not abnormal. Respondents were given the following options to reply: never, once, twice, 3–5 times, 6–10 times, and 11 or more times.

Table 10.1 shows replies to the question, in the predefined categories. It appears that 44 per cent of employees never took an absence during the year, whereas 56 per cent took some time off. Of those who did, few people had as many as six spells of absence. Accordingly, these replies were amalgamated in the analysis with the category 3–5 times absent, so that four distinct response categories were retained.

Potential Explanations of Absence

The potential influences on absence, which are considered in the analysis to be presented, are in four groups. These are: the employee's ability to attend, together with some personal background characteristics; pressure to attend; job-related factors, including personnel policies; and broad motivational factors, including organizational commitment. Each group will be briefly explained before the results are presented.

Ability to attend and personal background characteristics

An analysis of absence should take account of constraints which may simply prevent the individual from attending. Two main variables of this type were investigated. The first one was an indicator of the individual's health condition, based on the presence or absence of long-term illness. This indicator, which proved to be very important, was included in all the analyses.

The second type of variable measured family responsibilities, which are often assumed to undermine commitment to work, for women particularly. However, this assumption has not been well supported by evidence in the past (Bielby

1992), and the present study also found no indication that family circumstances were important for absence. Several different variables were tried, including whether there were any dependent children under the age of 5, and any dependent children at all. As they never made any difference to the results, even when women were considered separately, these variables were eventually dropped.

Because of differences in home responsibilities, women may face greater obstacles to regular attendance. Women may also have different health profiles from men, which may affect attendance. All analyses were performed separately for women and men in the sample, as well as for the whole combined sample, so as to get the clearest possible view of gender differences.

Other background characteristics of the individual were controlled or 'netted out' in the analysis, in the same way as in previous chapters. The background variables consisted of age group, the social class of current occupation, and the length of service in the present job.

Pressures to attend

How consistently employees attend may depend in part on how far they are required or pressured to attend, as well as on their personal motivation. The importance of organizational pressures to attend has been stressed in research by Edwards and Whitson (1993).

The present analysis included several items to explore this issue. One concerned the employee's perception of the level of importance attached by the supervisor to the avoidance of absence. If supervisors do not convey that regular attendance is expected of employees, absence may be more frequent. A further question asked whether the individual felt anxious about dismissal. Although this in part relates to wider issues of insecurity (covered in some of the job-related items discussed below), it seems plausible that a perceived threat of dismissal will specifically reduce the frequency of absence. Similar thinking underlies the third item in this group, the individual's perception of how easy or difficult it would be to get another job if the present one was lost.

Job-related factors and personnel policies

This was the largest group of variables in the analysis. The size of the organization was included: smaller organizations may foster lower absence rates because employees are more aware that others depend upon them. Also included was the sector in which the individual worked, divided into a 'social' sector (the public sector with a few additions) and a 'commercial' sector as in Chapter 9. Another characteristic included was the influence of trade unions (if present) on the workplace; this measure is discussed in Chapter 3.

Two further items concerned the type of employment contract in the job. One distinguished part-time from full-time workers (using a cut-off of thirty hours per week). The other distinguished those on short temporary contracts (less than one year), those on longer fixed-term contracts (one to three years), and those on permanent or open-ended contracts.

The analysis included several items about communication and participation: individuals' involvement in decisions affecting their jobs, the presence of 'quality circles', and the opportunity to take part in open, two-way communications with management.

Relationships with the supervisor may be important in a number of ways. The two items which, from previous chapters, seem to sum up the supervisory relationship most succinctly are whether the supervisor is seen as fair or biased, and whether the supervisor has a coaching role with subordinates. (A third item, specifically relating to control over absence, has already been discussed under 'pressure to attend'.)

Control and reward systems may affect absence, either directly by linking rewards to attendance, or indirectly by affecting general motivation and commitment. The analysis included the two composite variables for 'technical control' (work-pacing and incentive systems) and 'performance management', a measure which reflected internal progression, target-setting, appraisal, and merit pay.

The final items in this block concerned corporate welfare policies. One was whether working conditions were felt to be unsafe or unhealthy; the other was how the individual rated the available fringe benefits such as pensions and sick pay.

Motivation or commitment at work

Job-satisfaction measures have most often been used to assess motivation in studies of absenteeism. Two measures of this kind were used in the analyses here, satisfaction with pay and satisfaction with job security. But we focus more on the newer measures of organizational commitment. As in Chapter 9 (see there for further details), three types of commitment were distinguished for our analyses:

- 'Value commitment': formed from three items expressing the individual's identification with the organization and its values;
- 'Flexibility commitment': formed from two items expressing the individual's willingness to be extremely flexible over work and pay in the interests of the organization;
- 'Effort commitment': based on a single item expressing the individual's willingness to make exceptional efforts for the organization.

Finally, three other items were placed in this block which have played a large part in the previous chapters. Although not motivational items as those are usually thought of, they seem to sum up a great deal of the motivational changes which people were experiencing at work. These were task discretion (the usual index of discretion introduced in Chapter 2); upskilling in work, as also described in Chapter 2; and work strain, as described in Chapter 8.

Two Types of Analysis

There were two main versions of the analysis: one was designed to assess the impact of the organization's detailed personnel policies, and the other the impact of the broader factors of a 'motivational' type, which have just been outlined. Version 1 of the analysis omitted the motivational factors and focused on the organization's job-related policies and personnel practices. The results estimate the net impact of each policy or practice on employee absence, allowing for the impact of all the other policies and practices. The net impact of any personnel practice also includes any more diffuse effect it has on motivation. For example, supportive supervisory practices may reduce absence directly by increasing loyalty to the supervisor and the work group, and indirectly by contributing towards organizational commitment.

The Version 2 analysis focused on the motivational factors, consisting of task discretion, upskilling, work stress, satisfaction with pay and security, and organizational commitment. These factors were now added to the original Version 1 analysis. As already stressed, the organization's personnel policies such as the style of supervision, or communications, or reward and control systems, also affect motivation and through motivation, absence. But now we wish to assess the role of motivation *over and above* the specific policies and practices of the organization. The underlying assumption is that motivation can be stimulated in many kinds of subtle and elusive ways, ranging from a 'pat on the back' to an ethos of service to the community. There are also internal sources of motivation in the individual, perhaps springing from formative experiences many years before. It is impossible to include every such source of motivation in an analysis, but the broad motivational factors are intended to capture their joint effect. In this analysis, then, they represent the motivational 'added value' which cannot be reduced to personnel policies.

Results for Version 1 Analysis

The Version 1 analysis was performed first for men and women combined, and then for men and for women separately. The results which were significant in any of these analyses are summarized in Table 10.2.

Personal characteristics and ability to attend

The first panel of results in Table 10.2 shows personal characteristics which were strongly related to absence: sex, age, and health status. Health or ill-health is the chief indicator of 'ability to attend', and not surprisingly those with long-term illnesses were much more likely to report absences. This confirms how important it is to take account of ill-health when analysing absence statistics.

TABLE 10.2 *Effects of personal and job factors on frequency of absence, omitting the 'motivational' factors*

	All		Women		Men	
	Coeff.	Sig.	Coeff.	Sig.	Coeff.	Sig.
(A) *Ability to attend*						
Female	0.43	***	—		—	
Age 20–4	0.18	***	0.23	***	0.53	***
25–34	0.09	**	0.06	n.s.	0.46	**
35–44	−0.09	*	−0.08	n.s.	0.24	n.s.
45–54	−0.18	***	−0.21	***	0.20	n.s.
Long-term sickness	0.52	***	0.49	***	0.61	***
(B) *Pressure to attend*						
Supervisor attaches v. high						
importance to good attendance	−0.15	*	−0.24	*	−0.07	n.s.
Union influential	0.20	**	0.18	n.s.	0.27	**
Union v. influential	0.27	*	0.37	n.s.	0.30	n.s.
(C) *Job-related factors*						
Part-time contract	−0.26	***	−0.17	***	−0.35	***
Temporary (short-term) contract	−0.26	***	−0.59	***	0.01	n.s.
Adverse physical working conditions	0.17	***	0.20	***	0.11	n.s.
Supervisor biased	0.25	***	0.23	**	0.29	***
'Technical control'	0.03	n.s.	0.09	**	−0.02	n.s.
Open communications	−0.13	**	−0.08	n.s.	−0.17	**
Supervisor coaches	−0.10	*	−0.17	**	−0.02	n.s.

Note: Ordered probit analysis. Sample nos. are 2,837 (all), 1,367 (women), and 1,470 (men). Non-significant effects are not shown.

Another familiar result is that women on average had more frequent absence than men.[1] This difference in attendance rates is often assumed to reflect 'ability to attend', since women are expected to take the main responsibility for childcare and for looking after sick children. As already mentioned, however, we found no evidence of the age or number of children affecting absence, so childcare does not seem to be the key factor. Another possibility is that women are themselves more prone to short-term illnesses, for which there is some independent evidence.[2] We were only able to control for longer-term or persistent illness, so this seems a reasonable explanation.

[1] OECD (1991; see Table 6.6) reports for the UK in 1988 that the incidence rate of injury or illness absence for full-time male workers was 1.8% while for full-time female workers it was 2.4% and for part-time female workers it was 2.3%. Women had higher absence rates in all countries except Luxembourg.

[2] The General Household Survey 1992 (see Table 3.9) reports the following percentages consulting with their general practitioner in the fourteen days before survey: men aged 16–44, 9%, women aged 16–44 18%; men aged 45–64 13%, women aged 45–64 18%. The differences in long-standing illness rates between men and women, however, are slight (*ibid.*).

The remaining personal characteristic which strongly affected absence was age. Younger workers (those under 35, and especially those under 25) were more likely to be absent, while older workers were generally 'more dependable', with the lowest absence rates being recorded by the 45–54 age group. Surprisingly, perhaps, once age was controlled there was no independent influence from the employee's length of service with the organization.

Pressure to attend

The results (in panel (B) of Table 10.2) confirmed that external pressures to attend reduced absence, but only to a moderate degree. Where the supervisor placed high importance on attendance, there was a reduction in the likelihood of absence, but the influence was only a clear one in the case of men. Anxiety about dismissal might also be expected to reduce the absence rate, but the influence in this case was not significant. So the lead or example given by the supervisor was more effective in reducing absence than a disciplinary threat.

Another relevant finding concerned the influence of unions on the organization of work. Where unions had a medium or high level of influence, the frequency of absence was greater; this could be because a strong union limits management's ability to exert pressures to attend.

Job-related factors and personnel policies

Job-related factors formed the largest block of influences considered in this set of analyses. The significant results are shown in panel (C) of Table 10.2. The table does *not* show factors or policies which were non-significant in all analyses.

The employee's type of employment contract was an important influence on frequency of absence. *Part-time employees* were particularly unlikely to have frequent absence, and this was as true of the small proportion of male part-time workers as it was of women in part-time work. Workers on *short-term (up to one year) temporary contracts* were also less likely than permanent workers to take absence, but this applied only in the case of women on such contracts, not men. There was no difference in absence rates between permanent employees and those on fixed-term contracts of one to three years.

Since most part-time employees are women, and since short-term contracts reduced the frequency of absence only for women, this may seem to contradict the earlier finding about women being more often absent than men. In fact, however, it fits quite well if the reason for women's more frequent absence is more frequent short-term illnesses. These will arrive at any time, so *some* will occur on non-work days or (for temporary workers) in between spells of work. As a result, frequency of absence, over a fixed time such as a year, will appear to be less for the part-time workers or those on very short contracts. In addition, part-time workers or 'temps' may be more likely to fix visits to the doctor or dentist on non-work days.

We find more clear-cut evidence of how employers can influence absence when we turn to the more detailed personnel policies and practices. Nine types of policy or practice were evaluated and the results fell into three groups. Three types of practice were connected with higher absence, two types with lower absence, and four types had no clear effect one way or the other. Policies effective in reducing absence were, therefore, very much in the minority (recall, however, that at this stage we are not considering broader motivational factors like upskilling or task discretion).

It is perhaps an overlooked point that some inadequate personnel practices actually increase absence and so raise labour costs. Absence was particularly increased where employees felt that their working conditions were unhealthy or unsafe (nearly 30 per cent of employees felt this). The most obvious interpretation is that such working conditions lead to more frequent work-related injuries or illnesses. People who find their work physically unpleasant or dangerous may also take days off as a temporary escape.

The exercise of arbitrary supervisory power also seemed to increase absence. Employees who regarded their supervisors as favouring some staff over others—in other words, as biased—were more likely to be absent. It seems possible that employees take days off to escape the unpleasant feelings which such a relationship arouses.[3]

While both bad working conditions and biased supervision increased absence for men and women about equally, the third aggravating factor affected only women. This was the use of 'technical control' systems, involving work-pacing and incentive payment schemes. The existence of high absence in certain industries, like leather and footwear, which focus piecework payment systems on female workers has been noted in the past (Goodman *et al.* 1977). There are at least two ways of interpreting the link. One is as an escape from the work pressure created by individual incentives. The other, and more plausible, is that workers on piecework incentives can compensate for days off through short bursts of higher production, so there is less financial incentive for regular attendance.

Of the two practices which helped to reduce the frequency of absence, one was more effective with men and the other with women. Open two-way communications led to a reduction in absence for the sample as a whole, but in the separate analyses the effect was significant only for men. Conversely, absence was reduced where there was a supervisor who coached staff, but in the separate analyses this was significant only in the case of women.

It is also worth noting the personnel practices which proved to have no effect on absence. These included several which might have been expected to reduce

---

[3] Another possibility, however, is that the influence runs in the opposite direction: employees who are frequently absent may get adverse treatment from the supervisor, and then interpret this as bias.

absence through a motivational or 'loyalty' effect. Performance-management systems, despite their contribution to organizational commitment, did not affect absence. Nor did participation in work decisions, despite all the other connections this had with a positive experience of employment. Taking part in quality circles also made no difference to absence. Again, employees' favourable views of company welfare benefits such as occupational pensions neither reduced nor increased absence.

On the whole, then, what might be thought of as 'positive' personnel practices had limited effects on absence. Absence, it seems, could be reduced more by avoiding detrimental practices. However, we have not yet considered how far absence might be affected by employee motivation.

Absence and motivation

In the Steers and Rhodes model of absence, the central influence on absence is how individuals respond to their job situation, which forms the primary source of job satisfaction. Hence it is not enough to look at detailed personnel practices, one should look especially at broad indicators of satisfaction and commitment. Version 2 of the analysis is the appropriate framework for considering these broader factors. It assesses the 'added value' to the organization which motivation produces through lower absence. The relevant results are summarized in Table 10.3.

A first impression is how strikingly different the results are for men and women. The motivational factors which most influence men's absence are distinct from those which influence women's. A second impression is that, as happened with personnel practices, not all the significant effects are in the direction of reducing absence. Some motivational factors increase the frequency of absence.

TABLE 10.3 *Motivational factors related to frequency of absence (Version 2 analysis)*

	All		Women		Men	
	Coeff.	Sig.	Coeff.	Sig.	Coeff.	Sig.
Task discretion	−0.01	n.s.	0.01	n.s.	−0.04	**
Upskilling	0.06	***	0.02	n.s.	0.09	***
Work strain	0.03	***	0.05	***	0.01	n.s.
OC[a]						
values	−0.03	n.s.	−0.05	*	−0.00	n.s.
flexibility	0.01	n.s.	−0.00	n.s.	0.01	n.s.
effort	−0.10	**	0.01	n.s.	−0.21	***

[a] OC = Organizational commitment

Note: Ordered probit analysis. Sample nos. are 2,837 (all), 1,367 (women), and 1,470 (men).

Surprisingly, task discretion, such a key dimension of the experience of work, did not significantly reduce absence for the sample as a whole, nor for women considered separately. However, among men, those given high task discretion were found to be absent less often.

Even more surprising was to find that on the whole upskilling was connected with a *higher* level of absence. To make sense of this, recall that upskilling is associated with a higher degree of work pressure and responsibility, both of which increase work strain (see Chapters 2 and 8). There was a clear contrast between men and women on the questions concerning upskilling and stress. Although in both cases upskilling and stress tended to increase absence, in the case of women only the effect of work strain was statistically significant while in the case of men it was only upskilling which was significant. It seems that for men the pressures leading to absence are particularly concentrated in upskilling (and the associated change in responsibility), while for women absence is a response to wider, more varied kinds of stress.

Of the three types of organizational commitment, commitment to effort was the strongest influence in restricting absence, but this was significant only in the case of men. Conversely, commitment to the organization's values influenced women not to be absent from work, but did not have this influence on men. Flexibility commitment—a willingness to do any kind of work and even take a pay cut for the organization's sake—did not independently influence absence rates for either men or women. It seems, then, that men have low absence when they are determined to make exceptional work efforts, while women have low absence when they feel in harmony with the organization's values.

The most important point, though, is simply that organizational commitment (in one form or another) does reduce absence even over and above the many different kinds of personnel practices and policies which have been considered. So organizational commitment is more than 'mere words': it changes behaviour.

On the other hand, employees' satisfaction with their pay had no bearing on their frequency of absence, nor did their satisfaction or dissatisfaction with job security. So organizational commitment came through more strongly than satisfaction as an indicator of behaviour.

Conclusions about Absence

This section of the analysis was guided by the model proposed by Steers and Rhodes. The results confirmed a basic premiss of their model, namely that employee absence is a combination of ability to attend and motivation to attend. Long-term illness was a most important obstacle to attendance. In addition, a relatively high absence rate among women may reflect greater proneness to short-term illness. Part-time workers were less often absent and this tends to support that interpretation.

The policies and practices which influenced absence were substantially different between male and female employees. Men were less likely to be absent when they enjoyed high task discretion and open two-way communications, and when they were not exposed to adverse working conditions. Women were less likely to be absent when their supervisor had a coaching relationship with subordinates and communicated the importance of high attendance, while they were more likely to be absent if working under a technical control system. The only common factor for men and women was higher absence rates where the supervisor was seen as being biased.

These findings also indicate that high attendance rates will be achieved by removing factors which make the job unpleasant, such as poor supervision or bad working conditions, as much as by 'positive' policies such as communication and involvement.

High supervisory pressure for attendance led to reduced frequency of absence. However, this influence was fairly small, and the suggestion is that organizations can reduce absence rates only to a small degree by direct pressures of this kind. Moreover, though a substantial minority of employees feared dismissal, this had no effect on their frequency of absence.

The motivational aspects of the Steers and Rhodes model were quite well supported. Higher levels of organizational commitment led to lower frequency of absence, over and above the wide range of personnel policies and practices which were considered here. And while commitment reduces absence, work strain increases it. It was revealing to find, also, that upskilling actually contributed to increased frequency of absence. Since upskilling is closely connected with increased responsibility, the link between upskilling and absence seems likely to be stress-related.

QUALITY IN JOB PERFORMANCE

Job performance is perhaps the acid test for the motivational ideas of both organizational theorists and management practitioners. The former widely assume that greater organizational commitment contributes to better performance, as more committed employees should be more motivated to work well in their organization's behalf (Kalleberg and Marsden 1993). This assumption is probably shared by many practitioners, especially those adopting the HRM approach. Moreover, people with better skills are usually thought to be more capable of performing their work tasks well, irrespective of their organizational commitment. And performance is assumed to be affected by whether employees are able to exercise autonomy and discretion in their work, and whether feedback is available concerning results achieved. All these are factors encountered in the previous chapters, reflecting major recent tendencies in the organization of work.

But our review has also shown that more traditional ideas about control and reward systems still exert a powerful influence in the great majority of organizations. Management continues to rely heavily on control systems which pace and measure work, or on those which set targets and appraise performance against them, and on systems of payment which mesh with these controls. They do so even at the same time as they try to develop the newer emphasis on upskilling, task discretion, and employee commitment. A key question to be posed, therefore, is how strongly each approach influences the performance of individual employees. With the two approaches running side-by-side, there is an opportunity to test their relative strength.

The focus here is on the quality of work performance. Quality has been of particular competitive concern for management (see examples in Peters and Waterman 1982), and notions of 'total quality management' and 'customer care' have figured prominently in corporate strategies. The management of quality, whether of product or service, involves aspects of control, skill, and commitment, but it is far from obvious how to balance these, or reconcile their tensions (Hill 1991b; DuGay and Salaman 1992; Lawler 1994; Hackman and Wageman 1995).

Although we do not here analyse work effort, Chapter 3 has examined influences on work pressure, a concept which is closely connected with effort. At several points, results from the earlier analysis will be compared with those obtained here.

Measuring Job Performance

Survey measures of an employee's job performance have to be based on the respondent's self-reports. The limitations and advantages of self-reports have been discussed in general terms earlier in the chapter.

Self-report measures have been used previously in a number of studies of performance (e.g. Pruden and Rees 1972; Busch and Busch 1978; Darden *et al.* 1989). Such measures have been found to correlate highly with more objective measures. In a study comparing self-reported job performance ratings and ratings given by their superiors, Heneman (1974) found that the self-report measures had less halo error, restriction of range, and leniency than the supposedly more objective measures derived from management. In addition, more objective measures, such as supervisor's ratings and output measures, are useful only in specific settings (for instance, where tasks and outputs are well defined) and cannot be easily applied to a large and increasing range of loosely defined jobs, such as those which concern social interaction (Steers 1977b; Judge and Ferris 1993). Self-reports have the particular advantage that they can be obtained from workers in all types of jobs.

In the present survey, the self-reports of performance were of a relative or comparative type. The respondent was asked to compare her or his performance

TABLE 10.4 *Self-reported work quality (column %)*

Works much better	12
Somewhat better	36
Same	52
Worse	1
No.	3,275

Note: for question wording, see text.

with others doing similar work. The wording of the question was as follows: 'Compared with other people who do the same or similar kind of work that you do, how *well* would you say you do your job?' The possible replies were much worse, somewhat worse, about the same, somewhat better, and much better.

The tendency for people to give socially acceptable replies was one danger, so (as in the case of the question concerning absence) these questions were placed in the privately completed section of the interview. Even so, the replies were strongly biased towards the favourable end, with many people stating that they worked better, and only a handful stating that they worked less well than others (Table 10.4). A reasonable way of simplifying the replies was to divide them into two: those seeing themselves as working better than others, and the remainder.

Care has to be taken in interpreting a comparative type of question. Respondents giving a favourable judgement of their own performance may partly be saying that they are good workers, but they may also be implying that other workers are performing poorly. Conversely, when respondents say that their performance is just average, they may also believe that the average performance standards in their workplace are excellent.

Potential Influences on Quality of Performance

The information used to explain differences in quality of job performance was largely the same as in the analysis of absence, with a few additional items. The grouping of the items, to represent influences of different types, is also broadly similar to the previous analysis, so can be presented briefly.

A frequently used theoretical model assumes that job performance is affected by three major sets of variables: (i) ability to perform; (ii) task characteristics which facilitate or obstruct performance; and (iii) motivation to perform well (see Porter and Lawler 1968; Steers 1977*a*; Blumberg and Pringle 1982). To these we add external pressures to perform, such as threats or sanctions against poor work (as in the similar extension made to the model of absence, above).

There are no direct measures in the survey of 'ability to perform', but a variety of characteristics provide indicators. The individual's social class position is one of the most effective of these. The ability to perform well also depends upon experience, here represented both by age and by length of service with the present employer. Ill-health is included as something which may limit the capacity to perform well.

External pressures to perform are assessed in a similar way as for the analysis of absence. The importance attached by the supervisor to attendance is replaced by the importance the supervisor attaches to quality.

Job-related factors and personnel polices were represented in the analysis by the same wide range of variables as in the case of the analysis of absence, with one additional factor. This is 'feedback of performance': employees are assumed to work better if they understand the contribution of their work and if they are noticed when they do their work well. Two correlated questions covering these aspects were summed into a single measure of feedback.

Motivational factors were approached through the same broad measures, including organizational commitment, which have already been discussed in some detail in the analysis of absence.

As noted earlier, a person's view of their own performance will be coloured by their view of the general work standards in the organization. To allow for this, the analyses included an extra question which asked 'How efficiently is work carried out in this organization?' It was assumed that when employees feel that work is carried out very efficiently in their organization, they are less likely to describe themselves as working well compared with others. As with absence, a Version 1 analysis was carried out to assess the impact of job-related factors and personnel policies, and a Version 2 analysis to assess the additional impact, or 'added value', of the motivational factors. The main findings from Version 1 are summarized in Table 10.5, and those from Version 2 in Table 10.6.

Results for Quality of Performance

Personal characteristics reflecting ability or difficulty in working well had little bearing on self-reported quality of work. There were no clear differences by age, social class, or state of health. These negative findings were in striking contrast to the findings for absence.

However, there was a small overall difference by gender, with women being less likely on the whole to state that their work quality was above average. But women with longer periods of service with their employers did rate their own work quality above average, while this did not occur in the case of men (see panel (A) of results). Indeed, as we shall see, the results of the separate male and female analyses were generally very different, so the overall gender difference means little in itself.

Some indicators of external pressures on work quality are shown in panel (B) of the table. A striking feature of this block of results is that women are unaffected by any of these potential pressures. It is only men's results which are significant.

TABLE 10.5 *Factors related to high quality of job performance (Version 1 analysis)*

	All		Women		Men	
	Coeff.	Sig.	Coeff.	Sig.	Coeff.	Sig.
(A) *Ability*						
Female	0.83	*	—		—	
Length of service	1.07	*	1.15	**	1.01	n.s.
(B) *External pressure*						
Supervisor attaches high						
importance to quality in work	0.51	**	0.84	n.s.	0.38	***
v. high importance	0.54	**	0.65	n.s.	0.46	**
Union influential	0.84	n.s.	1.09	n.s.	0.64	*
Union v. influential	0.64	n.s.	2.20	n.s.	0.39	**
Size—under 10	0.73	*	1.01	n.s.	0.46	***
(C) *Job-related factors*						
Part-time contract	0.80	n.s.	0.73	*	1.42	n.s.
Temporary (short-term) contract	0.51	**	0.55	n.s.	0.44	**
Fixed-term contract	0.66	*	0.62	n.s.	0.69	n.s.
Performance management	1.12	***	1.17	***	1.08	*
Take part in work decisions	1.21	*	1.30	*	1.19	n.s.
Feedback	1.16	***	1.22	**	1.09	n.s.
Supervisor coaches	0.66	***	0.73	*	0.63	**

Note: Logistic regression—multiplicative effects on odds. Sample nos. are 2,837 (all), 1,367 (women), and 1,470 (men). Non-significant effects are not shown.

One result is particularly surprising. Where the supervisor attached a high or a very high importance to quality, men were *less* likely to say that they themselves produced above-average work. A possible explanation is that such supervisors may raise quality standards right across their work groups, making it harder for any one individual to feel above average. There are, of course, other interpretations, but this one fits with some other results. Female employees thinking their organization efficient were also less likely to consider their work quality above average. And both men and women who said that their supervisor helped staff learn to do the work better—a 'coaching' supervisor—were also less likely to think their own quality particularly good. If coaching by the supervisor raises general standards of work quality in the work group, this makes it more difficult for any one individual to shine relative to the others.

The result about the coaching supervisor really forms part of the next block of results, concerning the impacts of job-related factors and personnel policies, which is summarized in panel (C) of the table. In general these results are rather strong and clear. It is first of all apparent that the so-called 'flexible' or non-standard employment contracts are less likely to produce high-quality performance. Women on part-time contracts were much less likely to see themselves as above-average performers. The same tended to apply to both men and women on temporary or fixed-term contracts, although to a significant degree only in the case of men. Here is confirmation of one of the objections often made to flexible employment practices, namely a detrimental impact on quality (O'Reilly 1994).

Turning to control systems, one finds that performance management, involving features such as target-setting, appraisal systems, and merit pay, had particularly clear and positive affects on the quality of work. Employees, both female and male, who worked under a highly developed system of performance management were considerably more likely to report working at above-average quality. In contrast, 'technical control' (work-pacing and incentives) had no positive effect on quality although it was not detrimental either. It should be recalled, though, that in Chapter 3 technical control was shown to increase work pressure for both non-manual and manual workers, whereas performance management did so only for the non-manual group.

Employees who took part in work decisions, and felt that they got effective feedback, were also more likely to regard themselves as above-average workers in terms of quality. This was significantly so only among women, but men's results tended in the same direction.

Motivation and quality

Earlier chapters have argued that increases in skill and the widening of task discretion have played central roles in the changing experience of work. We have also shown that these changes were strongly related to increased work pressures and work demands. A missing piece of the picture so far is their relation to work quality.

As Table 10.6 shows, both task discretion and upskilling were positive influences on work quality. Task discretion enhanced work quality for both men and women, one of the few factors to do so. Upskilling was linked to work quality

TABLE 10.6 *Motivational influences on high quality of job performance (Version 2 analysis)*

	All		Women		Men	
	Coeff.	Sig.	Coeff.	Sig.	Coeff.	Sig.
Task discretion	1.05	**	1.05	*	1.05	**
Upskilling	1.08	*	1.00	n.s.	1.17	***
Work strain	1.04	*	1.05	*	1.02	n.s.
OC[a]						
values	1.00	n.s.	0.96	n.s.	1.03	n.s.
flexibility	1.01	n.s.	1.00	n.s.	1.04	n.s.

[a] OC = Organizational commitment.

Note: Logistic regression—multiplicative effects on odds. Sample nos. are 2,837 (all), 1,367 (women), and 1,470 (men).

only for men, not for women. This fits earlier findings about the lower degree of responsibility given to female employees when they are upskilled, and their more limited role in using new technology. It seems that upskilling has been given to women in a relatively restricted way, one which does not get reflected in a sense of high-quality performance at work. Conversely, another key indicator of change, work strain, was significantly linked to high-quality performance only for women, not men. It seems that women, more than men, have to reconcile themselves to experiencing a degree of stress if they are to pursue high-quality work.

The final results in this part of the analysis concern organizational commitment. One of the main reasons why firms seek higher levels of commitment among employees, rather than relying entirely on incentives and controls, has been to foster greater personal responsibility for quality. However, *commitment was unrelated to perceived quality of performance*. Non-significant results were obtained both for the separate measures of 'value commitment' and 'flexibility commitment', and (in a separate analysis, not shown) for a composite commitment measure which included 'effort commitment' as well as the other two facets.[4]

Conclusions about Quality of Job Performance

We introduced this part of the analysis as a kind of contest between motivational factors and systems of control. The result, however, has once again suggested that employers need both. This is similar to the analysis in Chapter 3, where both task discretion and control systems were found to increase work pressure.

But only certain motivational factors and certain forms of control are effective in supporting quality of performance. On the side of control, performance-management systems showed a positive impact on both men and women, while technical-control systems were ineffective. Other effective management practices were participation in work decisions and feedback of performance. A supervisor who emphasized quality, or one who coached staff to work better, seemed at first sight to have the opposite effect, but a plausible explanation is that such supervisors achieve a levelling up of performance which leaves less scope for individuals to excel.

On the side of broader motivational factors, task discretion proved to be particularly important, but upskilling was significant only for men. This fits earlier evidence that upskilling for women tended to be of a relatively restricted type.

The most striking negative result in this part of the analysis was the failure of organizational commitment to influence the quality of work performance. There are, of course, many potential explanations for this result. We are applying an unusually stringent test here, in which the effect of commitment is assessed net of the specific practices and policies, such as involvement in workplace decisions,

---

[4] 'Effort commitment' was not included as a separate item in the analysis, because its wording was too close to that of the question concerning quality of performance.

which help to develop commitment. Perhaps, though, the simplest explanation is that most or many employers have not yet projected quality as a key value of the organization. Unless the organization is itself manifestly committed to quality, there is no reason why commitment to the organization should lead the employee to raise the quality of her or his own performance.

LABOUR TURNOVER

Labour turnover, like absence, is an important element of labour costs. The turnover of employees leads to direct costs for recruitment and for training new employees. It also leads to indirect costs which may be substantial, for example through disruption of work while new people are being recruited. Organizations with high levels of labour turnover may need to carry surplus labour, or hire temporary workers, in order to maintain production or levels of service.

It has been suggested by industrial sociologists that greater organizational commitment contributes to greater employee identification with the goals and values of the organization and that this reduces the likelihood of voluntary job turnover (Mowday *et al.* 1982; Shore and Martin 1989; Huselid and Day 1991). A main argument for adopting a commitment-oriented management approach has been to reduce the costs of turnover by eliciting employee commitment (Brown *et al.* 1993).

A model to explain individual job turnover naturally has many similarities to the preceding models of absence and job performance. Motivation to stay, in the form of commitment to the organization, is again likely to be of interest, as are the various task and organizational characteristics which make an organization relatively attractive or unattractive. However, one might expect that economic factors loom larger in the decision to stay or leave. These would include satisfaction with present remuneration, and perceptions of how easy or difficult it will be to find an equally good job. Family and life-cycle characteristics have also often been assumed important in relation to job turnover, as in the case of young workers, a point which will shortly be discussed further.

Measuring Job Turnover

Since 'labour turnover' is used to describe aggregate movements out of jobs, the term 'job turnover' will be used here to refer to individual leaving decisions. There are many ways of measuring job and labour turnover (Price 1977). Longitudinal studies can use actual job changes after the initial survey. In a cross-sectional study, like the present one, it is more convenient to consider employees' voluntary *turnover intention*, i.e. whether or not the individual plans or expects to leave their present employer voluntarily.

The measure of leaving intentions is derived from two questions in the survey. The first question concerns whether the individual expects to leave *in the next year*, for any reason. The second question establishes the main reason for expecting to leave. The intention to leave is defined as expecting to leave either 'to work for another employer' or 'to work as self-employed'. Those with leaving intentions so defined comprised 13 per cent of the sample.

The obvious limitation of the measure is that, in practice, some people with the intention of leaving may not in fact leave, while others will leave without formulating an intention of doing so in advance. Extensive research has been conducted on the relationship between behavioural intentions and behaviour, and some research has also been conducted into the relation between leaving intentions and actual leaving decisions. The results of this research are broadly reassuring about the value of this kind of measure. Expressed intentions are good predictors of later behaviour, provided that the two are defined in a consistent way and the conditions which affect intentions do not change (Ajzen 1988). In addition, it may be disruptive for employers to have employees who wish to leave, even if they are unable to carry out this intention in practice.

Influences on Leaving Intentions

The factors used in this analysis to explain leaving intentions were the same as for the analysis of influences on absence presented earlier in the chapter. The relevant concepts are also similar, with motivation and job-related factors to the fore.

However, some individual characteristics are likely to be more important in the case of leaving intentions than in the case of absence. It has frequently been found, for instance, that turnover is higher, and leaving intentions more common, among younger workers than older workers. This is because younger workers are at a relatively earlier stage of their career and, because of fewer family commitments, are more able to take risks by changing employers (Webel and Bedeian 1989; Weisberg and Kirschenbaum 1991; Jenkins 1993). In addition, job change is used by younger workers as a way of testing the labour market and finding out what various employers are like. Other individual characteristics which have been found, in previous research, to increase job turnover or leaving intentions include higher-educational qualifications and higher-level occupations.

Another feature of job turnover is that it is likely to be particularly sensitive to good external opportunities, whereas for absence or job performance, these are merely a minor factor. In contrast to the external pressures which 'push' workers to attend or to work well, there are external 'pull' factors inducing employees to leave, such as higher wages elsewhere.

In analysing leaving intentions, the method was the same as for the two previous sections of the chapter. There were two versions of the analysis, the first to assess the effectiveness of a wide range of job-related factors and personnel practices on leaving intentions, the second to identify the added value provided by broad motivational factors, such as organizational commitment, task discretion, satisfaction with pay, and work strain.

The analysis was also repeated separately for men and for women: this was because family or life-cycle factors may affect the labour market mobility of men and women in different ways. As with the absence analysis, some preliminary work found no significant direct effects of marital status or dependent children on labour market turnover, and these variables were not included in the eventual analyses presented here. However, the separate analyses by gender helped to ensure that wider differences in mobility were taken into account. It turned out that the differences in findings, between men and women, were much more extensive than the similarities (see Hakim 1996*b*).

Findings about Leaving Intentions

Gender differences were particularly conspicuous in the individual factors which affected leaving intentions (Table 10.7, panel (A)). An intention of leaving in the near future was much more common among those aged under 35, and especially under 25, but this was almost entirely the result of leaving intentions *among young women*. Conversely, there were much higher leaving intentions among

TABLE 10.7 *Factors related to leaving intentions (Version 1 analysis)*

	All		Women		Men	
	Coeff.	Sig.	Coeff.	Sig.	Coeff.	Sig.
(A) *Personal characteristics*						
Female	1.09	n.s.	—		—	
Age 20–4	2.89	**	9.39	**	1.43	n.s.
25–34	2.10	*	9.12	**	0.82	n.s.
35–44	1.31	n.s.	4.35	*	0.73	n.s.
45–54	1.48	n.s.	4.35	*	0.96	n.s.
Professional/managerial	1.62	**	1.13	n.s.	2.51	***
Lower non-manual	1.43	*	1.11	n.s.	2.51	**
Length of service	0.93	*	1.01	n.s.	0.78	***
(B) *External pressures*						
Union slightly influential	0.57	***	0.64	n.s.	0.49	**
Union influential	0.46	**	0.49	*	0.40	**
Union v. influential	0.51	n.s.	1.32	n.s.	0.15	*
Get job fairly diff.	1.62	***	1.15	n.s.	2.39	***
Get job easy	2.44	***	1.54	*	3.86	***
Anxious dismissal	1.60	***	1.68	*	1.58	*
(C) *Job-related factors*						
Performance management	0.84	***	0.85	**	0.81	***
Take part in work decisions	0.75	*	0.98	n.s.	0.60	**
Supervisor coaches	0.65	**	0.44	***	0.97	n.s.
Communications	0.68	**	0.79	n.s.	0.63	*
Good benefits	0.54	***	0.55	*	0.48	**

Note: Logistic regression—multiplicative effects on odds. Sample nos. are 2,837 (all), 1,367 (women), and 1,470 (men). Non-significant effects are not shown.

non-manual than manual workers, and lower leaving intentions for those with longer service in the organization: however, these relationships applied *only to men*, not women.

The differences between men and women were less marked in terms of external pressures pushing individuals to stay or leave (panel (B) of Table 10.7). Both women and men were less likely to plan to leave if there was a fairly influential union in the workplace, and more likely to do so if they felt there was a reasonable chance of getting a job. In both these cases, it is true, men seemed more responsive to the influences. Interestingly, both men and women felt more likely to leave of their own accord where they feared dismissal. A 'hire-and-fire' atmosphere appears to raise labour costs by increasing voluntary turnover as well.

Of the three performance measures considered in this chapter, leaving intentions seemed the most sensitive to job-related policies and practices (panel (C) of Table 10.7). It is interesting to note first, though, that type of contract was not a significant influence on leaving intentions. Contrary to what is often assumed, neither a part-time nor a temporary job in itself led individuals to want to leave.

Men and women were about equally influenced by some of the policies, but in other cases there were marked gender differences. The presence of performance-management systems had a particularly clear impact, across both men and women, on reducing leaving intentions. And where employees saw and appreciated good fringe benefits, such as pensions, they were much less likely to want to leave; this applied equally to women and men.

Practices which had a gender-specific effect included open two-way communications, and providing the individual with opportunities to take part in decisions affecting work. These positive effects of participation only applied in a clear way to men, not women. Conversely, where the supervisor coached employees and helped them to learn, this substantially reduced leaving intentions among female employees, but did not influence men to a significant degree.

Motivation and leaving intentions

The additional influence of motivational factors on leaving intentions is summarized in Table 10.8. Here there was broad similarity between the results for men and women, but with one important exception. Feelings of high work strain made women significantly more likely to want to leave, but appeared to have no effect on men's leaving intentions. This gender difference underlines the greater impact of work strain on women which the previous analyses in this chapter have found.

For the first time in this chapter, satisfaction with pay emerged as an important influence. As expected, those who were satisfied were much less likely to be thinking about leaving than those who were dissatisfied, and this applied about equally to men and women. Pay satisfaction, it should be noted, was one of the variables included in the Version 2 analyses of absences and of work quality, but was non-significant in both cases.

TABLE 10.8 *Motivational effects on leaving intentions (Version 2 analysis)*

	All		Women		Men	
	Coeff.	Sig.	Coeff.	Sig.	Coeff.	Sig.
Pay satisfaction	0.80	***	0.77	***	0.79	***
Task discretion	1.04	n.s.	1.04	n.s.	1.03	n.s.
Upskilling	0.86	**	0.88	n.s.	0.87	n.s.
Work stress	1.04	*	1.12	***	0.98	n.s.
OC[a]						
values	0.87	**	0.89	*	0.81	**
flexibility	0.66	***	0.74	***	0.59	***
effort	0.84	n.s.	0.79	n.s.	0.86	n.s.

[a] OC = Organizational commitment

Note: Logistic regression—multiplicative effects on odds. Sample nos. are 2,837 (all), 1,367 (women), and 1,470 (men).

Task discretion, although it reduced absence among male employees, had no effect on leaving intentions. On the other hand, upskilling reduced leaving intentions even though it increased absence. The effect of upskilling may in part at least be an economic one: continuing training perhaps largely imparts skills which are specific to the present employer, and will only be fully recognized and rewarded by remaining there.

Finally, both men and women were much less likely to want to leave when they had high *organizational commitment*. 'Flexibility commitment' was particularly powerful, but this is hardly surprising, since both the items used for this measure refer explicitly to wanting to stay with the organization. In addition, high 'value commitment' also reduced the likelihood of wanting to leave. 'Effort commitment' also worked in the same direction, although its effect was not significant when the other two commitment measures were present.

Conclusions about Job Turnover

A major concern of this chapter has been the link between organizational commitment and various measures of labour cost and employee performance. This section of the chapter, which has found that job turnover is considerably reduced by a high level of organizational commitment, can be placed alongside the similar findings concerning frequency of absence. The analysis confirms that job turnover, like absence, is likely to be lower among employees with greater belief in the goals and values of their organizations. Organizational commitment can help to reduce labour costs.

Leaving intentions turned out to be sensitive to a wide range of organizational policies and practices, which can therefore be assumed to affect actual job turnover. Some of the practices reducing turnover were among those associated with the 'high commitment' management approach, notably participative relations in the workplace, and a supervisor who coaches team members. Performance-management systems were also effective in reducing leaving intentions, but these appear to represent a managerial 'control strategy' rather than the 'strategy of commitment' said to characterize the HRM approach (see the discussion in Chapter 3). In addition, leaving intentions were influenced by rather obvious economic considerations: by satisfaction with pay in the current job, by good fringe benefits, and by the perceived ease or difficulty of finding another job outside. Upskilling in the job, which was also effective in reducing leaving intentions, can be seen in part as a motivational practice and in part as an economic factor because of its bearing on future earnings potential inside the organization.

An important factor which *increased* leaving intentions—at least for female employees—was work strain. Work strain emerges from the whole set of analyses in this chapter as a particularly crucial influence on performance. Female employees who suffer from job strain will be more likely to think of changing employers. They will also be more likely to be absent. Employers face a complex issue of management here, because work strain is also associated with high levels of job performance. Can job performance be maintained if work strain is lowered, or can performance be raised without increasing it?

CONCLUSION: ORGANIZATIONAL POLICIES AND EFFICIENCY

This chapter has examined the influences on three aspects of employee behaviour which affect labour costs: absence, the quality of job performance, and job turnover (or leaving intentions). The findings can now be reviewed across all three types of outcome. The central question to be considered is how far organizations have identified policies and practices which are effective in managerial terms.

A striking finding was that no single policy or practice was significantly related to all three outcome measures. The most widely effective personnel policy appeared to be the use of performance-management systems. With their mix of internal progression, target-setting, appraisal, and merit pay, these systems strongly reduced job turnover and equally stimulated work quality. Moreover, the influences were significant for both female and male employees. It should also be recalled that, in Chapter 3, performance management was shown to increase work pressures, although only for non-manual workers. However, even performance-management systems had no influence on frequency of absence from work.

Other personnel policies had a positive impact only on one or two of the three outcome measures and then either only for men or only for women. Even though the results for men and for women often pointed in the same direction, it seemed that the strength of impact of each policy varied considerably by gender. In particular, open communications and participation in the work task had generally positive impacts for men while a supportive, coaching supervisor was more important for women.

A number of the policies and practices associated with the 'high commitment' management school are shown to produce results which management would see as desirable. But other approaches were equally effective in particular cases. The importance of performance-management systems has already been noted. When it came to controlling job turnover, the provision of ample fringe benefits was also important, and this exemplifies an older paternalistic tradition which lives on. The importance of working conditions for absence rates was another example which fits an old welfare tradition better than a new HRM perspective.

Important consequences also flowed from the use of 'flexible' employment contracts, but these consequences were ambiguous. Reduced absence rates among both temporary workers and part-time employees indicated one source of labour cost savings which may help to make these contracts attractive to employers. On the other hand, female part-time workers and male temporary workers felt less confident of their own work quality than workers on standard employment contracts. When added to the evidence from Chapter 6, about the low versatility of such workers, this suggests that the use of non-standard employment contracts may involve hidden costs.

Broad motivational factors, like the more specific personnel policies and practices, had a diversity of effects across the different kinds of outcomes.

- Organizational commitment reduced absence and job turnover but did not have an appreciable impact on the quality of job performance.
- Pay satisfaction reduced job turnover but otherwise had no positive effects.
- Task discretion did not affect job turnover but had favourable impacts on quality of work and, for men only, in reducing absence.
- Upskilling reduced labour turnover, and increased job performance, but it also increased absence rates for men, presumably because of the pressures resulting from the change in responsibility which went with skill.

These results leave little doubt that task discretion and upskilling, which have played such a large part in the account of earlier chapters, make a substantial difference to wider motivation and behaviour. Organizational commitment was also shown to make a difference to labour costs. But according to much popular management writing, there should also be a link between commitment and quality of performance, and this did not materialize here. Perhaps this negative finding reflects underlying weaknesses in organizational commitment, as suggested earlier in Chapter 9.

Work strain can be regarded as another indicator of motivation, albeit ambiguous since it can either be imposed or chosen. We found that work strain affected women's performance substantially, but men's hardly at all. For women it was linked positively with high work quality, but it also increased absence and job turnover. But while men were unaffected in this respect, the same kind of trade-off appeared, in their case, over upskilling, which raised the quality of their performance but increased their absence rates.

So there appears to be some tension between a high leverage on job performance on one hand, and maintaining high rates of attendance and labour force retention on the other. While a variety of policies and practices can stimulate higher performance, if work strain is also increased this may set a limit to the process. Here a fostering of organizational commitment may be attractive to the employer, in reducing the absence and turnover costs which a stressful drive for higher performance will induce.

Taken as a whole, this chapter confirms that employers are able to shape employees' behaviour and performance through personnel practices and through changes in the nature of work which affect motivation. But they are not able to do so in any simple way. Each policy produces only part of the desired results and has to be complemented by others, and some policies produce partly adverse consequences which must then be corrected in yet other ways. Determined pursuit by management of low labour costs and high work performance will not, according to these results, lead them towards a unified and coherent strategy, such as the 'high commitment strategy'. Rather, it will lead them towards increasingly complex sets of practices devised pragmatically to extract more from the employee. The question, which only experience can answer, is how far employees can continue to assimilate and reconcile the increasingly complex demands made of them.

Conclusion

The last decade has been marked by an extensive restructuring of British industry in the context of two recessions, a major technological revolution, and heightened international competitiveness. The central objectives of this research have been to assess the impact of these developments on the nature of work and on the experience of employment. Four main types of change have been of focal interest: the rise in skill levels, the implementation of new managerial policies with respect to employee relations, the increase of non-standard employment contracts, and the growth of labour market insecurity. While there has been no shortage of speculation about how widely such factors have affected the workforce, or about the depth of their impact, there has been remarkably little hard evidence. In the successive chapters of this book, we have sought step-by-step to build up a clearer picture. What assessment finally emerges of the extent and significance of these changes? How have they effected the characteristics of jobs? How have they effected the traditional forms of the employment relationship? What has been their impact upon the subjective quality of work experience and upon the way people perceive the organizations they work for? And, finally, how have they affected the major structural divisions in the workforce: those between men and women and between people in different occupational classes? Has there been a tendency towards the convergence or towards the polarization of employment conditions and work experiences?

THE CHANGING NATURE OF WORK

The Rise in Skill Levels

A good deal of the literature on the quality of work has emphasized the central importance of skill level. It is skill level that is seen as determining the ability of people to sustain and develop their creativity through work. Theories of the 'degradation of work' are ultimately theories of the tendency of industrial or capitalist development to undercut skills through an increasingly specialized division of labour which separates the conceptual aspects of work from the execution of tasks.

The evidence from this research with respect to skill trends is very consistent. The main direction of change in the last decade has been one of rising skill levels. Indeed, what is striking is just how pervasive the process of upskilling has been in recent years. A majority of all employees (63 per cent) reported that the skills required in their work had increased over the previous five years, compared to a mere 9 per cent who had experienced deskilling. A number of quite different tests confirmed the picture. Comparing the situation in the 1990s with surveys carried out in the previous decade, it was clear that employers were requiring higher qualifications for jobs of the same level and that employees were receiving more training in order to be able to carry them out. Moreover, employers attached sufficient value to the increase in skills for it to be clearly reflected in earnings even when age, sex, and occupational class had been taken into account.

While any one measure inevitably has its shortcomings, the cumulative nature of the evidence in favour of upskilling is impressive. Moreover, it was a process that affected both men and women and (albeit to different degrees) all social classes. It could not be explained away simply as a tendency for people to move upwards in the course of their careers. Among those that had remained in the same occupation as five years earlier, 62 per cent stated that the skill required in their job had increased and, even among those who had remained in exactly the same job as five years earlier, this was the case for 56 per cent. There can be little doubt that there has been a pervasive process of skill transformation affecting the working lives of a very substantial sector of the workforce.

Quite apart from its direct importance for the quality of work life, the rising level of skills was associated with other significant changes in the characteristics of work. To begin with, it was closely related to an improvement in the intrinsic interest of work. Both higher levels of skill and the experience of a recent enhancement in the skill requirements of the job were linked to greater variety in the nature of work and to greater possibilities for self-development through the work process itself. Further, they were intimately linked to the degree of discretion that people were allowed in making decisions about how to carry out the work. The evidence showed that there had been a significant increase over time in task discretion, a development that affected most occupational classes. Employers appear to have responded to the rising skill levels of work by devolving greater decision-making responsibility onto employees. This had the advantage of making better use of the knowledge of those directly involved in the work, while at the same time making it possible to cut back on the personnel employed in middle management and supervisory positions.

What factors underlay these trends in skill? The development that perhaps most caught the imagination of commentators of the changing economic scene was the spread of new computer-based technologies. The scenarios that were developed ranged from apocalyptic visions of the collapse of work to Arcadian images of a transformation of skills that would lead to the end of alienation in work.

There can be little doubt from the evidence examined here that the spread of new technologies indeed had been spectacular. Within little more than half a decade, the proportion of employees working with automated or computerized equipment had risen from only 39 per cent to over half the workforce (56 per cent). While it had swept most rapidly through the higher skilled categories of the workforce, no occupational class was unaffected. And if men were more likely to be users of new technologies, change over time had been even more rapid in women's work. In contrast to the false prophecies of earlier decades, those who argued that a pervasive transformation of the technical infrastructure of the world of work was now underway were clearly correct.

Further the evidence from the research points firmly in favour of the view that new technologies were associated with a rise in skill requirements. It was associated with higher demands by employers in terms of education and training even within occupational classes and those who had directly experienced the introduction of advanced technologies were more likely to report an increase in the skill demands of their job. Moreover, those who worked with advanced technologies were more likely to have seen their responsibilities for the work task increase than other employees. But this restructuring of work roles around new technologies was in no sense a deterministic outcome of the technologies themselves. Rather it reflected the interaction of new technical possibilities with managerial philosophies about appropriate methods of work organization. Most notably, employers appear to have responded in rather different ways to new technology depending on whether the workforce was male or female, increasing work responsibilities for men to a greater extent than for women.

The spread of advanced technology, then, does appear to have been associated with very positive changes in the nature of the work task and with higher levels of direct participation about relatively immediate work issues. However, in the wider context of the results, there must be doubts about the centrality that technology is often attributed in theories of change in the nature of work. It has already been seen that its implications for the design of jobs are contingent upon managerial choices about forms of work organization and that it is far from transforming the underlying employment relationship. But, in addition to this, it was only one of the factors that have contributed to the changing nature of work tasks and of the immediate work environment.

To begin with, the upward trend in skills appears to have been a much more widespread phenomenon than could be accounted for purely in terms of technological change. The overriding emphasis on the importance of technology for the nature of work derives from a historical period when the economic structure was dominated by manufacturing industry. The agenda of research has lagged well behind the transformations occurring in the real economy. One of the most important changes in the post-war era has been the shift in most advanced capitalist societies to economies based on the service industries. This has been accompanied by a profound shift in the underlying nature of work. The problematic of the alienation of the factory worker tied to the machine or the assembly line that dominated so much of the interwar and post-war literature is now beginning to look curiously dated.

In Britain in the early 1990s, only 6 per cent of employees were involved in assembly-line work, and only a further 11 per cent were primarily concerned with working with machines. Instead, the type of work that has grown in the wake of the expanding service economy is work that principally involves interacting with people. Our evidence indicates that just under half of the workforce (46 per cent) were predominantly involved in what might be termed people-work. This could take very different forms: it could involve caring responsibilities, selling products or services to people, or organizing people in their leisure and their work. But they pose radically different problems with respect to the experience of work to those of the repetitiveness and monotony of machine work. Moreover, it is clear that the performance requirements of people-work have been rising sharply in recent years, making this also a major contributor to the overall process of upskilling.

New Forms of Management

A second theme that has reappeared in many parts of the discussion is the nature and significance of managerial policies for regulating the workforce. The predominant post-war model, at least in the large-firm sector of British industry, had been based on the joint regulation of the conditions of employment through negotiation with trade unions. This was widely seen as implying a significant retreat in terms of traditional managerial prerogatives. Not only the terms of employment, but also the detailed organization of work, were increasingly subject to negotiation and agreement between management and shopfloor representatives. As shop-steward influence rose, many observers pointed to the decline of the power of supervision. Supervisors were reluctant to use disciplinary authority, when they were regularly bypassed by shop steward appeals to middle management. Rather management pinned its hopes on a policy of 'constitutionalism'. Where work roles were the outcome of joint regulation, it was more likely that they would be regarded as legitimate by the workforce. The economic and political conditions that had sustained this approach became increasingly precarious from the early 1980s.

But if the traditional system was clearly under attack, how exactly was it changing? A number of commentators pointed to the emergence of new systems of management, which emphasized more formalized, 'bureaucratic' procedures of performance assessment and career reward. The distinctive feature of this approach was that it involved a new emphasis on the direct relationship between management and the workforce, in contrast to the traditional pattern where relationships were mediated through the institutions of joint regulation. The importance of this was noted by writers from different intellectual perspectives: they were to be found in the ranks both of radical left critics of capitalist institutions and of authors of prescriptive guides to management practice. The latter found a common identity under the rubric of 'human resource management'. However, there was still relatively little evidence about the extent of such developments in Britain, let alone any empirical assessment of their implications.

Our evidence certainly indicated a hardening of employer attitudes to trade union influence. Whereas in the mid-1980s nearly two-thirds of employees (62 per cent) worked for employers that either actively encouraged or at least accepted trade unionism, by the time of the survey in the 1990s, this had fallen to less than 47 per cent. It was not just that there had been an increase in establishments where there were no unions present; the more hostile attitudes of employers were evident even if the comparison was confined to people in unionized establishments. There were good grounds for suspecting that employers were moving away from their reliance on joint regulation, with its implication of an arm's-length relationship with the shop floor, to more active forms of workforce management.

However, the research suggested that employers had followed quite diverse paths in seeking to strengthen their direct relationship with the workforce. Some had sought to increase the intensity of traditional forms of direct supervision, while others had adopted radically different strategies that involved giving greater discretion to employees themselves and shifting to more impersonal forms of control. The fact that these diverse forms of control were closely linked to the presence of unions certainly supports the view that they were an attempt by management to regain the initiative. But the pattern suggests a period of experiment with different types of policies by different types of employers, rather than the general endorsement of any specific new philosophy of management. But that said, there were clear signs of an increase in the types of practices commonly referred to as human resource management policies.

The literature on human resource management varies in its emphasis on 'individualistic' or 'collective' methods of developing a direct relationship between management and the workforce. In some versions, it is seen to involve encouraging collective participation at least in decisions about the relatively immediate work environment. However, the research found little evidence of a significant development of participative procedures. Only a small minority of British employees (32 per cent) felt that they could have any significant say over changes in work organization and only half felt that they could exercise any influence at all. There was no evidence at all that such participation had increased over the decade. Indeed, the proportion feeling that they could exercise some degree of influence may have declined somewhat since the mid-1980s. It was true that advanced technologies appeared to be conducive to higher levels of participation, but this has been counteracted by other shifts in managerial thinking that have tended to de-emphasize the importance of wider involvement of employees in organizational decision-making. The types of performance-management system that have been becoming more common are primarily those based on more individualized relationships between management and employees.

One of the most notable features about the development of these policies was that they were very much focused upon higher-level employees. Our estimates indicated that whereas just over half of managerial and professional and lower non-manual employees were under what might be defined as a performance-management system, this was the case for less than a third of manual workers. Manual workers, on the other hand, were more likely to be subject to technical control systems, in which the constraints on performance were embodied in the machinery itself or in payment systems based on measured output. The trends with respect to increased direct supervision were also strongly class stratified. In general, direct supervisory controls of everyday work performance were more likely to have decreased than to have increased for non-manual employees. However, among manual workers, direct supervision was more likely to have increased. In short, rather than the generalization of a new system of management across the workforce, our evidence suggests that employers were developing class-specific policies that tended to reinforce the differentiation between categories of employee.

In part, the adoption of performance-related policies can be seen as a method of extending managerial control to categories who had previously benefited from a relatively high-trust relationship. In that sense it represented an erosion of the privileged employment relationship enjoyed by employees such as professionals, with their traditional norms of autonomy. At the same time, it is clear that the growth of this type of control was linked to a policy of increasing the day-to-day responsibilities of employees for their work tasks. Whereas tighter supervisory control was associated with lower levels of task discretion, and technical control was neutral in this respect, there was a clear positive link between performance-management control systems and the level of task discretion. The introduction of performance management was not solely about the assertion of more detailed management control over employees that had been relatively autonomous in earlier decades. It was also linked to a wider policy of decentralizing work-task decision-making within organizations. This was, of course, inherently a highly ambivalent process. While more immediate forms of managerial control were lifted, longer-distance controls were strengthened.

Another factor that was closely linked to the spread of performance-management systems was advanced technology. Our evidence provided little support for the view advanced by some theorists that advanced technologies produce a working environment in which high-trust relationships make possible a reduction in systems of organizational control. There was no withering away of managerial control and, in some important respects, control was intensified. In certain sectors, employees working with advanced technology were more likely to be subject to forms of technical control, in which levels of work effort were determined

by the design of the technology itself and by payment systems based upon measured output. But, more generally, there was a very clear relationship between
advanced technology and the prevalence of systems of performance management. The introduction of new technological systems appears to have acted as
a stimulus for, and possibly as a facilitator of, the implementation of new philosophies of managerial control.

The persistence (albeit in modified form) of such organizational control systems reflects the fact that the relations between employees and management depend
on more than the quality of the immediate working environment. Although new
technologies were associated with upskilling and an improvement in the intrinsic interest of work, the terms of employment necessarily remain problematic
even where the quality of work is relatively high. There is no inherent reason
why people will be satisfied with, for instance, their pay or the scheduling of
their hours of work because they are involved in jobs that they find intrinsically
interesting. These are issues that require regulation at a higher organizational
level. However, there was no evidence that those who worked in more advanced
technical settings had greater collective control over their terms of employment.
Once organizational size had been taken into account, they were not more likely
than other workers to have collective representation either through elected works
councils or through trade unions. The commitment of employees to their organizations remained problematic and, consistent with this, there was no evidence that
advanced technology reduced the likelihood of industrial conflict. Given the persistence of conflicting interests, organizational control remained a central managerial concern.

While there are grounds for thinking that performance-management systems
are of growing importance, it is important to realize that they did not imply the
demise of the supervisor. A striking feature of the data was the pervasiveness
of supervisory activity. Not only were the great majority of employees supervised, but a remarkably high proportion also had some type of supervisory responsibility. In part, this omnipresence of the supervisor reflected the return of particular
types of employers to strategies of tighter direct supervision. But the supervisory process also remained important, albeit in rather different form, in the context of more sophisticated forms of control. The extension of an appraisal system
may well reinforce in the longer term the power of supervisors over those that
work for them, even though the relationship is no longer one of day-to-day direction of the person's work. These new control systems are sometimes defined
as 'impersonal', and it is suggested that one of their virtues is that they remove
the potential for interpersonal friction that was seen as one of the most damaging aspects of traditional supervisory control. However, it is unlikely that such
performance-management systems are impersonal in any very strict sense of the
term. Certainly, the procedures are more formalized and the short-term arbitrary
power of supervisors is curtailed. But the importance of the quality of the personal relationship between supervisor and supervised is likely to remain of funda-

mental importance. Where appraisal systems are linked to merit rewards, either through payment or promotion, there is also ample scope for dispute about the impartiality of assessments that have considerable implications for employees' fortunes both in the short and in the long term.

However, the most fundamental dilemma that may be associated with such reward policies could well be with the way they relate to another important structural trend in systems of work organization. It was noted earlier that there has been a widespread policy of enhancing the responsibilities of employees and decentralizing decision-making. One advantage of this for employers is that it has made it possible to cut back on the costly ranks of middle and lower management. There has been a general shift towards organizations with flatter structures. Yet there is an important way in which this type of development may conflict with the increased individualization of reward structures. One of the factors that makes a system of regular appraisal a powerful lever is that it is seen as linked to the opportunities that people have for career advancement. It is integrally related to the presence of a reasonably active internal labour market. But as the organizational pyramid becomes flatter, people's chances of receiving promotion are likely to diminish. One of the notable findings of our data was that people were now less likely to believe that their future career opportunities lay through their own organization than had been the case in the mid-1980s. If organizational career incentives decline, the control aspects of performance-management systems will become increasingly divorced from any real capacity to provide rewards. By raising expectations that are destined to remain unfilled, such systems could become a source of resentment rather than of increased social integration.

Non-Standard Contracts and the Polarization of the Workforce?

It has been seen that the more 'integrative' forms of management policy, which have been the primary focus of discussion about changes in patterns of management, were in practice heavily centred on non-manual employees, in particular professionals and managers. This was a development then that tended to confirm earlier lines of social division in the workforce. A number of writers, however, have gone further than this in suggesting that recent trends have made divisions between different categories of employee even sharper. One of the major factors behind this increased polarization of the workforce is seen to be the expansion of non-standard forms of employment contract, in particular part-time and temporary contracts. The picture drawn contrasted a relatively stable and integrated stratum of higher-skilled employees on the one hand, with an increasingly insecure and flexible stratum of low-skilled employees on the other.

The issue is certainly an important one in that just under a third of all employees were on non-standard contracts. Our evidence confirmed that part-time and temporary workers suffered from substantially poorer terms of employment. However, it did not support some of the conventional views of the nature of such work, and it underlined the distinctiveness of these different labour market segments. Indeed, our results pointed to the need to distinguish clearly not only between the conditions associated with part-time and temporary work, but between different types of temporary work.

A first point to note is that the various non-standard contract workers varied very much in terms of their skill distribution. Women in part-time work were certainly in markedly less skilled jobs, whatever indicator one takes. They were also less likely than either men or women in full-time work to have seen an increase in the skill requirements of their work. Indeed, they were disadvantaged on the more detailed skilled dimensions, even when allowance was made for the relatively low occupational class of their jobs. The low skill-level of the work was associated with little discretion over how the task was to be done and with work that was repetitive, lacking variety, and offering little in the way of opportunities for self-development. Although part-time employees in the public sector were rather better placed than their private sector equivalents, the overall pattern remains rather bleak.

However, this picture of the quality of the work cannot be generalized to all non-standard contract workers. Temporary workers on short-term contracts (less than a year) were much more evenly distributed over the skill structure, and were only slightly more likely to be unskilled than permanent employees. Those who were on medium-term contracts (one to three years), representing approximately half of all temporary workers, were at least as qualified as the permanent workforce. These medium-term 'contract' workers had shared fully in the process of upskilling, they were even more likely to have received training from their employer than permanent employees, and the quality of the jobs they did was just as high. These looked very much like 'entry' jobs into what was often quite high-quality work.

Second, the different types of non-standard contract offered specific types of flexibility. Neither female part-timers nor temporary workers were involved in jobs that could be defined as 'flexible' in any very general way. Indeed, there were several dimensions of flexibility on which both categories of non-standard worker were indistinguishable, or even less flexible, than permanent workers. Part-time employees were less likely than permanent employees to switch between different types of work on the job and there was no difference in this type of 'task flexibility' between temporary workers and permanent employees. Both part-time and temporary workers were less likely to be on incentive or merit payment systems that related pay to performance. Finally, part-time workers were less likely than permanent workers to put in extra hours of work and temporary workers were no more likely to do so. Although both types of non-standard contract clearly offered employers particular types of flexibility, they were in other respects relatively inflexible working arrangements.

Third, there were marked differences between the various types of non-standard contract in terms of the likely difficulty of being able to move out of that position in the labour market. Part-time employees were clearly heavily constrained in their opportunities for getting promotion into better work. They were much less likely than full-time employees to think that they had any type of career ladder they could move up and they were far less likely to think they had a reasonable chance of getting promotion to a better job within their organization. This absence of serious career opportunities also characterized the jobs of short-term temporary workers. However, those on medium-term contracts were much more optimistic about their career prospects and indeed were as likely as permanent employees to think that they would get promotion in their current organization.

Finally, the different types of non-standard contract had very different implications for job security. There was no evidence to support the common view that part-timers had lower job security. Certainly they were less likely to be protected by employment protection legislation and they perceived this fact quite clearly. Our estimates suggested that only 53 per cent of part-time workers would have been covered by employment protection, compared with 71 per cent of women working full-time and 74 per cent of men working full-time. But, in practice, female part-time employees felt just as secure in their jobs as permanent employees. Moreover, an analysis based on people's work histories of the relative risks of becoming unemployed for part-time and full-time employees suggested that they were quite correct in this perception. It seemed likely that the lack of formal security of their position was counterbalanced by the fact that they tended to work in the more protected service industries (and quite substantially in the public sector services). In contrast, both types of temporary worker did clearly suffer from much higher levels of job insecurity, particularly those on contracts of less than a year.

While those on non-standard contracts shared the fact that they were in some significant respects disadvantaged relative to permanent full-time employees, they were highly differentiated in the degree, and in the particular types, of disadvantage that they experienced. Simply placing these employees together into an undifferentiated category of the 'peripheral' or 'flexible' workforce is to obscure the very distinctive character of these labour market positions, whether in terms of skill, promotion chances, or security.

The group that did come very close to the notion of a 'peripheral' workforce was that of employees on short-term contracts of less than a year, although even these workers were more evenly distributed across the different skill categories than is often suggested. This clear peripheral group constituted 6 per cent of the workforce. Longer-term contract workers, although also less secure, were in a more complex position. They were as skilled as the permanent workforce and a significant proportion saw their jobs as opening the way to a career path. A substantial proportion appear to have been in the position of trainees, preparing to enter relatively good-quality jobs. Part-time workers were particularly disadvantaged in terms of skill and promotion opportunities, but they had a relatively high sense of job security. Given their level of effective job security, it is far from clear that they can be meaningfully classified as part of a flexible or peripheral workforce. Overall, the pattern of differentiation was considerably more complex than is often allowed and our analyses showed how important it is to look at people's experiences as well as at formal contractual provisions.

Finally, although both temporary and part-time workers experienced clear (if rather different) disadvantages in aspects of their employment conditions, there was no evidence of polarization across time. Not only had these categories grown very little over the decade as a share of the overall workforce, but there was no sign that their employment conditions had become worse relative to those on standard contracts. Indeed, a notable finding was that the relative skill position of part-time workers (although not the responsibility they were allowed in their work) had improved. This reflected an increase since the mid-1980s in the proportion employed in lower non-manual work and a decrease in the proportion in semi- and non-skilled manual work. There also appears to been some degree of convergence in the job insecurity of temporary and permanent workers (in part as a result of the declining security of 'permanent' workers).

Unemployment and Job Security

An even more severe form of differentiation of life chances than that resulting from the diversification of types of contract came from the growth of job insecurity and unemployment. This was clearly one of the most fundamental changes in the character of the labour market, affecting those with so-called 'permanent' contracts. Our charting of the overall careers of the people interviewed underlined the rise in vulnerability to unemployment since the 1970s for both men and women. Indeed, it was striking that among the most recent birth cohort in the survey, 40 per cent of men and nearly 30 per cent of women had experienced a spell of unemployment.

There were, however, important variations in the overall experience of employment instability. The deterioration of the labour market appears to have been particularly severe in its effects for men, particularly for young men. The position of men deteriorated with respect to all measures of stability. They were more likely to become unemployed, they were less likely to be in stable employment in the sense of being continuously in work, and they were less likely to have long-term employment with any given employer. For women, although there was a rise in the risk of unemployment, there was at the same time an *increase* in their overall employment stability. This apparent paradox results from the fact that they were less likely to withdraw from the labour market to have children and that they made greater use of maternity leave. Moreover, there was also an increase in the extent to which women remained in long-term employment with particular employers, indicating that they were securing more stable positions within organizations.

The factors affecting the risk of unemployment also differed in some important respects between men and women. Perhaps the most striking difference was the extent to which men were more heavily trapped by their previous educational and occupational backgrounds. Their degree of vulnerability was more strongly linked to family background, in particular the extent to which they came from families where the parents had encouraged education, and to the actual educational qualifications they had acquired. The risk for men was also more closely related to the occupational class of the jobs they had obtained (even taking account of early family circumstances). While there was evidence that in recent years an increased proportion of the unemployed were coming from higher occupational classes, it was nonetheless clear that it was still those in the manual working class that experienced the greatest risk of unemployment.

These differences have to be seen in the context of the very high level of gender segregation in the job market, with very different patterns in the development of job opportunities for men and for women. In the male job market, the contraction of jobs was above all in skilled and non-skilled manual work. Insofar as new job opportunities emerged, they tended to be in higher-level positions, which could be entered only through the acquisition of formal educational qualifications. There was then a very high educational penalty with respect to employment stability and transitions between contracting and expanding areas of employment were very difficult to make.

In the female job market, in contrast, there was a major expansion even of the number of non-skilled jobs (albeit often part-time) with the development of the service sector. The class of previous job was then much less decisive in determining the availability of job opportunities. This may also account in part for the lesser importance of education and family background for women's risk of unemployment. But, in addition to this, it must be remembered that in the service sector social skills were a considerably more important component of skill. Education is likely to be a weaker guide to the level of social skill than it is to the technical skills needed in jobs primarily concerned with the production of information or objects.

But despite these differences, there was one very important feature common to the experiences of job insecurity of men and women. For both, there was a clear process of entrapment whereby once people became unemployed they were much more vulnerable to further spells of unemployment. Given that education and early family circumstances had been taken into account, it seems unlikely that this reflects individual factors such as personality. Rather, the cumulative evidence showed that those coming out of unemployment were generally forced to take jobs in a much less secure sector of the labour market. They were excluded from more technologically advanced work environments (which were associated with greater job security), they were more likely to find themselves in temporary work, and they were more likely to be in jobs without trade union protection.

Moreover, it is clear that, even for those who did manage to stay in employment, the experience of unemployment had very severe long-term effects for their careers and for the quality of their work life. Those who had been unemployed were less likely to experience upward career mobility and they found themselves in jobs where there was less chance either to exercise or to develop their skills. Unemployment was not then a transitional experience, merely moving people from contracting to expanding areas of the economy. It left an enduring mark on people's work careers—greatly increasing the risk of future unemployment, making it more difficult to move to better jobs, and confining people to relatively poor-quality work. Further, the increasing risk of unemployment left its mark even on those who had not been directly affected, as was evident in the widespread worries about job security among those in employment and the fact that the greatest change in recent years in people's job preferences lay in the increased importance they attached to having a secure job.

THE EXPERIENCE OF WORK

What has been the impact of the changes that have marked the last decade on the quality of people's working lives? There are two different approaches that have been adopted in the literature on the quality of work. The first focuses on the objective character of the work situation. It has typically attached central importance to the opportunities that work provides for people to achieve self-development. In particular, it has emphasized the extent to which the skill level of jobs allows people to be involved in challenging and creative work, the degree of autonomy that they have in the work process, and finally their ability to participate in wider decisions that affect their employment conditions. The second approach concentrates on people's psychological response to their work situation, most typically their degree of satisfaction with their work. A more recent development has been to examine negative indicators of the psychological effects of work, such as its implications for psychological stress.

Both approaches have their strengths and their weaknesses. The 'objective' approach has the advantage that it is based on indicators that are unlikely to be heavily affected by the level of people's aspirations. People may say that they are content with their work, but this may merely reflect a lack of knowledge of other opportunities or a fatalistic acceptance that they are unlikely to be able to find anything better. Yet ultimately it would be difficult to accept that the objective quality of working life had improved if this conflicted significantly with people's own perceptions of their well-being. What type of improvement would it be if it left people feeling more miserable? The subjective approach has the advantage of ensuring that the analyst is not postulating an improvement that is meaningless to those affected. Clearly, a valid assessment of change in the quality of work life has to take account both of the types of objective change that are regarded as critical to self-realization and of people's psychological response to their work situation. Indeed, it is the links between the two which must be the central focus of research interest.

The single most commonly emphasized objective condition affecting the quality of work is that of skill level and it has been seen that skills developments have been in the direction that should have been conducive to a much more positive experience of work. The evidence is more ambivalent with respect to autonomy. There was certainly an increase in task discretion, but, as has been seen, greater task discretion did not mean that organizational controls were lifted. Rather there was a shift to different types of control that were potentially no less powerful than traditional forms of control in constraining employees to conform to organizational norms. It is an empirical issue whether the benefits deriving from greater task discretion persisted despite, or were largely cancelled out by, the emergence of new forms of performance control. Finally, there has been no improvement at all with respect to the third factor that has been seen as vital to the quality of work—direct participation in wider organizational decisions.

The trends with respect to the most commonly emphasized objective indicators of the quality of work are then far from convergent. It is primarily with respect to skill and task discretion that one can detect a significant improvement in the objective characteristics of work. It should also be remembered that most of the arguments about determinants simply take for granted that employees have high levels of security in their jobs. The notion of progressive self-realization through work fits poorly with an employment system that offers short-term jobs and little perspective for the future.

How have these changes in practice affected people's subjective experience in work? Our strategy in assessing this was to use both positive and negative indicators of job experience. The main positive indicator was job involvement. Many earlier studies had used measures of job satisfaction. These have been correctly criticized on the grounds that they may tell us little about the quality of work: people may be satisfied with undemanding and uninteresting work if they feel that the available alternatives are similar or even worse. The indicator of job involvement was designed to capture more clearly whether or not people found the work intrinsically interesting. Our negative indicators were of two types: a measure of work strain, on the one hand, reflecting the mental and physical fatigue involved in the work, and a measure of psychological distress, on the other, which referred to whether a person was vulnerable to anxiety or depression.

A first point to note is that the quality of work was something that was of central importance to people. In the 1960s, there had been forecasts that the long-term trend was towards a decline in the traditional work ethic and its replacement by an instrumental approach to work, in which a job would be seen simply as a way of generating the income needed to sustain family and leisure activities. However, there was no evidence to support this view. There had been no decline since the early 1980s in the extent to which people felt committed to employment irrespective of the financial benefits it brought. In terms of their priorities in choosing a job, the great majority of employees continued to attach a high level of importance to the intrinsic aspects of work. The view that people are income maximizers is not borne out by the data: high income was ranked only sixth in importance among job characteristics.

One important reason why the intrinsic qualities of work have not declined in importance is education. Higher levels of education generate higher expectations about the quality of work and educational levels have been rising. Another important factor, to which we will return, has been the changing importance of employment for women's lives.

The importance of the intrinsic aspects of work was strongly confirmed by the relationship between skill experiences and job involvement. The higher people's level of skill the more involved they were in their work and those that had experienced an increase in their skills had higher levels of involvement even within skill categories. Part of the reason why skill was so important was that it affected other aspects of the quality of work, such as the variety and the capacity for self-development in work. But even when these (and a wide range of other personal and organizational factors) were controlled for, an improvement in skills still remained a highly significant factor, suggesting that people enjoy the challenge of more complex work. And conversely, the experience of deskilling was powerfully associated with low levels of job involvement.

A very similar picture emerged for the effects of task discretion. Higher levels of task discretion were linked to higher job involvement. This might be expected given that there was a close link between skill and the scope people were given for making decisions on the job. As skills levels have increased, there has been a tendency for employees to be given greater say over how the job is to be done. But while part of the effect of task discretion can be attributed to skill in this way, the analysis showed it still retained an importance in its own right. Even at the same levels of skill, and with the same experiences of skill change, those who had more responsibility for deciding how to do their job felt more involved in their work.

The research provides strong support for arguments that the raising of skill levels and the granting of increased discretion to employees in work are key factors in improving the quality of work experience. But, at the same time, it suggested that the effects of such developments may be more ambivalent than is sometimes assumed. The negative side of the processes of upskilling and increased task discretion was that both were associated with a marked intensification of work effort. As people have had to learn and carry out more complex tasks, and take greater responsibility for everyday decision-making, the effort that they have had to put into their work has sharply increased. Indeed, employers are likely to have favoured such strategies in good part because they perceived them as having benefits in terms of higher levels of employee effort.

The implication of this was that higher skill levels, upskilling, and greater task discretion were all strongly related to the level of strain people experienced as a result of their work, in their mental tension, and physical fatigue. They were more likely to keep worrying about job problems when they got home, to find it difficult to unwind, and to feel exhausted at the end of the working day. This may have had a wider impact on the quality of people's lives. Although this could not be explored very far in our data, it was notable that work strain appeared to have negative effects on people's satisfaction with their family, social, and leisure lives.

If skill and task discretion had ambivalent effects for employee well-being, what were the implications of changing forms of managerial control? The analysis has focused in particular on two forms of control—technical control (in which the control of work performance was embodied in the technology of the work process or in the payment system) and performance-management control systems which relied on *post facto* systems of formalized appraisal and merit reward.

Much of the critique in the literature of the negative effect of managerial control systems on work experiences has focused on the implications of technical control, as exemplified for instance in the assembly-line production characteristics of the automobile industry. Our evidence fully supports the conclusions of earlier case-study work. Technical control systems were associated with lower job involvement and higher levels of work strain. These effects persisted even when a wide range of personal, task, and organizational characteristics had been taken into account. Technical control appeared then to increase work strain, even over and above its implications for the intensity of work effort.

It is, however, the implications of performance-management systems which are of particular interest, since these are becoming increasingly important in the context of higher skills and more decentralized forms of work organization. The effects of performance-related systems would appear to be markedly different from those based on technical control. Instead of reducing work motivation, they were related to higher levels of job involvement. It might be thought that this reflected the fact that such systems were particularly likely to be put into place where employees were allowed greater task discretion. But even when this was controlled for, the positive effect persisted. At the same time, they were associated with higher levels of work strain, mainly as a result of the fact that they were linked to higher levels of work effort.

Finally, the evidence strongly underlined the negative effects for people's work experiences of job insecurity. It was not formal contractual status—whether people were on 'temporary' or 'permanent' contracts—that was most important, but the *de facto* job insecurity that people experienced irrespective of contract status. Employees who felt that their jobs were under threat felt much lower levels of involvement in their work, they experienced higher levels of work strain, and, most crucially, they were more likely to display the more severe symptoms of psychological distress. It is notable that these effects still emerged at a high level of statistical significance even when occupational class, the characteristics of the work task, the level of work effort, and the forms of organizational control had been taken into account. While the changing character of work tasks led to an improvement in the quality of work over the decade, this was counterbalanced by the very negative consequences of heightened job insecurity.

ORGANIZATIONAL COMMITMENT AND PERFORMANCE

It has been argued that one of the most marked changes in work in the last decade has been the sharp rise in skill levels, together with the increased responsibility devolved on employees for everyday decision-making about how work is to be carried out. However, it has been frequently suggested in the literature that such developments place a much stronger pressure on employers to ensure high levels of commitment to the organization if they are to secure effective performance. Skill and responsibility transfer important power resources to employees and increase the dependence of management on the discretionary effort of employees. If organizations are to obtain higher levels of performance, they will need to ensure that employees identify with organizational goals and values. How extensive then is such organizational commitment? How successful were managerial policies designed to enhance commitment? And how far did the research support the view that organizational commitment was important for performance?

One of the most striking findings at the descriptive level was just how limited organizational commitment appeared to be among British employees. While there was little sign of explicit alienation, it was nonetheless notable that only 8 per cent of employees were strongly of the view that their own values and those of their organization were very similar, only 14 per cent that they were proud of their organization, and only 30 per cent that they felt loyalty to it. Moreover, only 28 per cent felt sufficiently attached to their organization to say that they would turn down another job if it offered higher pay.

Why has there been such a manifest failure to create stronger bonds? It might be that there are inherent limits to developing organizational commitment, given the tensions inherent in an employment relationship. It is also clear that the improved quality of work deriving from upskilling, while effective in increasing people's willingness to put in discretionary effort for their organization, did not lead in any automatic way to a greater commitment to organizational values. There was a positive association in the 'social' sector of employment, but it had no impact at all in the 'commercial' sector.

It did appear, however, that there were ways in which employers could enhance commitment. But the types of managerial policy that were effective in this respect were still of fairly limited coverage. For instance, performance-management systems were effective in increasing people's identification with organizational goals, but less than half of the workforce (42 per cent) were involved in them in any significant way, and these were predominantly people in higher occupational positions. Even more striking is the neglect by employers of the development of more participative practices that would increase employees' involvement in wider workplace decisions. This appeared to be one of the most powerful mechanisms for generating higher levels of commitment, with its influence persisting even when a wide range of other factors had been taken into account. But only a third of the workforce felt they could have a significant say over such decisions, and the proportion appears to have declined over the decade.

A fundamental difficulty confronting management's attempts to secure commitment was that this appeared to be strongly linked to what people perceived to be the nature of organizational values. And the values that counted were not so much 'market' values of efficiency, innovation, and reward as 'social' values of a more collective type. People felt committed to organizations when they regarded them as having a general approach that was caring about the welfare of employees and when they saw their activities as useful for society. An important point that emerged in this respect was the difference between the 'commercial' and the 'social' sectors. The importance of social values for commitment was vital for commitment in both sectors. However, it is clear that commercial sector firms had much more difficulty drawing on this source of commitment. The difference with respect to perceived usefulness to society was particularly great. Whereas 73 per cent of employees in the social sector were strongly of the view that their organization did something useful for society, this was the case for only 26 per cent in the commercial sector.

How important in practice was organizational commitment for employee performance? The evidence suggested that it was important in two principal respects. It had a significant effect in reducing levels of absenteeism. It also made an important independent contribution to reducing job turnover, as measured by people's intentions to leave the organization in the coming year. Both of these factors have very considerable implications for labour costs, so it is clear that arguments about the significance of organizational commitment do have to be taken seriously.

However, it should also be noted that the evidence did not provide any support for the view that organizational commitment added anything over and above other task and organizational characteristics with respect to the quality of work performance. This appeared to be much more powerfully influenced by more immediate job characteristics. In particular, employees with greater task discretion and employees who had experienced upskilling in recent years were higher on qualitative job performance. It was also notable that upskilling tended to reduce job turnover (presumably in part because the skills that were acquired were firm-specific).

The importance of participative practices again came out strongly from the analyses. Individual participation in decisions about work organization were strongly linked to higher work quality, as well as having the effect (at least for men) of reducing the likelihood of job turnover. Finally, it should be noted that performance-management systems also had a marked effect on job performance.

In contrast to those who have argued that they are primarily a matter of rhetoric, the empirical evidence again confirms the effectiveness of the incentive structure they provide. As with task characteristics, the main effect of these policies is a direct one; it does not depend upon their generating organizational commitment.

The importance of upskilling and task discretion for job performance is noteworthy, because it suggests that there is no inherent contradiction between improving the quality of work life for employees and sustaining high levels of performance. Indeed, more intrinsically worthwhile work may lead to a greater concern with quality.

However, there were clearly limits to the extent to which these processes could be developed without risking the emergence of negative effects. It was noted in the last section that both factors had the downside of increasing work strain. In the case of upskilling, this appears to have been sufficiently serious to have increased the likelihood of absence. Moreover, work strain had a significant impact for women on both absenteeism and the likelihood of leaving the job. There appears then to be a tension between policies that tend to stimulate higher performance and the problem of maintaining work strain within levels that do not have a serious impact on absence and turnover.

EMPLOYMENT CHANGE AND SOCIAL DIVISIONS IN THE LABOUR FORCE

The discussion up to this point has largely dwelt on the general trends with respect to work and employment. But this leaves the question of whether specific social groups may have been rather differently affected by these patterns of change. Given the size of the sample, there were limits to which these issues could be explored. In particular, the sample numbers were too small to examine with any rigour the very important issue of differences in experience linked to ethnicity, particularly given what is known about the sharply contrasting patterns that can be found between ethnic minorities. However, the research has attempted to examine the differential impact of change with respect to two major types of social division: gender and class. The most general issue is whether there has been any marked tendency for convergence or divergence in work experiences.

Gender and Employment

One of the most notable findings of the research was the very sharp rise in women's commitment to employment over the decade. Whereas in 1981, women had been quite clearly less committed to remain in employment irrespective of financial need, this difference had entirely disappeared by the 1990s. Moreover, this change did not reflect a sharp polarization of the trends in work motivation between women in full-time work, who were becoming more highly integrated into the workforce and women in part-time work whose commitment remained marginal. Rather the rise in employment commitment was even more marked among women part-timers than among full-timers, leading to increased homogeneity in women's attitudes to work. It is clear that the period witnessed a major and very broad cultural shift in the significance of employment in women's lives.

Yet, while women's commitment to work has become very similar to that of men, the results of the research underlined the continuing extent of gender inequality in employment and labour market experiences in the 1990s. A comparison of the measures of the skill position of men and women showed that women were far more likely to be in low-skilled occupations where there were no requirements for qualifications, training, and experience on the job. They were also less likely to have benefited from the opportunity to increase their skills on the job in recent years. These marked skill disadvantages were reinforced by the fact they were given less scope to take decisions about how to do the work. Women's skill disadvantage was not simply a result of their distribution by occupational class; rather, with the exception of those in professional and managerial positions, it was evident within each occupational class.

But what had been the pattern of change over time? Had the trend been towards growing polarization between men's and women's skill levels, as some have suggested, or has there been convergence? If one turns to comparison over time, there is substantial evidence that the gender gap for skill is being closed. The same trend emerged on each of the indicators of skill. The signs are that the process of convergence has occurred through cohort replacement. It was notable that the younger the age group, the smaller the skill gap. Indeed, in the youngest age group (those between 20 and 25) the gender gap had reversed: it was now women rather than men who were in the more highly qualified jobs. This parallels the increasing superiority of girls' educational attainment over the same decade.

Once more there was no evidence that this reflected any growth of differentiation between different types of women's work. It has been suggested that, while more highly qualified women in full-time work may have been able to improve their position, this has been counterbalanced by a deterioration in the work situation of those in less skilled positions, especially those in part-time work. It was certainly the case that women in part-time work scored lower on all of the skill indicators than other women and this remained true even if one took account of occupational class and age. However, taking the trend over time, the evidence indicates that the 'skill gap' for part-timers has narrowed *even more* than for other women. Indeed, much of the relative improvement in the skill position of women over the decade is attributable to the changing position of part-timers, who had been increasingly recruited into intermediate rather than non-skilled class positions.

But while women's skill position clearly improved, this was not the case with respect to the decision-making discretion they were given in the job. For men, the rise in levels of skill had been accompanied by greater responsibility on the job. But comparisons over the decade showed no relative improvement in

women's task discretion. However, in this respect, there were variations by occupational class. Among women in professional and managerial jobs, there was a
narrowing of the gap between men and women, whereas this was not the case
in other class positions. This provides some support for the view that the qualification lever has been particularly important in improving women's work status
in higher occupations. The discrepancy between the trends in skill and discretion
was particularly sharp among women in part-time work. Although part-timers
had seen a marked improvement in their skill position, there was even some
evidence of a decline in their task discretion. The difference in men's and women's
experiences of responsibility on the job was also evident with respect to new technology. New technology had the same effect for men and women in enhancing
skill levels, but it was only for men that it led to an increase in the scope for
taking decisions on the job. The evidence converges to suggest that employers
have continued to adopt gender-specific organizational policies that have operated
to women's disadvantage.

However, there was one major respect in which women appear to have been
at a relative advantage to men. Whereas men's employment stability has declined sharply over the decades, women's employment has become more stable.
Women have experienced an increase in their vulnerability to unemployment,
but this was less marked than for men. In terms of the continuity of their work
careers, it was overshadowed by the decline in the time they spent out of the
labour market and by the increased use of maternity leave. Moreover, for women
there has been a rise in the proportion with long-term relationships with particular
employers, whereas for men there has been a decline. The stability of women's
employment position was particularly brought out in the examination of part-
time work. Even though this type of employment had much poorer legal protection
than full-time work, in practice employment security appeared to be just as great.
The fact that women enjoyed greater protection from the sharp deterioration of the
labour market in the 1980s and early 1990s must be related in part to the high
degree of gender segregation of the occupational structure. Women's work was
heavily concentrated in the expanding service sector and they benefited from a
greater availability of job opportunities.

Overall, the pattern suggests that the decade saw a much deeper degree of
integration of women into the employment system. Their commitment to employment became indistinguishable from that of men, there was marked convergence
in skill levels, and they experienced higher levels of employment stability. These
patterns were not confined to a particular sector of women's work, but appear
to have been very wide-ranging. The main respect in which there was no evidence
of improvement was that of responsibility on the job. While employers have
made increasing use of women's skills, they have been reluctant to extend to
them the level of trust in decision-making accorded to male employees.

There was, moreover, a price that women paid for their increased integration
into employment. They were more likely than men to have suffered from work

strain in the sense of mental and physical fatigue. This may have reflected the increased demands on them in work or the problems of reconciling these work demands with their continued prime responsibility for child-rearing and domestic work. While they had a similar level of organizational commitment to men, they were more likely to be absent from work. It was, moreover, notable that work strain had a much stronger relationship to performance, absence, and leaving intentions for women than for men. All the signs point to the fact that the change in women's employment position has substantially increased the pressures they face and that there remain major difficulties in satisfactorily reconciling the demands of family and employment.

Class Convergence?

If the main trend with respect to gender divisions has been towards convergence, has there also been a tendency towards greater similarity in the employment relationship between classes?

It has been noted that the rise in skill levels was a very widespread phenomenon. A majority of those in non-manual and skilled manual class positions had benefited in recent years from skill development. At least 70 per cent of professionals and managers, lower non-manual employees, and technicians and supervisors had seen their skills increase, and this was the case for 64 per cent of skilled manual workers. But there was a major divide between those in higher and intermediate class positions and those in non-skilled manual work. Less than half (45 per cent) of non-skilled workers had seen an increase in the skill requirements of their work over the previous five years. Given the close links between skill and other intrinsic aspects of work, it is not surprising that there was also a sharp difference between classes in the degree of involvement that people felt with their work.

Moreover, once age and sex had been taken into account, there was no evidence that the relative position of the non-skilled had improved over time. The only class where there was firm evidence that skill levels had risen relative to those in professional and managerial work was that of lower non-manual workers, who had seen a significant increase over time in the qualifications required for their jobs. The broader picture is one of a general upward movement in skills, leaving relative class differentials much the same. Non-skilled workers continued to benefit far less from the opportunities for skill development, increasing the precariousness of their position in the labour market.

The move towards giving employees greater scope for taking decisions about tasks was more widespread across the class structure. All classes improved their position relative to professionals and managers with respect to decisions about how to carry out their normal daily tasks. There was, however, no change in relative class positions with respect to the more demanding criteria of task discretion—namely the ability to initiate new tasks—except in the case of skilled manual workers. Finally, in terms of wider forms of participation in workplace decisions, there remained a very sharp break between professionals and managers, who had relatively high levels of say over changes that affected their work, and all other employees who were largely excluded from such decisions.

It has been seen that the increase in task discretion did not mean a reduction in organizational controls. The system of control (involving both sanctions and rewards) has been held to be a crucial defining difference in class relationships. Those in lower-class positions are considered to be involved in an employment relationship regulated by the labour contract which involves 'a short-term and specific exchange of money for effort' (Erikson and Goldthorpe 1992: 41–2). Those in higher-class positions in contrast are thought to benefit from 'service' employment conditions, involving a longer-term and diffuse form of exchange of work for reward.

While the decade has seen significant developments in forms of managerial control, it is clear that these have been highly class stratified. It seems likely that professionals and managers and lower non-manual employees may have lost some of the autonomy they formerly enjoyed. But the forms of control that were imposed upon them still differed very considerably from those that characterized manual work. It was primarily with respect to these higher categories that management developed new forms of performance-management control. While these involved a distinct formalization of relationships, tighter monitoring, and an implicit reduction of trust, they were still premissed on a long-term relationship. The evaluation of performance was based on periodic assessment rather than daily monitoring and an important component of the reward structure was the prospect of longer-term career advancement. In contrast, manual workers were much more likely to be subject to technical control, which relied on machine-pacing and short-term pay incentives related to output. While there had been significant modifications in the forms of control within classes, these left intact the basic distinctions between classes.

Finally, an essential criterion of the type of employment relationship is the level of job security that people can expect. Here again the disadvantages of being in manual work, at least for men, stood out very clearly from the analyses of work-history experience. The major structural transformation of British industry had involved the decimation of large sectors of traditional manufacturing industry and it was those in manual work who bore the brunt of the costs in terms of vulnerability to unemployment. Moreover, given that those in non-skilled work were much less likely to benefit from ongoing improvement in their skills, they were in a far weaker position to cope with a rapidly changing technical environment.

There is some sign over the period of an increase in the proportion of the unemployed drawn from professional and managerial employees. But the predominance of manual workers in the ranks of the unemployed remained overwhelming (61 per cent for the period 1990–2). Moreover, those in professional and managerial work were far better protected by formal dismissal procedures and they were less likely to be confronted by the problem of obsolescence of skill when seeking new work. It was seen that when professionals and managers became unemployed they found new work by trading down in terms of job status. But for non-skilled manual workers there was much less scope for accepting work with lower skill; rather trading down had to take the form of accepting highly insecure work that led, as has been shown, to further cycles of unemployment.

Overall, then, there have been few signs over the past decade of a major reduction of class differentiation in the nature of employment relationships. The most significant improvement was in the spread of task discretion, which gave people greater responsibility in the way they carried out their everyday tasks. But with respect to skill development, the way work performance was controlled, and job security, the British employment structure still remained fundamentally divided by class. Insofar as the quality of employment had improved this had been largely to the benefit of those in higher and intermediate class positions. The most severe costs of change had been borne by those in non-skilled manual work.

However, there were some developments that in the longer-term may herald significant changes in this respect. If the advantage with respect to job security of those in professional and managerial work were to be further eroded, then the efficacy of performance-management systems in securing the commitment of these employees to organizational objectives may well decline. Moreover, it has been seen that it is above all those in professional and managerial work who have had to support the high levels of work strain that have accompanied the restructuring of work. Decreasing commitment, together with rising levels of pressure in work, could lead to the development of a more aggressive assertion of their interests as employees. This could generate a spiral of declining trust relations, more constraining forms of managerial control, and more conflictual relations with employers.

Such a scenario would clearly imply for the future an increased similarity in the nature of the employment relationship between professionals and managers and those in other class positions. Moreover, it must be remembered that, in the face of such developments, professional and managerial employees would be relatively well organized to defend their interests. In contrast to the manual working class, which witnessed a very severe erosion of trade union membership over the decade, this was not the case among professionals and managers. As a result, they are now among the most highly organized sectors of the workforce. Even when other factors have been taken into account, they are as likely to be unionized as skilled manual workers and they are more likely to be trade union members than those in non-skilled work. With their combination of resources of both skill and organization, professional and managerial employees may well play a central role in defining the employment relations of the coming decades.

TOWARDS A NEW MODEL OF EMPLOYMENT?

Finally, what are the implications of the different elements of our analysis for
the broader picture of change in work and employment relations? A number of
analysts have argued that the present era has been witnessing a profound
change in the nature of work and the employment relationship, representing a
rupture from the 'Taylorist' or 'Fordist' models that dominated the immediate
post-war decades. The new era would be one in which a reskilled workforce
would experience more satisfying work, higher levels of task discretion, greater
autonomy from managerial control, higher levels of employee participation, and
relatively secure employment. There would be an effective convergence on the
type of employment relationship that has characterized the professional.

Our research has certainly confirmed that the decade has seen major changes
in the nature of work in Britain and a marked restructuring of employment relationships. However, while these theories were in good part correct in some of
their key assumptions about the way in which the work task was changing, our
results cast considerable doubt upon their analyses of the change occurring in
the employment relationship.

The central development that such theories were correct in highlighting was
the rising levels of skills. This partly reflected managerial responses to new technologies and partly the demands for higher-quality performance in work demanding social skills. Moreover, the process of upskilling was associated with a major
improvement in the intrinsic interest of the work task. Work was becoming more
varied and opportunities for self-development greater. It is clear, too, that employers had tended to respond to rising skill levels by increasing the scope for
decision-making on the job (at least for the male employees). In general, these
developments imply some degree of reintegration of the conception and execution
of work and therefore represented a marked break with Taylorist conceptions
of work organization. It must be remembered that the period we were studying
was one of very rapid sectoral and technological change. It is an empirical question whether this widespread process of upskilling will continue with the same
momentum in the future. But certainly, in this respect, the past decade has seen
a significant step forward in the 'humanization' of work.

The evidence for a major change in the nature of employment relationships was
much less convincing. The first area in which our evidence conflicts sharply
with these 'optimistic' scenarios of change is with respect to the organizational
control of work performance. As part of their vision of the emancipation of work,
they posit that a key trend is a shift from 'control' to 'commitment' as the underlying mechanism for ensuring high levels of work performance. Our evidence
indicates that this provides a far too simple, and ultimately a misleading, description of the processes that have been occurring. Certainly, there is evidence that
management has been developing new policies for regulating the employment

relationship, which in part have the purpose of making employees more committed to their organizations. However, these were also systems designed to achieve more effective control of work performance, in a situation in which work processes had become more skilled, more complex, and less transparent. Managerial control of work performance had become more sophisticated, but control and its effects remained fundamental aspects of employee experience. Further, contrary to the assumptions of those who argue for a qualitative change in relationships, there was no evidence of an increase in the participation of employees in organizational decision-making. Rather it has been seen that the effective level of participation may have declined over the decade.

Taking the overall picture, the striking fact remained the low level of commitment of employees to their organizations. In part, this may have reflected the relatively limited development of policies that could secure such commitment. Performance-management systems were growing but still covered only a minority of the workforce. But it is notable that one of the most effective methods for securing commitment—involving employees in wider decisions in the workplace—had been largely neglected by employers. Given the significance of organizational commitment for both absenteeism and turnover, the issue of how to involve employees more deeply in the life of their organizations is likely to be of growing relevance in a steadily more skilled and technically complex economy.

A related reservation about the optimistic scenario is that this had been a period in which there had been a marked intensification of work effort. This contributed to higher levels of work strain, which at least in the case of women was sufficiently serious to affect absentee levels and intentions to quit the job. In part, this intensification of effort was integrally linked to the process of upskilling and greater task discretion. As jobs became more challenging they also became more mentally and physically demanding. While the quality of tasks was certainly improving for those who had jobs, the very developments that made this possible were integrally related to higher levels of work strain. It is also clear that the intensification of work effort was linked to the emergence of new methods for controlling work performance. While these appear to have been relatively effective in bringing about higher involvement in work, they were at the same time a major factor contributing to increased work effort. The problem of reconciling higher demands on skill and performance with tolerable levels of work strain is likely to be high on the agenda over the next decade.

Another crucial aspect of the employment relationship, where the trends fail to support the optimistic scenario, is that of job security. Since the 1970s there has been a marked increase in the risk of unemployment. It also has been seen that, among a significant proportion of those employed, there was a sense of job insecurity. These are developments that were not limited in their effects to some marginal category of the workforce. The increased risk of unemployment affected all occupational classes, even if the absolute levels remained very different from one class to another. Theories of a change in 'production regime' explicitly or implicitly assume that employees have a high degree of stability of employment, which enables them to develop skill flexibility and helps ensure a high degree of organizational commitment. However, the last two decades have seen the undercutting rather than the reinforcement of such long-term employment relationships.

Finally, despite the marked rise in skill levels and improvements in the quality of work, there persisted major differences in experiences between sectors of the workforce. Our evidence did not support some versions of the polarization thesis. For instance, while there were major differences in the employment conditions of 'regular' and 'non-standard' workers, there was no evidence that these grew greater over the decade and in some respects differences diminished. There was also no evidence of an overall sharpening of gender inequalities in employment. Rather it was a decade that saw a considerable enhancement of women's integration into the workforce, although with persisting gender differences in task discretion and material rewards.

However, the experiences of the workforce remained deeply divided along class lines. It was those in higher and intermediate class positions who primarily benefited from the positive changes to the quality of employment. They were much more likely to have experienced upskilling and hence more intrinsically interesting work. They certainly paid a price for this in terms of increased work effort and work strain. But the principal costs of change, namely the severe distress linked to unemployment, fell on the manual workers. The employment structure, then, continued to generate fundamental differences in people's life chances.

While it appears correct, then, that 'Taylorist' conceptions of work (to the extent that they prevailed historically) are now in the process of being superseded and that employers are placing a new emphasis on raising the skill levels of the workforce, the changes that have been taking place in the employment relationship are much less dramatic. The structures of control of work performance are being modified, but control remains pervasive and possibly more intense in the pressures it brings to bear on work effort. Far from converging on a 'professional' model of the employment relationship, the terms of employment remained fundamentally differentiated by class.

TECHNICAL APPENDIX

This appendix provides supplementary information about some of the main technical aspects of the questionnaire development, sample construction, fieldwork, and data analysis for the Employment in Britain Survey of employees.

THE INTERVIEW QUESTIONNAIRE FOR THE EMPLOYED

The interview for employed people can be visualized as in three parts:

(i) The work history, with which the interview commenced, and which took an average of fifteen minutes to complete.

(ii) The main interview covering current and recent experience of work; this also concluded with questions about socio-economic characteristics of the individual and of the household. The average time for the main interview was about one hour.

(iii) A self-completion questionnaire, completed by the respondent (without intervention by the interviewer). This contained item scales suitable for self-completion, as well as some items which were of a relatively sensitive or personal nature. This took on average about ten minutes to complete.

To permit more ground to be covered within the cost and time constraints of the study, it was decided that some questions would be administered to only half the sample. This was achieved through the self-completion module, which for the main survey was prepared in two versions with partly differing content. The use of the two versions was alternated by each interviewer. The interviewer asked the respondent to answer the self-completion questionnaire on completion of the main interview. All but 3 per cent of respondents returned usable self-completion interviews.

There were two chief principles guiding the construction of the questionnaire. The first was relevance for testing theories and predictions about changes in the employment relationship. The links between these and questionnaire measures have been discussed in the various chapters. The second principle was comparability with other research: the aim was to contribute to the assessment of changes over time and cross-national differences.

Four main British sources were identified. These were the Women in Employment Survey of 1980, conducted by the Department of Employment and the Office of Population Censuses and Surveys (Martin and Roberts 1984); the 1984 Survey of Class in Modern Britain (CMB) conducted by the University of Essex (Marshall *et al.* 1988); the 1986 Social Change and Economic Life Initiative (SCELI) of the Economic and Social Research Council; and the British Social Attitudes Survey of 1989 (BSA89) (Jowell *et al.* 1990). The 1980, 1984, and 1989 surveys were comparable with the Employment in Britain Survey, in being based on nationally representative samples, although the 1980 survey did not collect data on men. The 1986 survey was not based on a nationally representative sample (it covered six travel-to-work areas around large towns or cities), but had a similar sample size and yielded a social class distribution which closely matched the national profile.

Occupation and Class

An important aspect of the design of the interview schedule was the inclusion of a life-long retrospective work history. The potential of this was first shown in the Women in Employment Survey and it was also adopted (in a different format) in the SCELI Survey. The present survey uses a somewhat simplified version of the procedure used in SCELI, and covered all changes of employment status since initial entry to the labour market, except those which lasted for less than one month.

The extensive occupational data collected was coded to the Registrar-General's Standard Occupational Classification (SOC) 1990. It was then converted, using a pro-gramme developed by Ken Prandy, University of Cambridge, and Peter Elias, Univer-sity of Warwick, into 'Goldthorpe Classes' (otherwise known as EGP classes). The class schema aims to differentiate positions in terms of their employment relations (for details, see Erikson and Goldthorpe 1992: 35–47). The details of the mapping of occupational unit groups into classes is given in Goldthorpe and Heath (1992). It is the only class schema to date which has serious evidence for its criterion validity (Evans 1992, 1996). In the text, we sometimes use it as a *proxy* for broad skill levels. It should be noted that this is not part of the interpretation of the schema by its authors, but reflects our own empir-ical finding that it predicts very well a range of skill measures (indeed, substantially better than the Registrar-General's classes, which are stated to represent skill groupings).

For presentational purposes, the class schema is used here in an aggregated form. Since the study focuses on the employed only, classes IVa, IVb, and IVc are excluded. The collapsed schema that we have adopted is as follows:

Class labels in text	Goldthorpe classes
Professional/managerial	I + II
Lower non-manual	IIIa
Tehnician/supervisory	V
Skilled manual	VI
Semi- and non-skilled	IIIb + VIIa +VIIb

The labels attached to the classes are our own. Professional/managerial, for instance, is referred to by the authors of the schema as the 'service class'. It should be noted that, in the schema used, routine non-manual employees (IIIb sales and services) are placed with the non-skilled, as recommended by Erikson and Goldthorpe (1992: 44).

Other Measures

With respect to particular measures, our measure of psychological well-being (or its con-verse, psychological distress) is taken from one of the most widely used and best validated measures: the General Health Questionnaire (Goldberg 1972). A number of items were used in the 1986 SCELI survey as well as in several studies of unemployed samples. The twelve-item version of this scale was used here. The measure of work-related strain was a four-item scale developed by Warr (1990).

Occasional items were taken from other nationally comparable sources. A single item measuring the concept of non-financial employment commitment was based on the version used by the MRC/ESRC Social and Applied Psychology Unit, University of Sheffield, in a number of large-scale inquiries (Warr 1982; Jackson *et al.* 1983). Another item, measuring the receipt of vocational education and training, was taken from the 1986

Training in Britain Survey (Rigg 1989). From the French national survey *Etude des Conditions de Vie* 1986–7, several items were taken concerning work with automation and computers, and physical working conditions. Wider comparisons concerning physical working conditions were made possible by inclusion of a series of items from the 1991–2 European Survey of the Working Environment by the European Foundation for the Improvement of Living and Working Conditions.

The nearest previous parallel to the present inquiry, in terms of the breadth of information collected, although not the aims, was perhaps the Quality of Working Life surveys conducted in the USA during the 1970s (Quinn and Staines 1979). Unfortunately, the items used in these American surveys were not replicated in Britain in the same period; also, the surveys were discontinued in the USA after 1980. Nonetheless, items from the Quality of Working Life Surveys were sometimes adopted for topics where there was no comparable British source.

The concept of *organizational commitment* was one where comparative national information for Britain would have been particularly valuable, but proved non-existent. A set of eight items used in the work module of the American General Social Survey (GSS) of 1991 was adopted here. These items in turn were drawn from the longer Organizational Commitment Questionnaire (OCQ) of Mowday (see Mowday *et al.* 1982).

Two of the most widely used motivational concepts are those of the 'work ethic' (Atieh *et al.* 1987; Furnham 1987) and the 'locus of control' (Rotter 1966). The seven-item Australian Work Ethic Scale (Ho and Lloyd 1984) and Spector's (1988) sixteen-item Work Locus of Control Scale were each shortened to five items. It should be recognized, of course, that the shortening of scales may detract from the richness or complexity of the original concepts.

SAMPLE DESIGN AND CONSTRUCTION

The Employed Sample

The sampling frame for the survey of employed people was the Postal Address File (PAF). This is a computerized database of all the addresses (but without names of residents) currently recognized by the Post Office. It has become the most commonly used sample frame for social surveys, replacing the Electoral Register. The advantage of PAF is that the addresses are continuously updated. A sample drawn from the PAF is likely to be very little affected by mobility, and also avoids the bias which, in the case of the Electoral Register, may result from non-registration.

The sample was derived by a commonly used multi-stage procedure. First 150 postal sectors were drawn by stratified random sampling, with probability proportional to the number of addresses in the sector. The decision to take 150 primary sampling units was based on experience, and is generally considered to be adequate to generate a nationally representative sample for Britain. A postal sector is roughly similar in size to a ward, with an average of 2,500 households per sector. The stratification procedure was carried out by CACI Inc., making use of population statistics held on their database, and applying as stratification factors geographical region, population density, and social class composition.

A practical advantage in selecting the primary sampling units with probability proportional to size is that an equal probability sample is then derived by having the same number of interviews in each primary sampling unit (Moser and Kalton 1971). A random sample of sixty addresses was drawn within each selected postal sector, in order to obtain thirty-six addresses with eligible members, and, with an assumed 70 per cent response rate, the desired twenty-five interviews per sector.

The final stage in the construction of the employed sample was doorstep screening and selection. Eligible individuals were those at the preselected addresses who were aged 20–60 (inclusive) and in employment. Self-employed people, people on temporary sick leave, and women on maternity leave, were counted as employed. If only one person was eligible at the address, an interview was sought with that person. If more than one person was eligible, the person to be asked for an interview was selected by means of a Kish Grid (Kish 1949; Moser and Kalton 1971).

WEIGHTING OF THE EMPLOYED SAMPLE

The chief reason for reweighting the employed sample results from the fact that only one person is interviewed at each address with eligible people. To correct for this, the number of eligible people at the address was recorded for each individual interviewed, and this (adjusted so as not to inflate the total sample size) was used as the primary weighting factor.

The distribution of the employed sample was compared with data from the Labour Force Survey in terms of age, gender, and social class. The only appreciable bias in the sample concerned gender, where there was an overrepresentation of women. Accordingly, a second weighting factor was introduced, down-weighting women and up-weighting men to match the proportions found in the April 1992 Labour Force Survey, while holding sample size constant.

The Unemployed Sample

The issues concerning the construction and weighting of the unemployed sample were particularly complex. Here only an outline description, focusing on principles rather than details, is provided.

For reasons of cost, it was necessary to sample and interview unemployed people in the same areas (that is, the same postal sectors) as the employed sample. The sample was generated on the computers of the National Unemployment Benefit System (NUBS) in matched areas. NUBS records virtually all unemployed people who draw benefit or obtain National Insurance credits by virtue of unemployed status. This was a simple random sample from each area, of those aged 20–60 and with at least three months of current unemployment. To meet the requirements of confidentiality, all those listed were given an opportunity by the Employment Service to withdraw their names from the sample before it was issued.

It should be noted that, in the interval between sampling and interview, a proportion of the unemployed sample had ceased to be unemployed. Some were in jobs, others had become economically inactive. These, however, were not excluded from the survey. This was a sample of unemployed people at the time of sampling, not at the point of interview.

Reweighting was necessary for the unemployed sample because the primary sampling units had been selected with probability proportional to size (number of addresses), which is a good proxy for employment but a relatively poor proxy for unemployment. The method of reweighting was based on Moser and Kalton (1971: 114–15) and implemented with the aid of tables, produced by the Employment Service, showing the actual numbers unemployed in each area at the time of sampling, and tables of the numbers of addresses in each primary sampling unit.

THE FIELDWORK

The success of any survey depends crucially on the care and the expertise with which it is carried out in the field. This task was performed by Public Attitude Surveys (PAS) Research Limited.

Development of the questionnaires involved several stages including: (*a*) two pre-tests, consisting of a total of 120 interviews, ninety-one with employees and twenty-nine with unemployed people; these involved a number of experiments with subsamples; and (*b*) a pilot survey of 150 interviews, 100 with employees and fifty with unemployed people.

The main fieldwork was preceded by whole-day briefing meetings for the interviewers. There were ten such briefings, held in April–May 1992. Interviewers were provided with an instruction pack prior to the briefings, and were required to complete examples on sample selection and work-history recording before attending the briefing. The briefings were led by the directing staff of PAS, to a plan agreed with the authors of the survey. Particular attention was paid to instructing the interviewers in the rules for sampling, and the use of the Kish Grid; and to practising the more complex parts of the questionnaire, notably the completion of the work-history section. Stress was also laid on the correct procedure to follow for administering the self-completion questionnaire. At least one of the authors was present at each briefing.

Interviewing commenced immediately after the briefings, and continued during the period late April to August 1992. All interviewers had assignments including both employed and unemployed samples, and the samples (employed and unemployed) were interviewed in parallel.

Details of the response rates, and of the sources of sample attrition, are shown in Table A.1. The overall net response rate (that is, excluding non-eligible or otherwise

TABLE A.1 *Response to the surveys*

(a) *Employed sample*		(b) *Unemployed sample*	
Sample issued	10,332	Sample issued	1,720
Nobody aged 20–60	2,725	Unusable address or not	
Nobody employed	1,149	known at address	303
Non-residential property	387	Total eligible sample	1,417 (100%)
Non-existent address	182	Refused/no contact	414
Empty/demolished	480		
Total eligible sample	5,409 (100%)		
Refused	939		
No contact (4+ calls)	556		
Other non-interview	45		
Total interviews	3,869 (72%)	Total interviews	1,003 (71%)

TABLE A.2 *Regional distribution of response rates (cell %)*

	Employed	Unemployed
Scotland	76	73
North	71	76
North-West	75	70
Yorks and Humberside	69	66
West Midlands	73	66
East Midlands	74	77
East Anglia	64	76
South-West	64	78
Wales	71	72
Greater London	67	55
Rest of South-East	71	76

non-valid sample) was 72 per cent in the case of the employed sample, and 71 per cent in the case of the unemployed sample. Regional variations in the response rates are shown in Table A.2; the main outlier is the low unemployed response rate in Greater London, which has also been commonly encountered in other surveys.

ANALYSIS METHODS

A wide range of analysis methods, often involving multivariate statistical methods, have been drawn upon in preparing this book. It is not possible, in this appendix, to provide a detailed explanation of all these methods. Rather, the aim is to provide an outline of our approach. This is in two sections, the first a short discussion of some scoring and scaling methods used to summarize and group the questionnaire responses, while the second concerns the multivariate statistical methods which have been applied. The statistical section is prefaced by some explanation of the causal language and inferences involved in the analysis, and a postscript offers some guidance on how the statistical results in the text should be read and interpreted.

Summarizing and Grouping Questionnaire Items

An extensive interview survey generates a great volume of information, and this poses considerable problems of data reduction and summarization. The chapters contain many references to scores or indices obtained by combining information from two or more items in the questionnaire. It is known that a score or scale based on several correlated items provides more reliable information than any of the component items. The power of explanation may, in turn, be improved through the use of composite measures of improved reliability, as has been demonstrated through research on the links between attitudes, personality, and behaviour (Ajzen 1988).

The design of the survey questionnaire was influenced by these considerations. The aim was generally to represent the important topics or aspects not merely by a single item, but by a group of related items which might be expected to form a composite measure. However, there were practical limits on how far this process could be taken; these limits arose from the need to cover many topics within an interview of tolerable length. The end result, therefore, was that many topics were represented by a small number of questions—say, two to four—rather than by a substantial battery of items. The longest scale used in the questionnaire was the twelve-item version of the General Health Questionnaire (GHQ).

Before bringing together a group of items to form a composite measure, the items were tested by means of reliability analysis. The statistical measure used to summarize inter-item reliability was Cronbach's alpha (Cronbach 1951). Where the grouping of items was not self-evident, principal components factor analysis was often used to guide the selection and, in many cases, to derive factor scores which could be used directly as the measures (see Lawley and Maxwell 1963; Lewis-Beck 1994). The standard practice with factor analysis which we followed (unless otherwise stated in the text) was (i) to use principal components analysis with rotation to an orthogonal solution by the varimax method, (ii) to ignore factors with an eigenvalue of less than 1, and (iii) to derive factor scores by the factor regression method.

In general, where the grouping of items resulted from a factor analysis procedure, the practice in this research has been to use factor scores to measure the concepts derived. Where, however, the grouping has been a priori, and has not been based on factor analysis, the score has been obtained by a simple summation of the items after testing their reliability.

Causal Language, Theory, and Inference

The study has largely been concerned with analysis of changes over time and with statistical sources of variation in outcomes. It is customary to report and discuss this kind of analysis in terms of 'influences' or 'effects' upon the changes or outcomes. This can be (mis)read as implying that the data are being used to *identify* a set of causes. But sets of causal relationships cannot be *identified* through social survey inquiries, especially those of a cross-sectional nature (for discussion, see Cook and Campbell 1979: ch. 7). It needs to be stressed, therefore, that the purpose of the analysis here is not to construct causal models. The purpose is the limited one of testing a number of predictions about the changing nature of work organization and the employment relationship, and of exposing them to the possibility of falsification. These predictions may come either from existing theories or from our own interpretations of recent changes.

In short, causality is not inferred from the analysis, rather there is a causal hypothesis which forms part of a prior theory, guides us to look at particular relationships in the data, and is tested through the analysis. It might be possible to write the whole account of the research while always keeping this distinction explicit, and never applying causal language directly to the data or the results of an analysis: but it would also be tedious to do so. For example, a statement like 'variable A was found to have a positive influence on variable B' might be rewritten 'the positive association between variable A and variable B, implied by the hypothesized influence of variable X on variable Y, has not been falsified.' The use of causal language in the present text should, then, be understood as a convenient shorthand for connecting the analysis to predictions of a causal type.

Multivariate Statistical Analysis

Four main types of multivariate analysis have been used in the research: these correspond to differences in the dependent, or outcome, variable which is being analysed.

(i) Multiple regression analysis by the ordinary least squares (OLS) method

Ordinary least squares regression requires that the dependent variable should be measured on a continuous numerical scale, and that this should be approximately normally distributed (Wonnacott and Wonnacott 1990; Gujarati 1995). It has great practical advantages by comparison with other methods, especially that the results have a direct numerical interpretation. We have therefore used it for analyses involving summative scales (obtained by adding the responses to several questions), provided that the responses were well distributed.

(ii) Non-linear regression analysis of binary outcomes

Often in the social sciences the outcome variable is binary (yes–no): for example, whether or not a person has experienced unemployment in a certain period. It is not appropriate to apply ordinary regression to this case. Instead, either logistic regression analysis or probit analysis may be used (Maddala 1983; Gujarati 1995).

For a binary outcome, what one observes is the relative frequency with which 'yes' and 'no' occur. In logistic regression analysis, this is converted into the form of *odds*. The term odds is used as in betting: if the outcome occurs nine times out of ten, then the odds-on are 9 to 1. For computational convenience, it is actually the natural logarithm of this ratio, or log-odds, which is used. The analysis then estimates the effects of the other variables under consideration on this outcome measure.

Probit analysis is very similar to logistic regression analysis, but makes a different assumption about the statistical distribution from which the observed outcomes have been drawn. In essence, it assumes that there is an underlying normal distribution, and that the two observed values ('yes' or 'no') represent areas above and below a cutting-point on this distribution. The results of the probit analysis do not lend themselves to easy numerical interpretation, and we have generally preferred the logistic regression analysis for this reason.

(iii) Non-linear analysis of ordered outcomes

Sometimes the outcome variable has several ordered categories, but not sufficiently many to justify use of ordinary regression. An example would be responses to a single item on organizational commitment, with four possible replies indicating the strength of commitment. In this case, ordered probit analysis is very useful (McKelvey and Zavoina 1975). As the name implies, this is an extension of probit analysis to the case where the various replies are interpreted as areas defined by several cutting-points through a normal distribution. As with probit analysis, the estimates do not lend themselves to straightforward numerical interpretation.

(iv) Analyses with events in time as the outcome

Analysis concerning events in time are different from other kinds of analysis (Allison 1982). Such an analysis occurs only in Chapter 5 of this book. One complication is that observations of events are likely to be 'censored' because only a certain period is visible in each person's history. Obviously, we cannot observe future time. Another main complication introduced by analysing the same people over time is that observations at different times are quite likely to be correlated, because they reflect the persistent attributes of that individual or that individual's long-term circumstances. Unless these background variables are fully captured in the analysis, they can have a distorting effect on the results. This is also known as the problem of individual heterogeneity. Methods of 'panel data analysis' have been specifically developed to handle this problem, but they become extremely complex for non-linear regression models (Hsiao 1986).

The analysis in Chapter 5 aimed to avoid the complications of a full panel data analysis. Each individual's work history was broken into a series of five-year blocks, and the observations became each time-block-within-person, or person-period. The dependent variable was whether or not unemployment (or some other labour market pattern) had taken place within a person-period. Since this is a binary variable, the logistic regression method described above can be applied.

The main simplifying assumption here is that unobserved influences on the dependent variable are uncorrelated across person-periods. This is unlikely to be strictly true, but the choice of a fairly long time-block (five years) should considerably reduce any such correlations. The importance of unobserved variables is also reduced by incorporating a wide range of observed influences into the analysis, including employment variables from the previous period. To the extent that individual heterogeneity is still present, the significance of the coefficients in this type of analysis is likely to be overstated. The final precaution to be taken, then, is to use a more stringent significance test than is elsewhere customary: for instance, to pay attention to effects only if they are significant at the 99 per cent-confidence level or beyond.

Interpreting the Results of a Multivariate Analysis

Since we have considerably simplified the way in which tables of results from multivariate analyses are usually presented in academic journals, it is important to define the conventions which have been adopted.

Usually the multivariate analyses in this book are presented in the following form:

Table example (extract)
Effects of Technical Control and Supervision on Work Pressure

	Coefficient	Significance
Technical control	0.15	**
Increased supervision	0.47	***
Reduced supervision	0.54	***

Ordinary regression (OLS) model; no. = 1,753.

In most cases, as in the example, the table focuses on a limited number of relationships rather than presenting the whole analysis; in some cases, for presentational reasons, results from a single analysis are divided across two or more tables. The text contains details of the control variables which were used in each main analysis.

The selected relationships are specified in the title. The dependent or outcome variable in this example is work pressure. The explanatory variables are technical control and supervision, also shown in the left-hand column. The column headed 'coefficients' shows the estimated effects of each explanatory variable on the dependent or outcome variable. Coefficients can be presented either in standardized or undstandardized form. The practice followed here is always to use the unstandardized coefficients, often referred to by statisticians as the *b*-coefficients. The reason for this preference is that unstandardized coefficients are more comparable across different samples.

How the coefficients are to be interpreted depends on the type of dependent variable being analysed, and hence on the form of analysis applied. This is indicated by the footnote to the table, which states that the analysis is an 'ordinary regression (OLS) model'. As explained above, ordinary regression should only be used when the dependent variable is measured on a continuous numerical scale. As a result the coefficients take a straightforward numerical interpretation. The first coefficient shown above, 0.15, indicates that for every additional unit of 'technical control', there is an increase on the work pressure scale of 0.15. The effect will be the same whether one is considering the difference between 0 and 1 on technical control, or between, say, 4 and 5. An important point to note is that, for dichotomous variables, the coefficients estimate effects which are measured relative to an implied dummy variable which is not shown (the reference variable). The results above say that those with increased supervision on average experienced 0.47 more units of work pressure than those whose level of supervision had not changed.

The variables in the example are not commensurate, so there is no way of directly comparing the magnitude of their coefficients on the dependent variable. This is the usual situation in social investigations. The only way of assessing the importance of the coefficients is to consider their statistical significance. This is achieved by means of the *t*-statistic. A *t*-statistic of approximately 2 (that is, the coefficient is equal to twice its standard error) is generally regarded as just significant; the result would have arisen by chance only 1 in 20 times. This is also referred to as 'significance at the 95-per cent confidence level'.

To simplify the presentation, the tables of results in the text do not show the standard errors and *t*-statistics, but summarize the conclusion (derived from the *t*-statistic) in terms of a 'starring' convention. A single * indicates that the result is significantly different from zero at the 95-per cent confidence level; ** indicates significance at the 99-per cent confidence level; and *** indicates significance at the 99.9-per cent confidence level. The abbreviation 'n.s.' means not significantly different from zero.

When a logistic regression analysis is used, the results will be presented in just the same form as above but the coefficients must be interpreted differently. As explained earlier, the logistic regression analysis estimates the effects on the log-odds of a binary outcome. In our tables, these estimates are converted so as to show the *multiplicative* effect of each explanatory variable on the odds of the outcome. For example, if we are estimating the effect of gender on receiving training, an estimate (after conversion) of 1.5 (with female as the positive value of gender) means that being a woman increases the odds of getting training by one-and-a-half times, relative to a man. If on the other hand the estimated effect is less than 1, that means that the odds of getting training is lower for a woman than a man. An estimate of 1 means that there is no effect either way.

A multiplicative relationship, unlike in ordinary regression, is non-linear. This means that the effects will be somewhat different for people with different characteristics. In the example given in the previous paragraph, the coefficient of 1.5 would refer to men and women who were, in other respects, at the average or reference values for the survey. The coefficient would vary somewhat for people with other combinations of characteristics.

As explained earlier, the probit or ordered probit analysis produces results which, superficially, are similar to those from the ordinary regression analysis, but lack any simple numerical interpretation. With probit or ordered probit analysis, therefore, reliance has to be placed chiefly on the significance shown in the table.

Finally, it should be noted that the likelihood or log-likelihood statistics often used with non-linear models have not been presented here with model results of that type. We have come to the view, much influenced by discussion with statisticians, that such summary statistics are often misleading in the case of non-linear models. There is more justification, however, for retaining the adjusted R^2 measure in the case of ordinary regression, and we have done so for models of that type. Even here, however, the significance statistics on the explanatory variables provide a sounder, though essentially judgemental, guide to whether a useful analysis has been produced.

REFERENCES

Ajzen, I. (1988), *Attitudes, Personality and Behavior*. Milton Keynes: Open University Press.

Allen, N. J., and Meyer, J. P. (1990), 'The measurement and antecedents of affective, continuance and normative commitment to the organization', *Journal of Occupational Psychology*, 63: 1–18.

Allison, P. (1982), 'Discrete-time methods for the analysis of event histories', in S. Lienhardt (ed.), *Sociological Methodology*. New York: Jossey-Bass.

—— (1984), *Event History Analysis: Regression for Longitudinal Event Data*. London: Sage Publications.

Anderson, E. (1993), *Value in Ethics and Economics*. London: Harvard University Press.

Angle, H. L., and Perry, J. L. (1981), 'An Empirical Assessment of Organizational Commitment and Organizational Effectiveness', in *Administrative Science Quarterly*, 21: 1–14.

Armstrong, P. (1986), 'Class and Class Control at the Point of Production—Foremen', in P. Armstrong (ed.), *White Collar Workers, Trade Unions and Class*. London: Croom Helm.

Argyris, C. (1957), *Personality and Organization*. New York: Harper & Row.

Atieh, J., Brief, A., and Vollath, P. (1987), 'The Protestant Work Ethic-conservatism paradox: beliefs and values in work and life', *Personality and Individual Differences*, 8: 577–80.

Atkinson, J. (1984), 'Manpower Strategies for Flexible Organisations', *Personnel Management*, 15(8): 28–31.

—— (1985), *Flexibility, Uncertainty and Manpower Management*, Report 89. Brighton: Institute of Manpower Studies.

—— (1993), *New Forms of Work in the UK*. Dublin: European Foundation for the Improvement of Living and Working Conditions.

—— and Meager, N. (1986), *Changing Work Patterns: How Companies Achieve Flexibility to Meet New Needs*. London: NEDO.

Attewell, P. (1990), What Is Skill? *Work and Occupations*, 17(4): 422–88.

Banks, M. H., Clegg, C. W., Jackson, P. R., Kemp, N. J., Stafford, E. M., and Wall, T. D. (1980), 'The use of the General Health Questionnaire as an indicator of mental health in occupational studies, in *Journal of Occupational Psychology*, 53: 187–94.

Barbash, J. (1983), 'Which Work Ethic?', in J. Barbash, R. J. Lampman, S. A. Levitan, and G. Tyler (eds.), *The Work Ethic—A Critical Analysis*. Madison, Wis.: Industrial Relations Research Association: 223–61.

Barnett, C. (1986), *The Audit of War: The Illusion and Reality of Britain as a Great Nation*. London: Macmillan.

—— and Starkey, K. (1994), 'The emergence of flexible networks in the UK television industry', *British Journal of Management*, 5: 251–60.

Barron, R. D., and Norris, G. M. (1976), 'Sexual divisions and the dual labour market', in D. L. Barker and S. Allen (eds.), *Dependence and Exploitation in Work and Marriage*. London: Longman.

Batstone, E. (1988), 'The Frontier of Control', in D. Gallie (ed.), *Employment in Britain*. Oxford: Blackwell.

Beatson, M. (1995), *Labour Market Flexibility*. Employment Department, Research Series no. 48.

—— and Butcher, S. (1993), 'Union density across the employed workforce', *Employment Gazette*, 101(1): 673–89.

Beaumont, P. (1987), *The Decline of Trade Union Organization*. London: Croom Helm.

Beck, U. (1992), *Risk Society. Towards a New Modernity*. London: Sage Publications.

Beechey, V., and Perkins, T. (1987), *A Matter of Hours: Women, Part-time Work and the Labour Market*. Oxford: Polity Press.

Beer, M., Spector, B., Lawrence, P. R., Mill, Q. D., and Walton, R. E. (1984), *Managing Human Assets*. New York: Free Press.

Bell, D. (1974), *The Coming of Post-Industrial Society*. Heinemann: London.

Benson, P. G., Dickinson, T. L., and Neidt, C. O. (1987), The Relationship Between Organizational Size and Turnover: A Longitudinal Investigation, *Human Relations*, 40(1): 15–29.

Berger, P. A., Steinmuller, P., and Sopp, P. (1993), 'Differentiation of life courses? Changing patterns of labour market sequences in West Germany', *European Sociological Review*, 9(1): 43–61.

Berger, S., and Piore, M. (1980), *Dualism and Discontinuity in Industrialised Societies*. Cambridge: Cambridge University Press.

Berggren, C. (1993), 'Lean Production—the End of History?', *Work, Employment and Society*, 7: 163–88.

Bessant, J. (1991), *Managing Advanced Manufacturing Technology*. Oxford: NCC/ Blackwell.

Bielby, D. D. (1992), 'Commitment to Work and Family', *Annual Review of Sociology*, 18: 281–302.

Bielby, W., and Beilby, D. (1989), 'Family Ties: Balancing commitment to work and family in dual earner households', *American Sociological Review*, 54: 76–89.

Blackburn, R. M., and Mann, M. (1979), *The Working Class in the Labour Market*. Cambridge: Cambridge University Press.

Blanchflower, D., and Cubbin, J. (1986), 'Strike propensities of the British workplace', *Oxford Bulletin of Economics and Statistics*, 48: 19–39.

Blauner, R. (1964), *Alienation and Freedom: The Factory Worker and his Industry*. Chicago: University of Chicago Press.

Blegen, M. A., Mueller, C. W., and Price, J. L. (1988), 'Measurement of Kinship Responsibility for Organizational Research', *Journal of Applied Psychology*, 73(3): 402–9.

Blumberg, Melvin, and Pringle, Charles D. (1982), 'The Missing Opportunity in Organizational Research: Some Implications for a Theory of Work Performance', *Academy of Management Review*, 7: 560–9.

Blumberg, P. (1968), *Industrial Democracy: The Sociology of Participation*. London: Constable.

Blyton, P., and Turnbull, P. (eds.) (1992), *Reassessing Human Resource Management*. London: Sage Publications.

Bommer, W. H., Johnson, J. L., Rich, G. A., Podsakoff, V. M., and Mackenzie, S. B. (1995), 'On the interchangeability of objective and subjective measures of employee performance: A meta-analysis', *Personnel Psychology*, 48: 587–605.

Boyer, R. (1988), *The Search for Labour Market Flexibility*. Oxford: Clarendon Press.

Bradley, K., and Hill, S. (1983), ' "After Japan": the quality circle transplant and productive efficiency', *British Journal of Industrial Relations*, 21: 291–311.

—— —— (1987), 'Quality circles and managerial interests', *Industrial Relations*, 26: 68–82.

Brannen, J., Meszaros, G., Moss, P., and Poland, G. (1994), *Employment and Family Life: A Review of Research in the UK (1980–1994)*. London: Dept. for Education and Employment.

Brannen, P. (1983), *Authority and Participation in Industry*. London: Batsford Academic.

Braverman, H. (1974), *Labor and Monopoly Capital. The Degradation of Work in the Twentieth Century*. New York: Monthly Review Press.

Brooke, Paul P. (1986), 'Beyond Steers and Rhodes Model of Employee Attendance', *Academy of Management Review*, 11(2): 345–61.

—— and Price, James L. (1989), 'The Determinants of Employee Absenteeism: An Empirical Test of a Causal Model', *Journal of Occupational Psychology*, 62: 1–19.

—— Russell, D. W., and Price, J. L. (1988), 'Discriminant Validation of Measures of Job Satisfaction, Job Involvement, and Organizational Commitment', *Journal of Applied Psychology*, 73(2): 139–45.

Brown, C., Reich, M., and Stern, D. (1993), 'Becoming a high-performance work organization: the role of security, employee involvement and training', *International Journal of Human Resource Management*, 4(2): 247–75.

Brown, R., Curran, M., and Cousins, J. (1983), *Changing Attitudes to Employment?* Research Paper 40. London: Dept. of Employment.

Buchanan, D. A., and Boddy, D. (1983), *Organisations in the Computer Age*. Aldershot: Gower.

Bulmer, M. (ed.) (1975), *Working Class Images of Society*. London: Routledge & Kegan Paul.

Burawoy, M. (1983), 'Factory management regimes under advanced capitalism', *American Sociological Review*, 48(3): 587–605.

Busch, Paul, and Busch, Ronald F. (1978), 'Women Contrasted to Men in the Industrial Sales Force: Job Satisfaction, Values, Role Clarity, Performance and Propensity to Leave', *Journal of Marketing Research*, 15: 438–48.

Campanelli, P., and Thomas, R. (1994), 'Working Lives Developmental Research. Issues surrounding the collection of life-time work histories'. London: Employment Department.

Caplan, R. D., Cobb, S., French, J. R. P., Van Harrison, R., and Pineau, S. R. (1975), *Job Demands and Worker Health*. Washington: US Dept. of Health, Education and Welfare.

Casey, B., Metcalf, H., and Millward, N. (1997), *Employers' Use of Flexible Labour*, Report no. 837. London: Policy Studies Institute.

Central Statistical Office (1996), *Social Trends 1996*. London: HMSO.

Chadwick Jones, J. (1969), *Automation and Behaviour: A Social Psychological Study*. London: Wiley-Interscience.

Cheng, Yuan (1994), *Education and Class: Chinese in Britain and the U.S.A.* Aldershot: Avebury Press.

Child, J., and Loveridge, R. (1990), *Information Technology in European Services. Towards a Microelectronic Future*. Oxford: Blackwell.

Chinoy, E. (1955), *Automobile Workers and the American Dream*. Garden City, Mich.: Doubleday.

Clark, A. (1996), 'Job Satisfaction in Britain', *British Journal of Industrial Relations*, 34: 189–217.

Clark, J., McLoughlin, I., Rose, H., and King, R. (1988), *The Process of Technological Change: New Technology and Social Choice in the Workplace.* Cambridge: Cambridge University Press.

Clegg, H. A. (1979), *The Changing System of Industrial Relations in Great Britain.* Oxford: Blackwell.

Cobb, S., and Kasl, S. V. (1977), *Termination: The Consequences of Job Loss.* London: Batsford Academic.

Coch, L., and French, J. (1948), 'Overcoming resistance to change'. *Human Relations*, 1: 512–32.

Commission on the Future of Worker-Management Relations (1994), *Report and Recommendations.* Washington, DC: Dept. of Labor.

Cook, T. D., and Campbell, D. T. (1979), *Quasi-Experimentation: Design and Analysis Issues for Field Settings.* Chicago: Rand-McNally.

Crompton, R., and Jones, G. (1984), *White-Collar Proletariat.* London: Macmillan.

Cronbach, L. J. (1951), 'Coefficient alpha and the internal structure of tests', *Psychometrica*, 16: 297–334.

Dailey, R. C., and Kirk, D. J. (1992), 'Distributive and Procedural Justice as Antecedents of Job Satisfaction and Intent to Turnover', *Human Relations*, 45(3): 305–17.

Dale, A., and Bamford, C. (1988), 'Temporary Workers: Cause for Concern or Complacency? *Work, Employment and Society*, 2: 191–209.

Daniel, W. W. (1987), *Workplace Industrial Relations and Technological Change.* London: Frances Pinter.

—— and Millward, N. (1983), *Workplace Industrial Relations in Britain.* London: Heinemann.

Darden, William R., Hampton, Ronald, and Howell, Roy D. (1989), 'Career Versus Organizational Commitment: Antecedents and Consequences of Retail Sales People', *Journal of Retailing*, 65: 80–106.

Darnell, A. C., and Evans, J. L. (1990), *The Limits of Econometrics.* London: Edward Elgar.

Davies, R. B., and Pickles, A. R. (1985), 'Longitudinal versus cross-sectional methods for behavioural research: A first-round knockabout', *Environment and Planning*, 17: 1315–29.

Davis, James A., and Smith, Tom W. (1992), *The General Social Survey: A User's Guide.* Newbury Park, Calif.: Sage Publications.

DeCotiis, Thomas A., and Summers, Timothy P. (1987), 'A Path Analysis of a Model of the Antecedents and Consequences of Organizational Commitment', in *Human Relations*, 40(7): 445–70.

Delbridge, R. (1995), 'Surviving JIT: Control and resistance in a Japanese transplant', *Journal of Management Studies*, 32(6): 803–17.

De Tersac, G. (1992), *Autonomie dans le travail.* Paris: Presses Universitaires de France.

De Vaus, D., and McAllister, I. (1991), 'Gender and Work Organization: Values and Satisfaction in Western Europe', *Work and Occupations*, 18(1): 72–93.

Dex, S. (1985), *The Sexual Division of Work.* Brighton: Harvester Press.

—— (1988), *Women's Attitudes towards Work.* Basingstoke: Macmillan.

—— and McCulloch, A. (1995), *Flexible Employment in Britain: A Statistical Analysis*. Manchester: Equal Opportunities Commission.

Dore, Ronald (1973), *British Factory, Japanese Factory: The Origins of Diversity in Industrial Relations*. Berkeley and Los Angeles: University of California Press.

Downing, H. (1980), 'Word Processors and the oppression of women', in T. Forester, (ed.), *The Microelectronics Revolution*. Oxford: Blackwell.

DuGay, P., and Salaman, G. (1992), 'The Cult(ure) of the Customer', *Journal of Management Studies*, 29(6): 615–33.

Dugue, E., and Maillebouis, M. (1994), *Autour de la competence: Les notions, les pratiques, les enjeux-Bibliographie*. Paris: CNAM.

Duncan, Otis Dudley, and Blau, Peter (1967), *The American Occupational Structure*. New York: Wiley & Sons.

Durand, C. (1978), *Le travail enchaine*. Paris: Seuil.

Edwards, P. K. (1986), *Conflict at Work: A Materialist Analysis of Workplace Relations*. Oxford: Blackwell.

—— (1987), *Managing the Factory*. Oxford: Blackwell.

—— (1992), 'Industrial conflict: Themes and Issues in Recent Research', *British Journal of Industrial Relations*, 30(3): 361–404.

—— and Scullion, Hugh (1982), *The Social Organization of Industrial Conflict*. Oxford: Blackwell.

—— and Whitson, Colin (1989), 'Industrial Discipline, the Control of Attendance, and the Subordination of Labour: Towards an Integrated Analysis', *Work, Employment and Society*, 3(1): 1989.

—— —— (1993), *Attending to Work: The Management of Attendance and Shopfloor Order*. Oxford: Blackwell.

Edwards, R. C. (1979), *Contested Terrain: The Transformation of the Workforce in the Twentieth Century*. New York: Basic Books.

Elgar, T. (1991), 'Task Flexibility and the Intensification of Labour in UK Manufacturing in the 1980s', in A. Pollert (ed.), *Farewell to Flexibility?* Oxford: Blackwell.

Erez, M. (1993), 'Participation in Goal-Setting: A Motivational Approach', in W. M. Lafferty and E. Rosenstein (eds.), *International Handbook of Participation in Organizations*. Oxford: Oxford University Press.

Erikson, R., and Goldthorpe, J. H. (1992), *The Constant Flux. A Study of Class Mobility in Industrial Societies*. Oxford: Oxford University Press.

Estrin, S., Grout, P., and Wadhwani, S. (1987), 'Profit sharing and employee share ownership', *Economic Policy*, 4: 13–62.

Etzioni, A. (1975), *A Comparative Analysis of Complex Organizations*, rev. and enlarged edn. New York: Free Press.

Evans, G. (1992), 'Testing the Validity of the Goldthorpe Class Schema', *European Sociological Review*, 8: 211–32.

—— (1996), 'Putting Men and Women into Classes', *Sociology*, 30: 209–34.

Fernie, S., and Metcalf, D. (1995), 'Participation, contingent pay, representation and workplace performance: evidence from Great Britain', *British Journal of Industrial Relations*, 33(3): 379–415.

—— —— and Woodland, S. (1994), *Does HRM Boost Employee-Management Relations?* Working Paper 548, London School of Economics: Centre for Economic Performance.

Fitzgerald, R. (1988), *British Labour Management and Industrial Welfare, 1846–1939*. London: Croom Helm.

Flanders, A. (1964), *The Fawley Productivity Agreements: A Case Study of Management and Collective Bargaining*. London: Faber.

Florkowski, G. W. (1994), 'Employment Growth and Stability under Profit-Sharing: A Longitudinal Study', *British Journal of Industrial Relations*, 32(3): 303–18.

Forester, Ted (1980), *The Microelectronics Revolution*. Oxford: Blackwell.

Foulkes, F. K. (1980), *Personnel Policies in Large Non-Union Companies*. Englewood Cliffs, NJ: Prentice-Hall.

Fox, A. (1974), *Beyond Contract: Work, Trust and Power Relations*. London: Faber & Faber.

Francis, A. (1986), *New Technology at Work*. Oxford: Oxford University Press.

Freeman, C. (1986), 'The diffusion of innovations—microelectronics technology', in R. Roy, and W. Wield (eds.), *Product Design and Technological Innovation*. Milton Keynes: Open University Press.

French, J. R. P., Caplan R. D., and Van Harrison R. (1982), *The Mechanisms of Job Stress and Strain*. New York: Wiley.

Friedmann, G. (1946), *Problemes humains du machinisme industriel*. Paris: Gallimard.

Frohlich, D., and Pekruhl, U. (1996), *Direct Participation and Organisational Change. Fashionable but Misunderstood. An Analysis of Recent Research in Europe, Japan and the USA*. Luxembourg: Office for Official Publications of the European Communities.

Furnham, A. (1987), 'Predicting Protestant work ethic beliefs', *European Journal of Personality*, 1: 93–106.

—— (1990), *The Protestant Work Ethic. The Psychology of Work-Related Beliefs and Behaviours*. London: Routledge.

Gallie, D. (1978), *In Search of the New Working Class*. Cambridge: Cambridge University Press.

—— (1988), 'Introduction', in D. Gallie (ed.) *Employment in Britain*. Oxford: Blackwell.

—— (1991), 'Patterns of Skill Change: Upskilling, Deskilling or the Polarization of Skills?' *Work, Employment and Society*, 5(3): 319–51.

—— (1996), 'Skill, Gender and the Quality of Employment', in R. Crompton, D. Gallie, and K. Purcell (eds.), *Changing Forms of Employment. Organisations, Skills and Gender*. London: Routledge.

—— Cheng, Y., Tomlinson, M., and White, M. (1994a), 'The Employment Commitment of Unemployed People', in M. White (ed.), *Unemployment and Public Policy in a Changing Labour Market*. London: Policy Studies Institute.

—— Kalleberg, Arne L., and White, Michael (1994b), 'Organizational Commitment in Britain and the United States.' Unpublished paper, Oxford University and the University of North Carolina at Chapel Hill.

—— Marsh, C., and Vogler, C. (1993), *Social Change and the Experience of Unemployment*. Oxford: Oxford University Press.

—— Penn, R., and Rose, M. (eds.) (1996), *Trade Unionism in Recession*. Oxford: Oxford University Press.

—— and Vogler, C. (1993), 'Unemployment and Attitudes to Work', in D. Gallie, C. Marsh, and C. Vogler (eds.), *Social Change and the Experience of Unemployment*. Oxford: Oxford University Press.

—— and White, M. (1993), *Employee Commitment and the Skills Revolution*. London: Policy Studies Institute.

—— —— (1994), 'Employer Policies, Employee Contracts and Labour Market Structure', in J. Rubery and F. Wilkinson (eds.), *Employer Strategy and the Labour Market*. Oxford: Oxford University Press.

Geary, J. (1992), 'Employment Flexibility and Human Resource Management', *Work, Employment and Society*, 6: 251–70.

—— (1995), 'Work Practices: The Structure of Work', in P. Edwards (ed.), *Industrial Relations. Theory and Practice in Britain*. Oxford: Blackwell.

—— and Sisson, K. (1994), *Europe: Direct Participation in Organisational Change. Introducing the EPOC Project*, Working Paper WP/94/18/EN. Dublin: European Foundation for the Improvement of Living and Working Conditions.

General Household Survey 1992 (1994). London: HMSO.

Gerhart, B. (1990), 'Voluntary Turnover and Alternative Job Opportunities', *Journal of Applied Psychology*, 75(5): 467–76.

Gershuny, J., and Marsh, C. (1993), 'Unemployment in Work Histories', in D. Gallie, C. Marsh, and C. Vogler (eds.), *Social Change and the Experience of Unemployment*. Oxford: Oxford University Press.

Gill, C., Beaupain, T., Frohlich, D., and Krieger, H. (1993), *Workplace Involvement in Technological Innovation in the European Community, ii: Issues of Participation*. Luxembourg: Office for Official Publications of the European Communities.

Goldberg, D. P. (1972), *The Detection of Psychiatric Illness by Questionnaire*. Oxford: Oxford University Press.

—— (1978), *Manual of the General Health Questionnaire*. Windsor: National Foundation for Educational Research.

Goldthorpe, John (1980), *Social Mobility and Class Structure in Modern Britain*. Oxford: Clarendon Press.

—— and Heath, A. (1992), *Revised Class Schema 1992*. London and Oxford: Joint Unit for the Study of Social Trends, Working Paper 13.

—— and Hope, K. (1974), *The Social Grading of Occupations: A New Approach and Scale*. Oxford: Oxford University Press.

—— Lockwood, D., Becchofer, F., and Platt, J. (1968), *The Affluent Worker: Industrial Attitudes and Behaviour*. Cambridge: Cambridge University Press.

—— —— —— —— (1969), *The Affluent Worker in the Class Structure*. Cambridge: Cambridge University Press.

Goodman, J. F. B. *et al.* (1977), *Rule Making and Industrial Peace*. London: Croom Helm.

Greene, W. H. (1990), *Econometric Analysis*. London: Macmillan.

Guest, David (1992), 'Employee Commitment and Control', in J. Hartley (ed.), *Employment Relations*. Oxford: Blackwell.

—— Conway, N., Briner, R., and Dickman, M. (1996), 'The State of the Psychological Contract in Employment', *Issues in People Management*, 16. London: Institute of Personnel and Development.

Gujarati, D. N. (1995), *Basic Econometrics*, 3rd edn. New York: McGraw-Hill International.

Hackman, J. R., and Oldham, G. R. (1975), 'The Development of the Job Diagnostic Survey', *Journal of Applied Psychology*, 60: 159–70.

—— —— (1980), *Work Redesign*. Reading, Mass.: Addison-Wesley.

—— and Wageman, R. (1995), 'Total Quality Management: Empirical, Conceptual and Practical Issues', *Administrative Science Quarterly*, 40: 309–42.

Hakim, C. (1987), 'Trends in the Flexible Workforce', *Employment Gazette* (Nov.): 549–60.

—— (1990), 'Core and Periphery in Employers' Workforce Strategies: Evidence from the 1987 ELUS Survey', *Work, Employment and Society*, 4: 157–88.

—— (1996*a*), *Key Issues in Women's Work: Female Heterogeneity and the Polarisation of Women's Employment*. London: Athlone Press.

—— (1996*b*), 'Labour mobility and employment stability: rhetoric and reality on the sex differential in labour market behaviour', *European Sociological Review*, 12(1): 33–52.

Halaby, Charles N., and Weakliem, David L. (1989), 'Worker Control and Attachment to the Firm', *American Journal of Sociology*, 95(3): 549–91.

Harrison, R. (1972), 'How to describe your organization', *Harvard Business Review*, 72(5): 119–28.

—— (1987), 'Harnessing personal energy: how companies can inspire employees', *Organizational Dynamics*, 16(2): 5–20.

Harrop, A., and Moss, P. (1995), 'Trends in Parental Employment', *Work, Employment and Society*, 9: 421–44.

Heath, Anthony, Mills, Colin, and Roberts, Jane (1992), 'Towards Meritocracy: New Evidence for an Old Problem', in Colin Crouch and Anthony Heath (eds.), *Social Research and Social Reform*. Oxford: Clarendon Press.

Heckman, J., and Singer, B. (1984), 'Econometric duration analysis', *Journal of Econometrics*, 24: 63–132.

Hedges, B. (1994), 'Work in a Changing Climate', in R. Jowell, J. Curtice, L. Brook, and D. Ahrendt (eds.), *British Social Attitudes. The 11th Report*. Dartmouth Publishing Co.: Aldershot.

Heneman, Herbert G. III. (1974), 'Comparisons of Self- and Superior Ratings of Managerial Performance', *Journal of Applied Psychology*, 59: 638–42.

Hibbett, A. (1991), 'Employee involvement: a recent survey', *Employment Gazette*, 99(12): 659–64.

Hill, A. B. (1984), *A Short Textbook of Medical Statistics*, rev. edn. London: Hodder & Stoughton.

Hill, S. (1991*a*), 'Why Quality Circles failed but Total Quality Management might succeed', *British Journal of Industrial Relations*, 29(4): 541–68.

—— (1991*b*), 'How Do You Manage a Flexible Firm? The Total Quality Model', *Work, Employment and Society*, 5: 397–415.

Hirschhorn, L. (1984), *Beyond Mechanization: Work and Technology in a Post-industrial Age*. Cambridge, Mass.: MIT Press.

Ho, R., and Lloyd, J. (1984), 'Development of an Australian work ethic scale', *Australian Psychologist*, 19: 321–32.

House, J. S. (1981), *Work Stress and Social Support*. Reading, Mass.: Addison-Wesley.

Hsiao, C. (1986), *Analysis of Panel Data*. Cambridge: Cambridge University Press.

Hunter, L., and McInnes, J. (1991), *Employer Use Strategies—Case Studies*, Research Paper 87. Sheffield: Dept. of Employment.

Huselid, M. A., and Day, N. E. (1991), 'Organizational Commitment, Job Involvement, and Turnover: A Substantive and Methodological Analysis', *Journal of Applied Psychology*, 76(3): 380–91.

Hyman, R. (1991), 'Plus ça change? The Theory of Production and the Production of Theory', in A. Pollert (ed.), *Farewell to Flexibility?* Oxford: Blackwell.

Incomes Data Services (1992), *Performance Management*, IDS Study 518, Nov.

—— (1997), *Performance Management*, IDS Study 626, May.

Industrial Relations Review and Report (1987), 'Taking the Pulse of Sickness Absence', IRRR, 405, Dec.

Jackson, P. R., Stafford, E. M., Banks, M. H., and Warr, P. B. (1983), 'Unemployment and psychological distress in young people: the moderating role of employment commitment', *Journal of Applied Psychology*, 68(3): 525–35.

Jacobi, O., and Hassel, A. (1995), 'Germany: Does Direct Participation Threaten the "German Model"?' *European Foundation for the Improvement of Working Conditions*, Working Paper WP/95/EN: 1–44.

Jahoda, M. (1982), *Employment and Unemployment: A Social-Psychological Analysis*. Cambridge: Cambridge University Press.

Jamal, M. (1990), 'Relationship of Job Stress and Type-A Behaviour to Employees' Job Satisfaction, Organizational Commitment, Psychosomatic Health Problems, and Turnover Motivation', *Human Relations*, 43(8): 727–38.

Jenkins, J. M. (1993), 'Self-monitoring and Turnover: The Impact of Personality on Intent to Leave', *Journal of Organizational Behaviour*, 14: 83–91.

Johns, G., and Nicholson, N. (1982), 'The Meanings of Absence', in B. M. Staw and L. L. Cummings (eds.), *Research in Organizational Behaviour*. Greenwich, Conn.: JAI: 127–72.

Jowell, R., Witherspoon, S., and Brook, L. (eds.) (1990), *British Social Attitudes: the 7th report*. Aldershot: Gower.

Judge, T. A., and Ferris, G. R. (1993), 'Social Context of Performance Evaluation Decisions', *Academy of Management Review*, 36: 80–105.

Jurgens, U., Malsch, T., and Dohse, K. (1993), *Breaking from Taylorism. Changing Forms of Work in the Automobile Industry*. Cambridge: Cambridge University Press.

Kalleberg, A. L., and Berg, I. (1987), *Work and Industry: Structures, Markets and Processes*. New York: Plenum.

—— and Leicht, K. T. (1986), 'Jobs and Skills: A Multivariate Structural Approach', *Social Science Research*, 15: 269–96.

—— and Losocco, K. A. (1983), 'Aging, values and rewards: explaining age differences in job satisfaction', *American Sociological Review*, 48: 78–90.

—— and Marsden, P. V. (1993), 'Organizational Commitment and Job Performance in the U.S. Labour Force', in Richard L. Simpson and Ida Harper Simpson (eds.), *Research in the Sociology of Work, V: The Meaning of Work*. Greenwich, Conn.: JAI Press.

Kanfer, R., Crosby, John V., and Brandt, David M. (1988), 'Investigating Behavioral Antecedents of Turnover at Three Job Tenure Levels', *Journal of Applied Psychology*, 73(2): 331–5.

Kaplinsky, R. (1984), *Automation: The Technology and the Society*. London: Longman.

Karasek, R., and Theorell, R. (1990), *Healthy Work. Stress, Productivity and the Reconstruction of Work Life*. New York: Basic Books.

Kelvin, P. (1980), 'Social Psychology 2001: The social psychological bases and implications of structural unemployment', in R. Gilmour and S. Duck (eds.), *The Development of Social Psychology*. New York: Academic Press.

Kenney, M., and Florida, R. (1993), *Beyond Mass Production: The Japanese System and its Transfer to the U.S.* Oxford: Oxford University Press.

Kern, H., and Schumann, M. (1987), 'Limits of the Division of Labour. New Production and Employment Concepts in West German Industry', *Economic and Industrial Democracy*, 8: 151–70.

—— —— (1992), 'New Concepts of Production and the Emergence of the Systems Controller', in P. S. Adler (ed.), *Technology and the Future of Work*. New York: Oxford University Press.

Kerr, C., Dunlop, J. T., Harbison, F., and Myers, C. A. (1960), *Industrialism and Industrial Man*. Cambridge, Mass.: Harvard University Press.

Kiesler, Charles A. (1971), *The Psychology of Commitment: Experiments Linking Behaviour to Belief*. New York: Academic Press.

Kim, Jae-On, and Mueller, Charles W. (1978), *Introduction to Factor Analysis: What It Is and How to Do It*. Beverly Hills, Calif.: Sage Publications.

Kish, L. (1949), 'A procedure for objective respondent selection within the household', *Journal of the American Statistical Association*, 44: 380–7.

Koch, J. L., and Steers, R. M. (1978), 'Job Attachment, Satisfaction and Turnover among Public Employees', *Journal of Vocational Behaviour*, 12: 119–28.

Kochan, T. A., Katz, H. A., and McKersie, R. B. (1994), *The Transformation of American Industrial Relations*, rev. edn. New York: Basic Books.

—— and Osterman, P. (1994), *The Mutual Gains Enterprise*, Cambridge, Mass.: Harvard University Press.

Kohn, A. (1993), *Punished by Rewards*. New York: Houghton Mifflin.

Kohn, M. L., and Schooler, C. (1983), *Work and Personality*, Norwood, N.J.: Ablex Publishing Corporation.

Landsberger, H. A. (1958), *Hawthorne Revisited: 'Management and the Worker'. Its Critics and Developments in Human Relations in Industry*. Ithaca, NY: Cornell University Press.

Lane, C. (1988), 'New technology and clerical work', in D. Gallie (ed.), *Employment in Britain*. Oxford: Blackwell.

Lane, R. E. (1991), *The Market Experience*. Cambridge: Cambridge University Press.

Lawler III, E. E. (1989), 'Participative Management in the United States: Three Classics Revisited', in C. J. Lammers and G. Szell (eds.), *International Handbook of Participation in Organisations, i. Organizational Democracy: Taking Stock*. Oxford: Oxford University Press.

—— (1994), 'TQM and Employee Involvement: Are they compatible?' *Academy of Management Journal*, 8(1): 68–76.

Lawley, D. N., and Maxwell, A. E. (1963), *Factor Analysis as a Statistical Method*. London: Butterworths.

Levitan, S. A., and Johnson, C. A. (1983), 'The Survival of Work', in J. Barbash, R. J. Lampman, S. A. Levitan, and G. Tyler (eds.), *The Work Ethic—A Critical Analysis*, 1–25. Madison, Wis.: Industrial Relations Research Association.

Lewis-Beck, M. S. (ed.) (1994), *Factor Analysis and Related Techniques*. London: Sage Publications.

Likert, R. (1961), *New Patterns of Management*. New York: McGraw-Hill.

Linhart, D. (1981), *L'appel de la sirene. L'accoutumance au travail*. Paris: Le Sycamore.

Lincoln, James R., and Kalleberg, Arne L. (1990), *Culture, Control and Commitment: A Study of Work Organization and Work Attitudes in the United States and Japan*. Cambridge: Cambridge University Press.

Lipset, Seymour Martin. (1991), 'American Exceptionalism Reaffirmed', in Byron Shafer (ed.), *Is America Different: A New Look at American Exceptionalism*. Oxford: Clarendon Press.

Lissenburgh, S. (1996), *Value for Money: The Costs and Benefits of Giving Part-Time Workers Equal Rights*. London: Trades Union Congress.

Littek, W., and Heisig, U. (1991), 'Competence, Control and Work Design', *Work and Occupations*, 18(1): 4–28.

Locke, E. A. (1976), 'The Nature and Causes of Job Satisfaction', in M. D. Dunnette (ed.), *Handbook of Industrial and Organizational Psychology*. Chicago: Rand McNally.

—— Latham, Gary P., and Erez, Miriam (1988), 'The Determinants of Goal Commitment', *Academy of Management Review*, (13): 23–39.

Lorence, Jon. (1987), 'A Test of "Gender" and "Job" Models of Sex Differences in Job Involvement', *Social Forces*, 66: 121–42.

—— and Mortimer, J. T. (1985), 'Job involvement through the life course: a panel study of three age groups', *American Sociological Review*, 50: 618–38.

Loscocco, K. A. (1989), 'The Interplay of Personal and Job Characteristics in Determining Work Commitment', *Social Science Research*, 18: 370–94.

—— and Kalleberg, A. L. (1988), 'Age and Meaning of Work in the United States and Japan', *Social Forces*, 67(2): 337–56.

Lundy, O., and Cowling, A. (1996), *Strategic Human Resource Management*. London: Routledge.

Lupton, T. (1963), *On the Shop Floor*. Oxford: Pergamon.

McGee, G. W., and Ford, R. C. (1987), 'Two (or More?) Dimensions of Organizational Commitment: Reexamination of the Affective and Continuance Commitment Scales', *Journal of Applied Psychology*, 72(4): 638–42.

McGregor, A., and Sproull, A. (1991), *Employer Labour Use Strategies: Analysis of a National Survey*, Research Paper 83. Sheffield: Dept. of Employment.

McGregor, D. (1960), *The Human Side of the Enterprise*. New York: McGraw-Hill.

McKelvey, R., and Zavoina, W. (1975), 'A statistical model for the analysis of ordinal level dependent variables', *Journal of Mathematical Sociology*, 4: 103–20.

McKersie, R. B. (1987), 'The transformation of American industrial relations: the abridged story', *Journal of Management Studies*, 24(5): 434–40.

McLoughlin, I., and Clark, J. (1994), *Technological Change at Work*, 2nd edn. Buckingham: Open University Press.

—— and Gourlay, S. (1994), *Enterprise Without Unions: Industrial Relations in the Non-union Firm*. Buckingham: Open University Press.

Maddala, G. S. (1983), *Limited Dependent and Qualitative Variables in Econometrics*. Cambridge: Cambridge University Press.

Mallet, S. (1969), *La nouvelle classe ouvriere*. Paris: Seuil.

Mangione, T. W. (1973), 'Turnover: Some Psychological and Demographic Correlates', in R. P. Quinn and T. W. Mangione (eds.), *The 1969–1970 Survey of Working Conditions*. Ann Arbor: University of Michigan, Survey Research Centre.

Mann, F. C., and Hoffman, L. R. (1960), *Automation and the Worker*. New York: R. Hall & Co.

March, J. G., and Simon, H. A. (1958), *Organizations*. New York: Wiley.

Marchington, M., Goodman, J., Wilkinson, S., and Ackers, P. (1992), *New Developments in Employee Involvement*, Research Series 2. London: Employment Department.

Marcus, P. M., and Smith, C. B. (1985), 'Absenteeism in an Organizational Context', *Work and Occupations*, 12(3): 251–68.

Marmot, M. G., Smith, G. D., Stansfield, S., Patel, C., North, F., Head, J., White, I., Brunner, E., and Feeny, A. (1991), 'Health Inequality among British civil servants: The Whitehall II Study', *Lancet*, 337: 1387–94.

Marsh, J. G., and Simon, H. A. (1958), *Organizations*. New York: Wiley.

Marshall, G., Newby, N., Rose, D., and Vogler, C. (1988), *Social Class in Modern Britain*. London: Hutchinson.

Marshall, T. H. (1964), 'Citizenship and Social Class', in *Class, Citizenzenship and Social Development*. Chicago: University of Chicago Press.

Martin, D. (1994), *Democratie industrielle. La participation directe dans les entreprises*. Paris: Presses Universitaires de France.

Martin, J., and Roberts, C. (1984), *Women and Employment: A Lifetime Perspective*. London: HMSO.

—— and Miller, G. A. (1986), 'Job Satisfaction and Absenteeism: Organizational, Individual, and Job Related Correlates', *Work and Occupations*, 13(1): 32–46.

Martin, R. (1984), 'New technology and industrial relations in Fleet Street; new technology will make it possible for managers to manage', in M. Warner (ed.), *Microprocessors, Manpower and Society*. Aldershot: Gower.

—— (1988), 'Technological change and manual work', in D. Gallie (ed.) *Employment in Britain*. Oxford: Blackwell.

Mayo, E. (1932), *Human Problems of an Industrial Civilisation*. New York: Macmillan.

—— (1949), *The Social Problems of an Industrial Civilisation*. London: Routledge & Kegan Paul.

Meyer, J. P., and Allen, N. J. (1984), 'Testing the "side-bet" theory of organizational commitment: Influences of work positions and family roles', *Journal of Applied Psychology*, 69: 372–8.

Miller, J. G., and Wheeler, K. G. (1992), 'Unravelling the Mysteries of Gender Differences in Intention to Leave the Organization', *Journal of Organizational Behaviour*, 13: 465–78.

Millward, N. (1994), *The New Industrial Relations*. London: Policy Studies Institute.

—— and Stevens, M. (1986), *British Workplace Industrial Relations 1980–1984*. Aldershot: Gower.

—— —— Smart, D., and Hawes, W. R. (1992), *Workplace Industrial Relations in Transition*. Aldershot: Dartmouth Publishing.

Mobley, W. H., Griffeth, R. W., Hand, H. H., and Meglino, B. M. (1979), 'Review and Conceptual Analysis of the Employee Turnover Process', *Psychological Bulletin*, 86: 493–522.

Morse, N., and Reimer, E. (1956), 'The Experimental Change of a Major Organizational Variable', *Journal of Abnormal and Social Psychology*, 52: 120.

Moser, C. A., and Kalton, G. (1971), '*Survey Methods in Social Investigation*', 2nd. edn. London: Heinemann Educational Books.

Mouzelis, N. P. (1967), *Organizations and Bureaucracy*. London: Routledge & Kegan Paul.

Mowday, Richard T., and Steers, R. M. (1979), 'The Measurement of Organizational Commitment', *Journal of Vocational Behaviour*, 14: 224–47.

—— Porter, Lyman W., Steers, Richard M. (1982), *Employee-Organization Linkages: The Psychology of Commitment, Absenteeism and Turnover*. New York: Academic Press.

Mueller, C. W., Wallace, J. E., and Price, J. L. (1992), 'Employee Commitment: Resolving Some Issues', *Work and Occupations*, 19(3): 211–36.

Naville, P. (1963), *Vers l'automatisme social? Problèmes du travail et de l'automation*. Paris: Gallimard.

—— Barrier, C., Grossin, W., Lahalle, D., Legotien, H., Moisy, B., Palierene, J., and Wackerman, G. (1961), *L'automation et le travail humain*. Paris: CNRS.

Nichols, T. (1991), 'Labour Intensification, Work Injuries and the Measurement of the Percentage Utilisation of Labour (PUL)', *British Journal of Industrial Relations*, 29(4): 569–92.

Nicholson, N., and Goodge, P. M. (1976), 'The Influence of Social, Organizational and Biographical Factors on Female Absence', *Journal of Management Studies*, 13: 234–54.

—— and Johns, G. (1985), 'The Absence Culture and the Psychological Contract: Who's in Control of Absence?' *Academy of Management Review*, 10(3): 397–407.

Northcott, J., and Walling, A. (1988), *The Impact of Microelectronics: Diffusion, Benefits and Problems in British Industry.* London: Policy Studies Institute.

O'Reilly, J. (1994), 'Part-Time Work and Employment Regulation: Britain and France in the Context of Europe', in M. White (ed.), *Unemployment and Public Policy in a Changing Labour Market.* London: Policy Studies Institute.

OECD (1991), *Employment Outlook 1991.* Paris: Organization for Economic Co-operation and Development.

Osterman, P. (1994), 'How common is workplace transformation and who adopts it?' *Industrial and Labor Relations Review*, 47(2): 173–88.

—— (1995), 'Skill, Training and Work Organization in American Establishments', *Industrial Relations*, 34(2): 125–46.

Parkes, K. R. (1982), 'Occupational stress among student nurses: A natural experiment', *Journal of Applied Psychology*, 67: 784–96.

Parsons, D., and Marshall, V. (1995), *Skills, Qualifications and Utilisation: A Research Review*, Research Series 67. London: Dept. for Education and Employment.

Patchen, M. (1970), *Participation, Achievement and Involvement on the Job.* New Jersey: Prentice-Hall.

Payne, J. (1990), *Adult Of-the-Job Skills Training: An Evaluation Study.* Sheffield: Training Agency Research and Development Series, 57.

—— (1991), *Women, Training and the Skills Shortage.* London: Policy Studies Institute.

—— and Payne, C. (1993), 'Unemployment and peripheral work', *Work, Employment and Society*, 7: 513–34.

Penn, R., Rose, M., and Rubery, J. (1994), *Skill and Occupational Change.* Oxford: Oxford University Press.

Peters, T. J., and Waterman, R. H. (1982), *In Search of Excellence: Lessons from America's Best-Run Companies.* New York: Harper & Row.

Pfeffer, J., and Lawler, J. (1980), 'The Effect of Job Alternatives, Extrinsic Rewards, and Commitment on Satisfaction with the Organization', *Administrative Science Quarterly*, 25: 38–56.

Pheysey, D. C. (1993), *Organizational Cultures: Types and Transformations.* London: Routledge, 1993.

Piore, M. J., and Sabel, C. F. (1984), *The Second Industrial Divide. Possibilities for Prosperity.* New York: Basic Books.

Polanyi, K. (1957), *The Great Transformation.* Boston: Beacon Press.

Porter, L. W., and Lawler, E. E. III. (1968), *Managerial Attitudes and Performance.* Homewood, Ill.: Irwin.

—— and Steers, R. M. (1973), 'Organizational, Work and Personal Factors in Employee Turnover and Absenteeism', *Psychological Bulletin*, 80: 151–76.

—— Steers, R. M., Mowday, R. T., and Boulian, P. V. (1974), 'Organizational Commitment, Job Satisfaction, and Turnover Among Psychiatric Technicians', *Journal of Applied Psychology*, 59: 603–9.

Price, J. L. (1977), *The Study of Turnover*. Ames, Ia.: Iowa University Press.

—— and Mueller, C. W. (1986), *Absenteeism and Turnover among Hospital Employees*. Greenwich, Conn.: JAI Press.

Procter, S. J., Rowlinson, M., McArdle, L., Hassard, J., and Forrester, P. (1994), 'Flexibility, Politics and Strategy: In Defence of the Model of the Flexible Firm', *Work, Employment and Society*, 8: 221–42.

Pruden, Henry O., and Rees, Richard M. (1972), 'Interorganization Role-Set Relations and the Performance and Satisfaction of Industrial Salesmen', *Administrative Science Quarterly*, 17: 601–9.

Purcell, J. (1991), 'The rediscovery of the management prerogative: the management of labour relations in the 1980s', *Oxford Review of Economic Policy*, 7(1): 33–43.

Pycio, Peter (1992), 'Job Performance and Absenteeism: A Review and Meta-Analysis', *Human Relations*, 45(2): 193–220.

Quinn, R. P., and Staines, G. L. (1979), *The 1977 Quality of Employment Survey*. Ann Arbor: Institute for Social Research.

Randall, D. M., and Cote, J. A. (1991), 'Interrelationships of Work Commitment Constructs', *Work and Occupations*, 18(2): 194–211.

Regalia, I. (1995), *Direct Participation: How the Social Partners view it. A comparative study of fifteen European countries*, Working Paper WP/95/73/EN. Dublin: European Foundation for the Improvement of Living and Working Conditions.

Rigg, M. (1989), *Training in Britain: A Study of Funding, Activity and Attitudes—Individuals' Perspectives*. London: HMSO.

Roethlisberger, F. J., and Dickson, W. J. (1939), *Management and the Worker*. Cambridge, Mass.: Harvard University Press.

Rose, D., Marshall, G., Newby, H., and Vogler, C. (1987), 'Goodbye to Supervisors?' *Work, Employment and Society*, 1(1): 7–24.

Rose, M. (1985), *Reworking the Work Ethic: Economic Values in Social-Cultural Politics*. London: Batsford Academic.

—— (1988), 'Attachment to Work and Social Values', in D. Gallie (ed.), *Employment in Britain*. Oxford: Blackwell.

Rotter, J. B. (1966), 'Generalized expectancies for internal versus external control of reinforcement', *Psychological Monographs: General and Applied*, vol. 80(1), whole no. 609.

Rousselet, J. (1974), *L'allergie au travail*. Paris: Seuil.

Salancik, G. R. (1977), 'Commitment and the Control of Organizational Belief', in B. M. Staw and G. R. Salancik (eds.), *New Directions in Organizational Behaviour*. Chicago: St Clair Press.

Selznick, P. (1957), *Leadership in Administration: A Sociological Interpretation*. New York: Harper & Row.

Sewell, G., and Wilkinson, B. (1992), 'Empowerment or Emasculation? Shopfloor Surveillance in a Total Quality Organization', in P. Blyton and P. Turnbull (eds.), *Reassessing Human Resource Management*. London: Sage Publications.

Shaiken, H. (1984), *Work Transformed: Automation and Labour in the Computer Age*. New York: Holt, Rinehart, and Winston.

Shore, L. M., and Martin, H. J. (1989), 'Job Satisfaction and Organizational Commitment in Relation to Work Performance and Turnover Intentions', *Human Relations*, 42(7): 625–38.

Sims, H. P., Szilagyi, A. D., and Keller, R. T. (1976), 'The Measurement of Job Characteristics', *Academy of Management Journal*, 19: 195–212.

Smith, C. (1987), *Technical Workers: Class, Labour and Trade Unionism.* London: Macmillan.

Smith, F. J. (1977), 'Work Attitudes as Predictors of Specific Day Attendance', *Journal of Applied Psychology*, 62: 16–19.

Smith, I. (1992), 'Reward Management and HRM', in P. Blyton and P. Turnbull (eds.), *Reassessing Human Resource Management.* London: Sage Publications.

Spector, P. E. (1986), 'Perceived Control by Employees: A Meta-Analysis of Studies Concerning Autonomy and Participation at Work', *Human Relations*, 39(11): 1005–16.

—— (1988), 'Development of the Work Locus of Control Scale', *Journal of Occupational Psychology*, 61: 335–40.

—— and Jex, S. M. (1991), 'Relations of Job Characteristics from Multiple Data Sources with Employee Affect, Absence, Turnover Intentions, and Health', *Journal of Applied Psychology*, 76(1): 46–53.

Spenner, K. (1990), 'Meanings, Methods and Measures', *Work and Occupations*, 17(4): 399–421.

Staw, Barry M. (1984), 'Organizational Behaviour: A Review and Reformulation of the Field's Outcome Variables', *Annual Review of Psychology*, 35: 627–66.

Steel, R. P., and Griffeth, R. W. (1989), 'The Elusive Relationship between Perceived Opportunity and Turnover Behaviour: A Methodological or Conceptual Artifact?', *Journal of Applied Psychology*, 74(6): 846–54.

Steers, Richard M. (1977a), *Organizational Effectiveness: A Behavioral View.* Santa Monica, Calif.: Goodyear Publishing Co.

—— (1977b), 'Antecedents and Outcomes of Organizational Commitment', *Administrative Science Quarterly*, 22: 46–56.

—— and Rhodes, Susan R. (1978), 'Major Influences on Employee Attendance: A Process Model', *Journal of Applied Psychology*, 63(4): 391–407.

Steinberg, R. J. (1990), 'Social Construction of Skill: Gender, Power and Comparable Worth', *Work and Occupations*, 17(4): 449–82.

Stewart, P., and Garrahan, P. (1995), 'Employee responses to new management techniques in the auto industry', *Work, Employment and Society*, 9: 515–36.

Storey, J. (1992), *Developments in the Management of Human Resources.* Oxford: Blackwell.

—— (ed.) (1995), *Human Resource Management: A Critical Text.* London: Routledge.

—— et al. (1993), *Human Resource Management Practices in Leicestershire: A Trends Monitor.* Loughborough University Business School.

Strauss, G. (1963), 'Some notes on power-equalization', in H. J. Leavitt (ed.), *The Social Science of Organizations: Four Perspectives.* Englewood Cliffs, NJ: Prentice-Hall, Inc.

Sussman, M. B., and Cogswell, B. E. (1971), 'Family Influences on Job Movement', *Human Relations*, 24: 477–87.

Tam, M. (1997), *Part-Time Employment: A Bridge or a Trap?* Aldershot: Avebury.

Tannenbaum, A. S., Kavcic, B., Rosner, M., Vianello, M., and Weiser, G. (1974), *Hierarchy in Organizations.* San Francisco: Jossey-Bass.

Temin, Peter (1990), 'Free Land and Federalism: American Economic Exceptionalism', in Byron Shafer (ed.), *Is America Different: A New Look at American Exceptionalism.* Oxford: Clarendon Press.

Thompson, P. (1989), *The Nature of Work*, 2nd edn. London: Macmillan.

Touraine, A. (1955), *L'evolution du travail ouvrier aux Usines Renault.* Paris: CNRS.

—— Durand, C., Pécaut, D., and Willener, A. (1965), *Workers' Attitudes to Technical Change.* Paris: OECD.

Treiman, Donald (1970), 'Industrialization and Social Stratification', in E. O. Lauman (ed.), *Social Stratification: Research and Theory in the 1970s*. Indianapolis: Bobbs Merrill.

Trist, E. L., Higgin, G. W., Murray, H., and Pollock, A. B. (1963), *Organisational Choice*. London: Tavistock.

Undy, R., Fosh, P., Morris, H., Smith, P., and Martin, R. (1996), *Managing the Unions. The Impact of Legislation on Trade Unions' Behaviour*. Oxford: Clarendon Press.

Useem, M. (1993), 'Management Commitment and Company Policies on Education and Training', *Human Resource Management*, 32(4): 411–34.

Vallas, S. P. (1990), 'The Concept of Skill', *Work and Occupations*, 17(4): 379–98.

Vroom, V. (1964), *Work and Motivation*. New York: Wiley.

Wajcman, J. (1991), 'Patriarchy, Technology and Conceptions of Skill', *Work and Occupations*, 18(1): 29–45.

Walker, C. R., and Guest, R. H. (1952), *The Man on the Assembly Line*. Cambridge, Mass.: Harvard University Press.

Wall, T. B., Kemp, N. J., Jackson, P. R., and Clegg, C. W. (1986), 'Outcomes of autonomous workgroups: A long-term field experiment', *Academy of Management Journal*, 29: 280–304.

Wall, T. D., and Clegg, C. W. (1981), 'A longitudinal field study of group work redesign', *Journal of Occupational Behaviour*, 2: 31–49.

—— —— Davies, R. T., Kemp, N. J., and Mueller, W. S. (1987), 'Advanced manufacturing technology and work simplification: an empirical study', *Journal of Occupational Behaviour*, 8: 233–50.

—— Corbett, J. M., Martin, R., Clegg, C. W., and Jackson, P. R. (1990), 'Advanced Manufacturing Technology, Work Design and Performance: A Change Study', *Journal of Applied Psychology*, 75: 691–7.

Walton, R. E. (1972), 'How to counter alienation in the plant', *Harvard Business Review*, 72(6): 70–81.

—— (1985a), 'From control to commitment in the workplace', *Harvard Business Review*, 85(2): 77–84.

—— (1985b), 'Towards a strategy of eliciting employee commitment based on policies of mutuality', in R. E. Walton and J. R. Lawrence (eds.), *HRM: Trends and Challenges*. Boston: Harvard Business School.

—— (1987), *Innovating to Compete*. London: Jossey-Bass.

Warr, P. (1982), 'A national study of non-financial employment commitment', *Journal of Occupational Psychology*, 55: 297–312.

—— (1987), *Work, Unemployment and Mental Health*. Oxford: Clarendon Press.

—— (1990), 'The measurement of well-being and other aspects of mental health', *Journal of Occupational Psychology*, 63: 193–210.

—— (1991), 'Mental Health, well-being and job satisfaction', in B. Hesketh and A. Adams (eds.) *Psychological Perspectives on Occupational Health and Rehabilitation*, 143–65. Sydney, Australia: Harcourt Brace Jovanovich Ltd.

—— Jackson, P., and Banks, M. (1988), 'Unemployment and mental health: some British studies', *Journal of Social Issues*, 44(4): 47–68.

Waterman, R. H. Jr., Waterman, J. A., and Collard, B. A. (1994), 'Toward a Career-Resilient Workforce', *Harvard Business Review*, July–Aug.: 87–95.

Watson, G. (1994), 'The Flexible Workforce and Patterns of Working Hours in the UK', *Employment Gazette*, July: 239–48.

Webel, J. D., and Bedeian, A. G. (1989), 'Intended Turnover as a Function of Age and Job Performance', *Journal of Organizational Behaviour*, 10: 275–81.

Weber, M. (1947), *The Theory of Social and Economic Organization*. New York: Free Press.

Wedderburn, D., and Crompton, R. (1972), *Workers' Attitudes and Technology*. Cambridge: Cambridge University Press.

Weisberg, J., and Kirschenbaum, A. (1991), 'Employee Turnover Intentions: Implications from a National Sample', *International Journal of Human Resource Management*, 2(3): 359–75.

Weitzman, M. L. (1985), 'Profit sharing as macroeconomic policy', *American Economic Review*, 75: 937–53.

Whelan, C., Hannan, D. F., and Creighton, S. (1991), *Unemployment, Poverty and Psychological Distress*. Dublin: Economic and Social Research Institute.

White, M. (1983), *Long-term Unemployment and Labour Markets*. London: Policy Studies Institute.

—— Gallie, D., Cheng, Y., and Tomlinson, M. (1996), *By Choice or Circumstance? Individuals in Unemployment*. Manuscript report: Policy Studies Institute.

—— (1991), *Against Unemployment*. London: Policy Studies Institute.

White R. K., and Lippitt, R. (1960), *Autocracy and Democracy: An Experimental Inquiry*. New York: Harper.

Whittington, R., McNulty, T., and Whipp, R. (1994), 'Market-driven change in professional services', *Journal of Management Studies*, 31: 829–45.

Wilkinson, A., and Willmott, H. (eds.) (1995), *Making Quality Critical*. London: Routledge.

Wilkinson, B. (1993), *The Shop Floor Politics of New Technology*. London: Heinemann.

Willmott, H. (1993), 'Strength is ignorance; slavery is freedom: managing culture in modern organizations', *Journal of Management Studies*, 30(4): 515–52.

Womack, J. P., Jones, D. T., and Roos, D. (1990), *The Machine that Changed the World*. New York: Rawson Associates, Macmillan Publishing Co.

Wonnacott, T. H., and Wonnacott, R. J. (1990), *Introductory Statistics*, 5th edn. New York: Wiley.

Wood, S. (1993), 'The Japanization of Fordism', *Economic and Industrial Democracy*, 14: 535–55.

—— (1996), 'High commitment management and payment systems', *Journal of Management Studies*, 33(1): 53–77.

—— and Albanese, M. T. (1995), 'Can We Speak of High Commitment Management on the Shop Floor?' *Journal of Management Studies*, 32(2): 215–47.

Woodward, J. (ed.) (1970), *Industrial Organization: Behaviour and Control*. Oxford: Oxford University Press.

Zuboff, S. (1983), 'The Work Ethic and Work Organisation', in J. Barbash, R. J. Lampman, S. A. Levitan, and G. Tyler (eds.), *The Work Ethic—A Critical Analysis*: 153–81. Madison, Wis.: Industrial Relations Research Association.

—— (1988), *In the Age of the Smart Machine: The Future of Work and Power*. New York: Basic Books.

NAME INDEX

SUBJECT INDEX